# Structural Defects Reference Manual for Low-Rise Buildings

# Structural Defects Reference Manual for Low-Rise Buildings

## Michael F. Atkinson
CEng, FIStructE, MICE

London and New York

First published 2000

by E & FN Spon
11 New Fetter Lane, London EC4P 4EE

Simultaneously published in the USA and Canada
by E & FN Spon, an imprint of Routledge
29 West 35th Street, New York, NY 10001

*E & FN Spon is an imprint of the Taylor & Francis Group*

© 2000 Michael F. Atkinson

Typeset in 10/12pt Sabon by Mathematical Composition Setters Ltd, Salisbury, Wiltshire
Printed and bound in Great Britain by St Edmundsbury Press, Bury St Edmunds, Suffolk

*British Library Cataloguing in Publication Data*
A catalogue record for this book is available from the British Library

*Library of Congress Cataloging in Publication Data*
A catalogue record for this book has been requested

ISBN 0-419-25790-X

# Contents

# Preface

Diagnosis of major structural damage in a building holds several fascinations for me. One is the process of problem solving; a second is the engineering challenge; and a third is the ability to learn from mistakes, be they design errors or bad workmanship. The majority of learning acquired from a building that has failed comes from hindsight. Everyone knows hindsight can be very exacting, but the secret is to learn from the errors entailed in the failures and devise systems that will reduce the frequency of such cases. This publication is not intended to be an answer to all defects likely to be encountered, but it should provide the basis for what is in effect a scientific examination, namely the structural investigation report. I hope it will be of use to building students, engineers, surveyors, loss adjusters, and other building-related professionals.

It is not written to assist in the carrying out of structural surveys for valuations or purchasing of a building, but for those situations where the structural defect may have become apparent at the point of buying or selling of a property. The defect already exists, and before a remediation strategy can be implemented, the cause of the structural damage must be identified.

Seldom is it found that structural failures are identical, similar though they may be; the details of each are dictated by many factors. It therefore follows that the remediation work will vary in each case, not only in type but also in extent when related to the age and the life expectancy of the building.

While this book deals primarily with domestic housing construction, much of the information is relevant to the design and construction of other types of low-rise building. The factual case histories are drawn from my extensive experience in dealing with major damage insurance claims.

Throughout the book please read 'he or she' for 'he'.

I am grateful to the following organisations for permission to reproduce diagrams and other data from their publications:

- The National House Building Council (NHBC)
- Extracts from British Standards and Codes of Practice are reproduced by permission of the British Standards Institution, 2 Park St, London W1, from whom copies of the complete Standards and Codes of Practice can be obtained.
- Construction Research Communications for the concrete mix design tables from BRE Digest 363.

I also thank the editorial staff at E & FN Spon for their assistance, and my wife for her administrative help and support.

I dedicate this book to our son, Robert Nirmal, who tragically died in May 1999. His encouragement will be long remembered.

M.F. Atkinson

# Chapter 1
# Background

## 1.1 Introduction

The aim of this publication is to assist those specialists – surveyors, engineers, builders and other interested parties – in identifying structural defects in a building, determining the cause of the defect, and – if required – recommending appropriate remedial measures to correct the problem.

This manual is intended to cover the range of low-rise conventionally constructed buildings up to four storeys in height, but the principles outlined in the various chapters can also be applied to other types of building, such as high-rise flats, factories and office buildings.

Disregarding wear and tear and the breakdown of inappropriate building materials, structural defects can result from

● excessive foundation settlement or differential settlement;
● subsidence caused by active mining or the collapse of old mine workings;
● subsidence as a result of trees drying out clay soils below a foundation;
● heave on foundations due to re-hydration of clay soils following tree removal;
● failure to provide for expansion of masonry in long terrace blocks;
● inadequate design of the foundations and super-structure;
● bad workmanship;
● the use of materials not fit for the purpose;
● spontaneous expansion of fill materials;
● overloading or early loading of structural elements;
● cutting or notching of existing structural elements;
● deep excavations for drains too close to the building;
● subsidence due to solution features in chalk, gypsum and salt beds;
● subsidence due to damaged drains;
● instability of sloping ground, failure of retaining structures;
● contaminated ground;
● geological faults;
● inclement weather, frost and wind;
● severe exposure.

To the layperson, cracking of a building as a result of foundation movement is often referred to as subsidence, when often the causation is excessive or differential settlement. It is worthwhile therefore to define the difference between subsidence and settlement.

● **Subsidence:** This occurs when the foundations of a building move up or down owing to the loss of support of the underlying strata below the foundation, or as a result of volumetric expansion changes in the soils or sub-base materials. Examples are mining subsidence, clay soils drying out owing to dry weather or trees, and clay heave caused by the re-hydration of clay soils when trees are removed from below a building plot. Tables 1.1 and 1.2 give examples of various subsidence modes.
● **Settlement:** this is defined as movement in the structure caused by the weight of the building compressing the ground upon which the building foundations rest. All buildings on compressible soils settle, and it rests with building designers to use their skill and professional judgement to provide foundations that will ensure that any settlements, total or differential, will be within an acceptable magnitude for the type of building being constructed.

For example, strip foundations built over peat beds generally suffer from excessive settlement, especially if the peat beds are thick and at shallow depth. Dwellings built on top of made-up ground are prone to excessive settlements as a result of the weight of the building, but such made-up ground can result in a phenomenon referred to as collapse settlement of the filled ground at depth. Often this is caused by water inundation.

All building in the United Kingdom takes place within a complex framework of constraints and guidance: for example, the Approved Documents that deal with the Building Regulation requirements, most of which are mandatory.

For the private housing sector, the major standard-setting body is the National House Building Council. This organisation was set up in 1936 by a group of builders with the object of improving the reputation of

**Table 1.1** Naturally occurring subsidences

| Type of subsidence feature | Cause and result | Failure mode |
|---|---|---|
| Solution features in limestone, chalk and gypsum strata. Can also occur in natural salt formations | Results in underground cavities, vertical sinkholes and breccia pipe features | Generally a conical collapse, which causes large surface hollows |
| Shrinkage and heave of clay soils | Can result from very prolonged dry weather, and from the influence of trees | Often cyclical, and usually takes place over a long period |
| Variable groundwater levels | Granular soils can become buoyant and lose strength, and cohesive soils can soften | The effects can be slow, depending on the speed of the drawdown |
| Frost heave on cohesive soils | Can result in differential movements | Often localised around the perimeter of the building, and slow to occur |
| Washout of fines from below foundations | Can be caused by natural flows of water or flooding of rivers | Can take place over a long period, but the damage can be sudden |
| Unstable slopes, land slip | Such large-scale movements can remove support to buildings | Can occur rapidly; often caused by a rise in pore water pressures |

**Table 1.2** Induced subsidence movement

| Type of subsidence feature | Cause and result | Failure mode |
|---|---|---|
| Long-wall mining for coal | If coal seams are relatively shallow, damage can be caused by the compression and tensile strains in the ground mass | Cracking of masonry; can be sudden if coal seam is shallow Local to large-scale damage can result |
| Mining for minerals such as salt, potash, metallic ores, fireclay, gypsum, stone flags, chalk, limestone | Collapses from this type of mining generally result from an unstable roof over the workings, which are often old and unrecorded | Crown holes at the surface; loss of support can cause severe damage to buildings |
| Made-up ground | Poorly compacted infills, very voided and settlement of fill under its own weight can result in excessive movements when loaded | Slow to sudden damage, depending on the load applied by the foundations |
| Inundation of infilled ground | Generally caused by the re-establishment of previous water table. Can be caused by water from burst pipes or drains, or some forms of vibro ground improvement works | Can occur quickly and can be ongoing and repetitive as the water table rises up through successive layers of loose fill |
| Loss of lateral support to a foundation | Generally results from drains being excavated too close to, and deeper than, the building foundation | Damage to a building can be sudden. Movements will be ongoing until the excavation is stabilised |
| Removal of soil below a foundation | Often caused by leaking or broken drains, water mains etc. | Damage can occur rapidly and at some distance from the water source |
| Consolidation of soft ground | Can be caused by adding fills over soft wet clays. This results in pore water being squeezed out, with subsequent settlement of the ground | Slow process, but can result in long-term subsidence damage |
| Shrinkage of cohesive soils | Desiccation as a result of external heat source, e.g. brick kilns | Long-term damage |
| Heave of cohesive soils | Can result from the effects of coldstore buildings | Long-term damage |
| Swelling of cohesive soils | Caused by the removal of trees and vegetation | Damage dependent on the plasticity index of the clays, and the water demand of the trees |
| Movements as a result of vibration of granular soils | Can be caused by moving traffic, pile driving, vibro-compaction ground improvement works | Localised situations can result in localised damage to masonry structures |

the housebuilding industry at a time when 'jerry building' was causing public concern. The NHBC has developed practical guidance, embraced in its technical standards and other publications.

Other bodies, such as TRADA, BSI and BRE, also provide comprehensive guidance on building construction.

Scotland has its own national Building Regulations in the form of the Building Standards (Scotland) Regulations, and in London there are the London Building (Construction) Bye-Laws, administered by the London borough surveyors.

Despite all these regulations, guidance and advice, structural defects continue to occur in buildings as a result of

● inadequate design;
● poor choice of materials;
● bad workmanship.

Of the many structural surveys undertaken by the author, the majority of the defects resulted from bad workmanship and inadequate supervision during the construction of the building. Many of the major structural damage claims accepted by the NHBC had resulted from inadequate foundations that had been approved by the local authority building control. Unfortunately, under the present legal system the local authority building control can be held to account only if any persons are injured or die as a result of the defective work. Since the NHBC became an Approved Inspector in 1984, which allowed it to carry out building control for builders, there has been a significant reduction in the number of major structural damage claims. This is due to the increased amount of statutory inspection of the significant parts of a building, such as foundation excavations, ground floor sub-base preparation, and first floor joists. Despite the occasional claims that hit the headlines, there is no doubt that the NHBC has played a significant role in improving building standards, and all at no cost to the public purse.

In 1998 the NHBC carried out a major strategic review of its Buildmark policy, and in April 1999 launched a revised policy, which for the first time gave house owners insurance cover against any problems arising from contaminated land, plus many other new initiatives.

Figure 1.1 shows the cost of major structural damage claims to the NHBC warranty for the period from October 1983 to September 1984.

The cost of claims paid by NHBC for brickwork and mortar failures from October 1985 to September 1986 is shown in Table 1.3. The total figure for the North West Region excluded remedial works amounting to £112 500 paid for by brick manufacturers who had supplied bricks not fit for the exposure conditions. The claims in 1983/1984 were £260 000, and in 1984/1985 were £170 000.

When selecting a facing brick for a site, if there are doubts about suitability, obtain written confirmation from the brick manufacturer that the brick is suitable for the location, both geographically and in the structure. By taking this action you can place the responsibility for any failure on the manufacturer provided you follow his advice.

During the financial year 1997/98 NHBC paid out £4 million arising from the failure of builders to meet their

*Table 1.3* Cost of brick failure claims to NHBC, October 1985 to September 1986

| NHBC region | Above dampcourse | | Below dampcourse | | Total cost (£) |
|---|---|---|---|---|---|
| | No. of claims | Cost (£) | No. of claims | Cost (£) | |
| North East | 28 | 153 980 | 8 | 15 160 | 169 140 |
| North West | 6 | 26 860 | 5 | 7 600 | 34 460ᵃ |
| Scotland | 3 | 3 200 | 22 | 7 670 | 10 870 |
| South East | 0 | – | 2 | 2 400 | 2 400 |
| South West | 8 | 6 400 | 6 | 10 100 | 16 500 |
| West | 0 | – | 7 | 8 100 | 8 100 |
| Central | 2 | 6 700 | 0 | – | 6 700 |
| East Midlands | 1 | 1 320 | 4 | 1 800 | 3 120 |
| East | 2 | 3 160 | 1 | 500 | 3 660 |
| South | 1 | 2 400 | 0 | – | 2 400 |
| Greater London | 0 | – | 0 | – | – |
| Total | 51 | 204 020 | 55 | 53 330 | 257 350 |

ᵃ In the North West brick manufacturers made a significant contribution towards the cost of replacing bricks not fit for the exposure conditions.

obligations in the first two years after legal completion. A further £9 million was spent on the repair of homes affected by major structural damage incurred in years 3–10 of the Buildmark policy. To put these figures into context, the annual construction cost for new homes covered by NHBC is approximately £7000 million.

## 1.2 Objectives

Most structural survey work generally involves working with various specialists. Figure 1.2 outlines the specialist contributions that can assist in the determination of a major structural damage claim and a final report on suitable remedial action. A structural survey should always be carried out by a chartered surveyor, or a chartered structural or civil engineer, and either one should be an expert in this field. He or she is entrusted with the investigation of the cause of the defect. It is fundamental that the cause of the damage be determined so that the correct remedial solution can be advised. The investigation must be carried out in the most economic fashion, and the final report should be fair and unbiased, and one that enables the client to take further appropriate action.

If he or she is required to produce a remediation scheme, it is essential to understand how the building and supporting strata will be affected by the remedial works. The recommended design solution must be able to be carried out without causing further deterioration of the building fabric, and safety must be a major consideration, followed by economy and practicality.

Structural surveys on a defective property may require the fixing of tell-tales and the establishment of a levelling grid so that long-term monitoring can be carried out.

For a little extra cost, a large amount of useful information can be gained by excavating several trial pits and carrying out some hand auger borings. In addition it is recommended that an initial desk study be

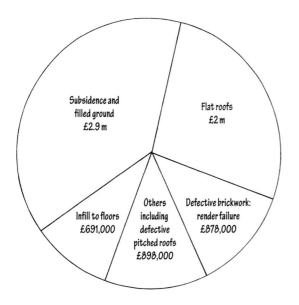

*Fig. 1.1* NHBC claims costs for 1983–84. *Note:* These claims are almost entirely from houses built in the 1970s. Houses built in the 1980s have a better record.

Subsidence and filled ground
£2.9 m

Flat roofs
£2 m

Infill to floors
£691,000

Others including defective pitched roofs
£898,000

Defective brickwork: render failure
£878,000

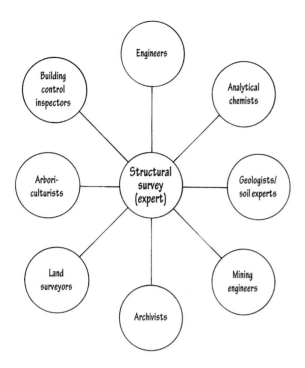

*Fig. 1.2* Specialist contributions to a structural survey.

carried out, checking such basic items as the geology, mining situation, and industrial history of the site. This information may be useful in planning the next stage of the investigation, and in deciding whether further specialist assistance is likely to be required.

Figure 1.2 shows many specialists in the building field, but it is unlikely that all of these will be called upon. While experts generally have an appreciation of these specialisations and should use these skills to further the investigation, they should be aware of their own limitations, and should recognise when they need separate specialist advice.

Determination of a cause of damage is essentially a problem-solving exercise. During the course of the investigation, likely initial theories will be discarded as the works are opened up. For example, if soft, wet ground is found under a foundation, it is prudent to have the drains checked and tested. If they are found to be in good order then the drains can be eliminated from the checklist.

Experts engaged to advise on the cause of major structural damage should have a detailed knowledge of building construction, and be experienced in investigating and appraising damaged buildings.

## 1.3 Household insurance and warranty insurance claims

Subsidence damage to a property is generally covered under the building owner's insurance policy. Most good insurance policies will cover subsidence, heave and landslip. They do not cover defects arising out of poor-quality building, inappropriate and/or inadequate building materials, inadequate design, or any damage caused by the owners' failure to carry out regular maintenance

to the building fabric. Subsidence cover for private housing only became available around about 1971, but insurance cover for heave on ground floors or foundations came into being in 1980.

In essence the building insurance policy is supposed to indemnify the building owner against loss from damage arising from an insured event or cause. It means that the insurer will be placing the insured party back in the same financial position in which they would have been but for the occurrence of the insured event or causation. This is generally done by appointing a chartered loss adjuster to investigate the claim, and by doing the following:

● paying for the cost of any remediation;
● agreeing with the building owner a cash settlement to cover the reduction in the market value of the building, i.e. the difference between the market value had the building not suffered any damage and the reduced market value of the building taking the structural damage into account.

### 1.3.1 Establishing a structural defects claim

When the owner of a property discovers structural damage to their property they should always inform their household insurance company or insurance warranty provider as soon as possible. Any failure to inform them quickly could result in a claim being turned down if the investigating surveyor considers the property has deteriorated over the period from discovering the defect to reporting it: for example, leaking drains can cause rapid deterioration in soils below a foundation. Prompt notification enables the insurer to set in motion works that would prevent the defect from getting worse.

When dealing with a damage claim arising out of subsidence caused by other parties, or by trees in a neighbour's land, it is important that the owners of the trees or their insurers are put on notice as quickly as possible. This will enable the tree owners' insurers to fully investigate the damage and any potential for further damage.

### 1.3.2 Ground floors

Most household insurance policies specifically exclude subsidence of ground floor slabs placed directly on the ground. The exception is where the damage to any foundations may have resulted in damage to the ground floor slab, or where the floor is left with an excessive slope.

In situations where floor slabs heave as a result of spontaneous expansion of sub-base fills, this could be referred to the insurers as a valid causation. Serious cases of ground floor subsidence are rare, and this is due to the NHBC's introducing a mandatory requirement for a fully suspended ground floor if the depth of upfill exceeded 600 mm.

### 1.3.3 Handling insurance claims for structural damage

When a structural damage claim is made by a building owner to his or her insurance provider, it is essential that the claim be fully investigated so that the extent of the damage can be assessed, and so that it can be determined whether or not the damage could become progressive: for example, subsidence of foundations on clay soils due to trees that are not yet fully mature.

The following parties are likely to be involved during the resolution of the structural damage claim:

- the policyholder – generally the building owner or mortgage provider;
- the insurance company – this includes insurance warranty providers such as NHBC;
- consulting engineers, chartered structural or civil engineers or chartered surveyor;
- chartered loss adjuster;
- qualified arboriculturist if there is a tree problem.

Where appropriate the insurance company may retain the services of an individual expert or a firm of consulting engineers to provide them with technical assistance.

### 1.3.4 Procedure

Insurance companies have not always dealt with structural defect claims in the same way. Over the last decade there has been a growing uniformity in the way such claims have been handled, mainly because of the millions of pounds spent on underpinning foundations as a result of extremely dry weather.

If an insurance company turns a claim down, building owners may have to employ their own expert to advise them as to the cause of the defect. The costs of this expert may have to be borne by the insured party, unless they can show that the insurers were wrong to reject their claim.

On other occasions, the defects may have come to light during a valuation survey carried out by a prospective purchaser.

Some subsidence or heave claims can take several years to resolve, especially if long-term monitoring is being carried out. It is important to recognise that all remedial solutions should be such that the property is put back into a saleable condition as soon as possible.

The remediation repairs should be fully documented in a schedule of works, and priced by at least three firms. The remediation works should be fully supervised during critical stages to ensure that they are properly carried out in accordance with the contract drawings.

### 1.3.5 Buildings covered by insurance warranties

The largest provider of warranty insurance for housing and other related buildings in the UK is the National House Building Council. The NHBC is also an independent standard-setting organisation and consumer protection body.

Its governing body comprises representatives from organisations interested in improving and raising the standards of housebuilding. Its board includes people who represent building societies, consumer groups, housebuilders, mortgage lenders, local authorities, housing associations and other professional bodies. The NHBC does not serve any one particular organisation or group.

The NHBC is an authorised insurance company with substantial gross reserves currently at £770 million. This provides the security for 1.7 million householders currently protected by Buildmark cover. Since the scheme was set up in 1936 over 5 million homes have been covered by NHBC.

There are other firms who offer warranty insurance on housing, such as Zurich Mutual and HAPM, but these firms do not have a large share of the market, and they do not provide the type of inspection service that NHBC offers.

#### The NHBC scheme

This operates in a fairly straightforward way. Builders who wish to join the NHBC scheme have to apply, pay a joining fee, and be vetted to ensure that they will be responsible members. The NHBC maintains a list of current builder and developer members. Builder members have both a technical and financial assessment; developers are assessed only for their financial state. This vetting is carried out both when they apply to join and afterwards on an ongoing basis.

The NHBC prescribes minimum building standards, referred to as the Technical Requirements, with which all members must comply when constructing a dwelling.

The NHBC has a large body of building inspectors and inspection managers covering builders' sites, and these are supported by qualified engineers operating from the regional offices. These field staff carry out spot check inspections of all dwellings registered and under construction, to ensure that, as far as can be seen, the builder has complied with the Technical Requirements, and this will allow the 10 year Buildmark insurance policy to be issued on completion.

The scheme obliges vendors of dwellings to offer the current prescribed form of House Purchaser's Agreement to house purchasers: this contains warranties relating to standards of workmanship and materials, and to the remedying of defects. It also provides an insurance scheme to indemnify house purchasers if there has been non-compliance with the Technical Requirements. The NHBC promulgates rules from time to time relating to the above matters, and takes disciplinary action against builders or developers who are in breach of the rules. The Rules, the ten-year Insurance Certificate and the House Purchaser's Agreement together constitute the NHBC insurance warranty scheme.

#### Conditions relating to NHBC membership

Builders' applications are subjected to vetting from both a technical and a financial point of view. Developers are subjected only to financial investigation, but are obliged under the NHBC rules to employ NHBC-approved builders as main contractors. The financial status of builders and developers is investigated by the taking-up of references, and by obtaining credit status reports.

Technical competence is assessed from an interview with the applicant builder, who must demonstrate adequate knowledge of the NHBC minimum technical standards. The builder is further assessed by on-site inspections of any dwelling under construction. Normally, a builder applicant is not fully accepted into the NHBC scheme until the NHBC has inspected a dwelling from commencement to completion of construction, and only if the builder's standards are seen to be in line with the Technical Requirements. If the builder's standards are below NHBC Technical Requirements, and show no signs of improvement, the application will be turned down.

The scheme operates using a premium rating system. This operates as an effective no claims bonus for good builders, with bad builders, or builders previously unknown to NHBC, paying increased premiums for each property registered.

### Discipline

The NHBC Disciplinary Sub-Committee deals with disciplinary matters. These include failure to build in accordance with NHBC Technical Requirements, failure to remedy reported defects, and other breaches of the rules.

### Technical requirements

Rule 10 obliges all builders to build in compliance with NHBC minimum building standards. These minimum building standards are contained in Volumes 1 and 2 of the *NHBC Standards*, a copy of which every builder receives.

In essence, the Technical Requirements set down minimum standards to ensure soundly constructed houses, and they incorporate the requirements of the statutory building regulations as outlined in the Approved Documents.

They cover all trades involved in housebuilding, and include some design matters. Special provisions cover timber frame design and construction: such constructions must be certified by an NHBC approved timber frame certifier, and a timber frame certificate HB 153 B must be duly signed and be available on site for inspection by the NHBC field staff.

### Hazardous ground conditions

NHBC has the power to require a builder to provide a full structural engineer's report and foundation proposals for buildings on hazardous ground for appraisal prior to commencement on site. The builder is obliged to provide NHBC with two months' prior notification of intention to build on such a site. This does not require the

submission of working details at this stage, but is intended to alert their field staff that such a development is imminent. The builder must register properties three weeks prior to commencement on site, and – provided the consulting engineer's foundation proposals and site investigation reports are submitted at the same time – appraisal should be completed before work starts on the foundations.

Under the NHBC Rules the builder must employ a chartered structural or civil engineer to deal with a hazardous site, and must have full professional insurance indemnities for an appropriate amount. These engineers are also required to carry out some site supervision, especially if there are specialist ground improvement works, piling and contamination issues.

If there is a dispute between a builder and NHBC as to whether the Technical Requirements have been complied with, NHBC can refuse to issue its 10 Year Insurance Certificate, and the builder has the right to independent arbitration in accordance with Rule 33.

Table 1.4 indicates the general effects when subsidence or settlement damage occurs to a building.

### NHBC Buildmark cover from 1 April 1999

Following a major strategic review, the NHBC initiated major changes to its Buildmark policy, and the Buildmark cover is clearly defined in a comprehensive document. The period of the warranty insurance cover on the dwelling is now 10 years from the date of legal completion with the purchaser.

*Table 1.4* The general effects of subsidence/settlement

| Effects of subsidence/settlement | Repercussions |
|---|---|
| *Buildings* | |
| Structural damage to floors and walls | Remediation works are generally required to stabilise the building |
| Possible demolition | Occupants will need to move out until the insurance claim is settled |
| Building may require temporary works to make it safe | This may be possible with the occupants still in the building |
| Insurance claim may have to be lodged | The owners may be liable for an excess on the policy |
| Property may become blighted | The building value could drop |
| *General* | |
| Danger to the public | Site could be sterilised. Any crown hole or subsidence hollow would require to be made safe |
| Damage and disruption to services. This could result in secondary effects | Services will need to be repaired quickly, especially water and sewage, if secondary damage is to be prevented |
| Personal injury and loss | Insurance claim |
| Loss or reduction of business | Insurance claim |

The Buildmark cover is divided into four main sections:

1   insurance cover before completion of the dwelling;
2   insurance cover during the first 2 years after legal completion of the dwelling;
3   insurance cover in years 3–10 after legal completion, including contaminated land cover;

4   Additional insurance cover in years 3–10 if NHBC subsidiary company carried out the building control function on the dwelling, acting as an approved inspector.

To enable people to fully understand the NHBC insurance document, which follows on pp. 8–14, a list of definitions is included; words with special meanings are printed in **bold** type and defined below.

*Definitions*

In the document, for clarity, 'you' and 'your' means the First Owner or a later Owner. NHBC and 'we' means the National House Building Council.

**Builder**   The Company or person named on the *Buildmark Offer* document who is responsible for the building or conversion of the *Home*.

**Buildmark**   The document containing the cover provided by NHBC and the *Builder*.

**Buildmark Offer**   The form which contains the offer of cover under *Buildmark* made by NHBC and the *Builder*.

**Common Parts**   Any of the following for which you are legally obliged to share responsibility for cost and upkeep with the *Owners* of other *Homes*:

(a)   The parts of a building containing a flat or maisonette.
(b)   Any garage, permanent outbuilding, retaining wall, boundary wall, external hand rail or balustrade, path, drive, garden area or paved area.
(c)   Any drainage system serving your *Home*.

Items in (b) are only included if they are newly built, or have been worked on by the *Builder* during conversion, at the date of the *Insurance Certificate*.

**Complete, Completion**   Legal completion of the first sale of the *Home* or, in the case of a *Home* being constructed under a building contract, the date when NHBC agrees that the *Home* substantially complies with *NHBC Requirements*.

**Contract**   A legal binding agreement or missive between the *First Owner* and the *Builder* for the purchase, building or conversion of the *Home*.

**Cost**   The cost we would have had to pay if we had arranged for the work to be done.

**Damage**   Physical damage to the *Home* caused by a *Defect*.

**Defect**   A breach of any mandatory *NHBC Requirement* by the *Builder* or anyone employed by him or acting for him. Failure to follow the guidance supporting the *NHBC Requirements* does not in itself amount to a *Defect*, as there may be other ways that the required performance can be achieved.

**First Owner, Owner**   The *First Owner* named on the *Buildmark Offer* and any later *Owner*. You must be (or have contracted to be) the freehold owner of the *Home*, or have a lease of at least 20 years (21 years in Scotland) of the *Home*.

**Owner**   Includes a mortgagee or heritable creditor in possession of the *Home*.

**Home(s)**   The house, bungalow, flat or maisonette referred to in the *Buildmark Offer*, together with any of the following which are included in the original *Contract*:

(a)   Any *Common Parts*;
(b)   The drainage system serving your *Home* for which you are responsible;
(c)   Any new heating system, air conditioning, smoke alarms, waste disposal units or water softening equipment newly installed at the date of the *Insurance Certificate*;
(d)   Any garage, permanent outbuilding, retaining wall, boundary wall, external handrail or balustrade, path, drive, garden area or paved area newly built, or worked on by the *Builder* during conversion, at the date of the *Insurance Certificate*.

**Home**   Does not include any fence, temporary structure, swimming pool, lift, or any electrical, electronic or mechanical equipment (whether built in or not) except the items listed in (c) above or which are necessary to comply with the Building Regulations. In Scotland it does not include any road, footpath or footway.

**Increased**   The figure of £500 was set in April 1999 and will be increased on 1 April each year in line with the Royal Institution of Chartered Surveyors' House Rebuilding Cost Index. The figure which applies to a claim is the one which was in force when the claim was first notified to NHBC.

**Insurance Certificate**   The certificate we issue on *Completion*, which brings Sections 2, 3 and 4 of this cover into operation.

**NHBC Requirements**   The mandatory *Requirements* we publish in the NHBC Standards which are in force either:

(a)   when the concreting of the foundations of a newly built *Home* or, if applicable, the *Common Parts* is begun; or
(b)   when conversion work affecting the *Home* or *Common Parts* is started.

**Original Purchase Price**   The amount notified to NHBC in the *Buildmark Offer*.

## Section 1 Cover before completion

### *Period of cover*

This part of the cover starts on the date you enter into the **Contract** and ends on the date shown on the Insurance Certificate.

### *Financial limits*

We will either pay for those losses and costs in the [left-hand] panel below or, at our option, arrange for the necessary work to be carried out at our expense. We will not pay for those losses and costs in the [right-hand] panel.

We will pay up to a total of £10 000 or 10% of the **Original Purchase Price**, whichever is greater.

| *What NHBC will pay* | *What NHBC will not pay* |
|---|---|
| If, due to his insolvency or fraud, the **Builder** does not start the **Home**, we will repay you the amount which you paid him under the **Contract** and which you cannot recover. | Anything not included in the original **Contract** with the **Builder**. |
| If, due to his insolvency or fraud, the **Builder** starts but fails to complete the **Home**, we will repay you the amount above, or pay you the extra cost above the **Original Purchase Price** for work necessary to complete the **Home** substantially in accordance with **NHBC Requirements**. | The cost of any work done by others without NHBC's written authorisation. |
| The **Cost** of the work that we have instructed the **Builder** to do, and which he fails to complete in accordance with **NHBC Technical Requirements** within the time set by NHBC. | Anything for which you have held back a sum of money. If you have done so, we will be entitled to deduct this amount from the sum that we would otherwise pay. If we carry out the work, you must pay us the amount before work starts. |

### *How to make a claim – what you must do*

Contact the NHBC office where your **Home** is located as soon as you believe the **Builder** might not complete the **Home** in accordance with the **Contract**.

Send us any evidence you have that the **Builder** is insolvent, or has acted fraudulently.

Get our written agreement before you take any action to have work carried out on your **Home** by anyone except the **Builder**.

If we ask for them, send us copies of any correspondence, contracts, plans, quotations, receipts and any other documents or information relating to your **Home**.

## Section 2 The first 2 years after completion

### The builder's obligations

This part of the cover tells you what the **Builder** must do if you give him written notice of **Defects** or **Damage** in your **Home**. This notice must be given as soon as possible within the period of cover.

The **Builder** must take the actions shown in the [left-hand] panel below, but he does not have to take action to deal with any of the items in the [right-hand] panel.

### Period of cover

This lasts for 2 years from the date of the Insurance Certificate.
There are special provisions for Common Parts.

| What the builder is liable for | What the builder is not liable for |
|---|---|
| Within a reasonable time and at his own expense, to put right any Defect or Damage to your Home or its Common Parts which is notified to him in writing within this period of the cover. | Wear and tear |
| | Deterioration caused by neglect or failure to carry out normal maintenance. |
| Any reasonable costs you incur, by prior agreement with the Builder, for removal, storage and appropriate alternative accommodation if it is necessary for you or anyone normally living in the Home to move out so that work can be done. | Dampness, condensation or shrinkage not caused by a Defect. |
| | Anything excluded by an endorsement by NHBC on the Insurance Certificate. |
| If he is given written notice of Defects or Damage within this period of cover, the Builder remains liable as above, even after this period of cover ends. | Anything caused by alterations or extensions to your Home after the date of the Insurance Certificate. |
| | Any Defect or Damage resulting from his compliance with written instructions given by or on behalf of the First Owner in respect of design, materials or workmanship. |
| | Any cost or expense greater than that necessary to carry out a workmanlike repair of the Defect or Damage. |
| | Any items falling outside the definition of Home. |
| | If you are not the First Owner, anything which you knew about when you acquired the Home and which resulted in a reduction in the purchase price you paid or which was taken into account in any other arrangement. |
| | Defects in multiple glazing panes in converted properties unless they were newly installed at the time of conversion. |

## The NHBC insurance

This part of the cover only applies if the Builder does not meet his obligations under Section 2.
We will either pay for the items in the [left-hand] panel below or, at our option, arrange for the necessary work to be carried out at our expense. We will not pay for items in the [right-hand] panel.

### Period of cover

This lasts for 2 years from the date of the Insurance Certificate.
There are special provisions for Common Parts.

### Financial limits

The most we will pay for all claims relating to the **Home** under Sections 2 and 3 together is the **Original Purchase Price** as shown on the **Insurance Certificate** up to a maximum of:

£500 000 for a newly built **Home**; or
£250 000 for a converted **Home**
(up to a total of £1 million for all the **Homes** in a continuous structure).

The financial limit will rise by 12% compound per year from the date of the **Insurance Certificate**. If we accept a claim, the financial limit will be reduced by the cost of each claim. The balance will then continue to rise by 12% compound per year.

The most we will pay for alternative accommodation, removals and storage is 10% of the financial limit at the time of the claim.

| What NHBC will pay | What NHBC will not pay |
|---|---|
| Any arbitration award or court judgment which you obtain against the **Builder** relating to obligations under Section 2 which he has failed to honour.<br><br>The **Cost** of any work contained in a Resolution Service report which is accepted by you and which the **Builder** does not complete or arrange to complete within the time set.<br><br>If the **Builder** is insolvent, the **Cost** of any work which he would otherwise have been liable for under Section 2. | Anything for which you have held back a sum of money. If you have done so, we will be entitled to deduct this amount from the sum that we would otherwise pay. If we carry out the work, you must pay us the amount before work starts.<br><br>Anything listed in the General Exclusions. |

### What you must do if you think there is a problem with your home

1   It is important to inspect your **Home** before and after you move in. The **Builder** is responsible for investigating any complaints and for putting right **Defects** or **Damage**.
2   Write to the **Builder** informing him of any items requiring attention as soon as you notice them. You must keep copies of all your letters as you may need these later to prove that problems were reported in the first 2 years.
3   If the **Builder** does not deal with your complaint to your satisfaction contact the NHBC office for the area where your **Home** is located. We will usually offer our Resolution Service.
4   Tell us if the **Builder** is insolvent and give us the opportunity to inspect your **Home**.
5   If we ask for them, send us copies of any correspondence, contracts, plans, quotations, receipts and any other documents or information relating to your **Home**.

### The Resolution Service

If there is a disagreement about the **Builder**'s obligations, we will usually try to resolve matters under our Resolution Service.

When we offer our Resolution Service, we will investigate any **Defects** or **Damage** which you have complained to the **Builder** about and which he has not put right within a reasonable time. We may need to visit your **Home**. We will then issue a report informing both you and the **Builder** of any work that he must carry out to fulfil his obligations under this Section.

The **Builder** must carry out the work within a reasonable period of time, which will be set by NHBC. You must allow the **Builder** reasonable access during normal working hours to carry out the work.

If the **Builder** does not carry out the work within the time set and has not agreed a programme with you to complete the work, we will, at our option, pay the **Cost** of the work detailed in our report or arrange for the work to be done.

If you disagree with our Resolution Service report there are other ways of resolving your dispute with the **Builder**. These are explained in the complaints and disputes procedures listed later. Please note that the Insurance Ombudsman Bureau cannot assist if you disagree with our Resolution Service report, as they can only deal with complaints regarding our insurance cover.

We have no liability under this Section unless we have issued a Resolution Service report which you have accepted or unless the **Builder** is insolvent or has failed to honour an arbitration award or court judgment.

### *Important note*

We will normally offer our Resolution Service. However, we can only help with disputes about **Defects** or **Damage**. We will not be able to help if you have a dispute about such matters as financial or contractual issues or boundary disputes. In these circumstances we will suggest you consider another type of dispute resolution procedure.

## NHBC regional offices

**Scotland**
42 Colinton Road, Edinburgh EH 10 5BT
Tel: (0131) 313 1001
Fax: (0131) 313 1211

**North**
Buildmark House, George Cayley Drive,
Clifton Moor, York YO30 4GG
Tel: (01904) 691666
Fax: (01904) 690474

**South East**
Buildmark House, London Road, Sevenoaks, Kent
TN13 1DE
Tel: (01732) 740177
Fax: (01732) 740978

**South West**
6–7 Clevedon Triangle Centre, Clevedon,
Somerset BS21 6HX
Tel: (01275) 337500
Fax: (01275) 337501

**Northern Ireland and Isle of Man**
Holyrood Court, 59 Malone Street, Belfast BT9 6SA
Tel: (028) 9068 3131
Fax: (028) 9068 3258

**East**
Buildmark House, Boycott Avenue, Oldbrook,
Milton Keynes, Bucks MK6 2RN
Tel: (01908) 691888
Fax: (01908) 678575

**West and Wales**
Buildmark House, 1–3 Roman Way Business
Centre, Droitwich, Worcs WR9 9AJ
Tel: (01905) 795111
Fax: (01905) 795116

**Amersham**
Buildmark House, Chiltern Avenue, Amersham,
Bucks HP6 5AP
Tel: (01494) 434477
Fax: (01494) 735201

## Section 3 Cover in years 3 to 10

Under this part of the cover, you must tell NHBC of your claim as soon as possible within the period of cover.

We will either pay for the items in the [left-hand] panel or, at our option, arrange for the necessary work to be carried out at our expense. We will not pay for the items in the [right-hand] panel.

There are special provisions for **Common Parts**.

### *Period of cover*

This starts 2 years after the date shown on the **Insurance Certificate** and ends 10 years after the date shown on the **Insurance Certificate**.

### Financial limits

The most we will pay for all claims relating to your **Home** under Sections 2 and 3 together is the **Original Purchase Price** as shown on the **Insurance Certificate** up to a maximum of:

£500 000 for a newly built **Home**; or
£250 000 for a converted **Home**
(up to a total of £1 million for all the **Homes** in a continuous structure).

The financial limit will rise by 12% compound per year from the date of the **Insurance Certificate**. If we accept a claim, the financial limit will be reduced by the cost of each claim. The balance will then continue to rise by 12% compound per year.

    The most we will pay for alternative accommodation, removals and storage is 10% of the financial limit at the time of the claim.

### What NHBC will pay

A  The full **Cost**, if it is more than £500 as **Increased**, of putting right any actual physical **Damage** to the **Home** caused by a **Defect** in the following parts of the house, bungalow, maisonette or flat and its garage or other permanent outbuilding, or its **Common Parts**:
  * Foundations
  * Load-bearing walls
  * External render and external vertical tile hanging
  * Load-bearing parts of the roof
  * Tile and slate coverings to pitched roofs
  * Load-bearing parts of the floors
  * Floor decking and screeds, where these fail to support loads
  * Retaining walls necessary for the structural stability of the house, bungalow, flat or maisonette, its garage or other permanent outbuilding
  * Multiple glazing panes to external windows and doors
  * Below-ground drainage for which you are responsible

B  The **Cost** of putting right any **Defect** in a flue or chimney which causes a present or imminent danger to the physical health and safety of anyone normally living in the **Home**

C  Any reasonable costs you incur by prior agreement with us for removal, storage and appropriate alternative accommodation if it is necessary for you or anyone normally living in the **Home** to move out so that work can be done.

### What NHBC will not pay

Any claim under **A** where the **Cost** of repair is £500 or less, as **Increased**.

**Damage** caused by shrinkage, thermal movement or movement between different types of materials.

**Damage** which is purely cosmetic, such as minor cracking, spalling or mortar erosion to brickwork, which does not impair the structural stability or weather tightness of the **Home** or which only affects decorations.

Any **Defect** in existing multiple glazing panes in converted properties unless they were newly installed at the time of conversion.

Anything which was or could have been reported to the **Builder** under Section 2. For these claims, please see Section 2 of the policy.

### How to make a claim – what you must do

Contact the NHBC office for the area where your **Home** is located as soon as the damage has been noticed. Give us the opportunity to inspect your **Home** before any work is done.

If we ask for them, send us copies of any correspondence, contracts, plans, quotations, receipts and any other documents or information relating to your **Home**.

We might ask you to pay a fee before investigating your claim. This will be refunded if your claim is valid or if we think it was reasonable for the claim to have been made.

# Section 4 Additional cover in years 3 to 10 if NHBC's subsidiary did the building control

This part of the cover only applies if NHBC Building Control Services Limited has done the building control. The **Insurance Certificate** will show if this applies to your policy. It only applies in England and Wales.

We will either pay for the items in the [left-hand] panel or, at our option, arrange for the necessary work to be carried out at our expense. We will not pay for the items in the [right-hand] panel.

## *Period of cover*

This starts 2 years after the date shown on the **Insurance Certificate** and Building Control Final Certificate and ends 10 years after the date shown on the **Insurance Certificate** and Building Control Final Certificate.

There are special provisions for **Common Parts**.

## *Financial limits*

The financial limit for a claim under this section is the original cost of the work covered by the NHBC Building Control Services Limited Final Certificate.

The financial limit will rise by 12% compound per year from the date of the **Insurance Certificate**. If we accept a claim, the financial limit will be reduced by the cost of each claim. The balance will then continue to rise by 12% compound per year.

The most we will pay for alternative accommodation, removals and storage is 10% of the financial limit at the time of the claim.

| *What NHBC will pay* | *What NHBC will not pay* |
|---|---|
| Repairs needed where there is a present or imminent danger to the physical health and safety of the occupants of the **Home** because the **Home** does not comply with the requirements of the Building Regulations that applied to the work at the time of construction or conversion in relation to the following: | Anything which we will pay for under another Section of this policy |
| | For claims that were referred to the **Builder** in the first 2 years, please see Section 2 of the policy. |
| • Structure<br>• Fire safety<br>• Site preparation and resistance to moisture<br>• Hygiene<br>• Drainage and waste disposal<br>• Heat-producing appliances<br>• Protection from falling, collision and impact<br>• Glazing – safety in relation to impact | |
| Any reasonable costs you incur, by prior agreement with us, for removal, storage and appropriate alternative accommodation if it is necessary for you or anyone normally living in the **Home** to move out so that work can be done. | |

# General conditions for claims to NHBC

1   If we accept any claim for which you could recover compensation from some other person, you must, at our expense, do whatever we may reasonably require:

(a) to recover compensation from that person for our benefit; or

(b) to enable us to enforce any rights you may have to that compensation by taking over your claim against that other person or in any other way.

2   You must take all reasonable steps to reduce damage. We will not pay for any work or other costs which result solely from your failure to do this.

## Important note

It is illegal to make a fraudulent claim.

# General exclusions for claims to NHBC

## NHBC will not be liable for

a   Any cost, loss or liability which is provided for by legislation or which is covered by any other insurance policy.

b   Anything excluded by an endorsement by NHBC on the **Insurance Certificate**.

c   Anything affecting or caused by alterations or extensions to the **Home** carried out after the date of the **Insurance Certificate**.

d   Any **Defect** or **Damage** resulting from compliance by the **Builder** with written instructions given by or on behalf of the **First Owner** in respect of design, materials or workmanship.

e   Wear and tear

f   **Damage** caused by neglect or failure to carry out normal maintenance.

g   Dampness, condensation or shrinkage not caused by a **Defect**.

h   Any **Defect** or **Damage** caused by the installation or presence of a swimming pool or lift.

i   Loss of enjoyment, inconvenience, distress or any other consequential loss affecting you or any loss of value to your **Home**.

j   Any professional fees except those reasonably incurred with our specific written consent. (Note – we may in our absolute discretion waive this exclusion if we accept a claim which we had at first rejected.)

k   Costs or expenses greater than would have been paid or incurred by a reasonable person in the position of the **Owner** spending his or her own money.

l   Costs which have already been taken into account by NHBC or by the **Builder** when making payment to or carrying out work for a previous **Owner**.

m   Costs which are attributable to your unreasonable delay in pursuing a claim.

n   If you are not the **First Owner**, anything which you knew about when you bought the **Home** and which resulted in a reduction in the purchase price you paid or which was taken into account in any other arrangement.

o   Replacement of any undamaged item solely because another item of the same nature, design or colour has to be replaced and the original items cannot be matched.

## Points to remember

- Damage to a building can result from poor forms of construction and from subsidence, and these situations can prove difficult to differentiate.
- Subsidence or heave caused by existing trees or trees removed can be very complex claims to deal with. Often they can be resolved only by long-term monitoring followed by appropriate remedial works.
- The natural geology may have a bearing on any subsidence. Check for geological faults; check whether there is any active mining under or adjacent to the building.
- Deep excavations can allow the ingress of water into soluble deposits such as gypsum or chalk. Check for the presence of soakaways around the building.
- Subsidence damage caused by solution features in chalks, gypsum or limestone pose a serious problem. Because the solution feature is generally an erratic incident and is relatively small in area it is difficult to locate accurately, and can be missed by even a thorough site investigation. The damage to the building is often the first indication of the problem.

- Behaviour of clay soils: Differential movement is characteristic of defects associated with clay shrinkage. The water content in the ground adjacent to the main external walls may drop, whereas the clays beneath the floor of the building internally remain unaffected. The external walls tend to settle, resulting in cracks to the masonry and in out-of-plumb walls.
- The bearing capacity of ground that has been made up by uncontrolled end-tipping is often poor and usually very variable. The made ground may consist of mixed materials, often organic, which means that they can vary greatly in volume as their water content changes. Materials such as peat and soft alluvium placed as backfill are very compressible, and settle under their own weight by a considerable magnitude. Placing any building over such ground requires extreme caution and good, sound engineering judgement, if large differential settlements are to be avoided. Often the only sensible option is a piled foundation.

## References

### National House-Building Council

*Good House Building Newsletter*, No. 6 (1984).
*Buildmark Document* (April 1999).

### Institution of Structural Engineers

*Subsidence of Low-Rise Buildings* (1994).

Powell, M.J.V. (ed.) *House Builders Reference Book*, Newnes Butterworth (1979).

# Chapter 2
# Surveys

## 2.1 Introduction

In determining the cause of major structural failure of a building a systematic and thorough approach should always be used in the diagnostic process. The aim of the initial survey should be first and foremost to establish the stability of the building. If damage has occurred that has the potential for a partial or complete collapse of the building then the occupants must be moved out immediately until the building has been made safe.

It is most important not to prejudge or allow oneself to be influenced by other parties or reports. For this reason an investigator should always inspect a dwelling and try to form his or her own opinions before reading other reports. Any discussions that may be necessary with other professionals should be held following the initial inspections.

The most common causes of structural failure, which can occur singularly or in a combination, generally result from the following:

- Settlement: a failure to accurately quantify the loadbearing capacity of the soils on which the building rests. Such a failure can result in cracking and distortions to the main walls. If the loadbearing capacity of the underlying soils varies considerably, greater movements generally reflect the differential situation.
- Subsidence: the effects of ground below the foundations of the building moving away from the foundation as a result of external influences such as mineral extraction, dewatering, solution features in chalk or gypsum strata.
- Poor structural design and inadequate detailing.
- Substandard materials and a failure to recognise or observe the particular requirements for proprietary materials.
- Bad workmanship.

Detailed examination of structural failures generally reveals a combination of the above factors as having contributed to the overall failure evident.

The usual causes of structural failures are inadequate foundations and poor ground conditions, followed by poor design and bad workmanship.

## 2.2 The survey report

A prerequisite for the diagnosis of structural damage to a building is a thorough, detailed and well-documented survey report carried out to a set pattern. The importance of this survey report cannot be overemphasised. It is often the only record of what may have been a long and complex survey, and it must contain all the information relevant to the survey, written up in a clear, concise manner with reasoned and logical explanations for the defect causations. It should also conclude with recommendations for putting right the defects and, if possible, an estimate of likely costs in carrying out the works.

Decisions must be made prior to carrying out the survey in order to define the objectives: e.g. where trial pits are required, what equipment will be needed, what the range of sampling will be. During the course of the survey a constant review should be carried out to ensure that the necessary information required is being obtained and is adequate to achieve the main objectives. In the review investigators should ask whether or not they are obtaining as much information as they might do. If they consider that the information obtained is insufficient then they should recommend that additional work be carried out.

Careful observations and additional monitoring or sampling will produce definite benefits. Detailed notes should be made, with relevant sketches of the damaged areas. In-situ tests may be required to determine the soil strengths, and these should be well documented to enable the various possible causes to be eliminated.

It may not be patently obvious whether the building is heaving or subsiding, and therefore it may be necessary to instigate a full plumb and level survey of the building as the start of a long-term monitoring process, together with the installation of tell-tales to record crack movements.

The diagnosis of the causes of structural failure is not always straightforward. Each defect is unique, and may be the result of a combination of factors. Before a solution can be provided the causation must be determined. A correct diagnosis will enable one to prepare a suitable remediation scheme that is both practical and economic, and is correct first time. Owners soon lose confidence when remediation works fail.

The normal order of producing a survey report is to collect all the information and prepare a draft report. This draft report should be reviewed, edited and rewritten where required. The investigator should then carry out a critical appraisal of the report to ensure that the original objectives have been met, and that these have been recorded and stated in a clear, precise and logical manner.

Following on from this stage the Investigator can then polish the report and make any adjustments to improve its presentation and make it more user friendly.

## 2.3 Survey equipment

To carry out a comprehensive investigation, an Investigator must be properly equipped. Table 2.1 shows the equipment checklist for carrying out structural surveys. Not all of the equipment listed will be required on every survey, but on certain jobs additional items will be required. It is a good practice to tick those items required and cross them on the checklist when assembled.

### 2.3.1 Note-taking

A good-quality A4 lined pad and folding clipboard for sketches, together with a shorthand pad, is the preferred equipment. For making notes a pencil is preferable to a ballpoint pen, especially in wet weather.

### 2.3.2 Measuring equipment

For measuring, a 3 m or 5 m flexible steel tape is essential, together with a 2 m folding rule for use indoors. A 300 mm steel ruler graduated in millimetres is useful for measuring in confined spaces.

*Table 2.1* Structural survey equipment checklist

| | | | |
|---|---|---|---|
| ☐ | Folding clipboard | Sectional ladder | ☐ |
| ☐ | Sample jars and sample bags | Hand auger equipment | ☐ |
| ☐ | A4 lined pads | Water level | ☐ |
| ☐ | Compass | Torch and batteries | ☐ |
| ☐ | 3 m tape | Hand lamp | ☐ |
| ☐ | 30 m tape | 300 mm steel rule | ☐ |
| ☐ | String line and blocks | Scale rule | ☐ |
| ☐ | Plumb bob and line | Small trowel | ☐ |
| ☐ | Transparent plastic ruler | Penknife | ☐ |
| ☐ | Dumpy level, tripod and staff | Masking tape | ☐ |
| ☐ | Builder's 1.0 m level | Spade | ☐ |
| ☐ | 3 m straightedge | Electric drill | ☐ |
| ☐ | Bradawl | Screwdriver | ☐ |
| ☐ | Pocket penetrometer | Vernier calipers | ☐ |
| ☐ | Shear vane apparatus | Overalls | ☐ |
| ☐ | 35 mm camera and films | Pencils | ☐ |
| ☐ | Avongard tell-tales | Pocket level | ☐ |
| ☐ | Hammer and chisel | | |

A similar ruler in transparent plastic is very useful for measuring crack widths. For measuring up external walls, or distances to adjacent buildings, a 30 m linen or plastic tape is useful.

Where Avongard measuring devices have been installed for long-term monitoring, these should be measured with extending vernier calipers (Fig. 2.1).

A dumpy level, tripod and extending staff are useful when carrying out plumb and level surveys and ground surveys (Fig. 2.2).

For checking the relative levels of dampcourses and floors a water level is very useful and accurate, and it can be used by one person.

A 1.0 m builder's spirit level, 3 m timber straightedge and a plumb-bob and line are also very useful for checking floors and walls for level and plumbness.

### 2.3.3 Probing equipment

For testing the soundness of existing timbers, checking for wood rot and beetle infestation, scraping mortar joints and probing the depths of cracks, a sharp metal probe is essential. A bradawl or sharp-ended screwdriver will suffice.

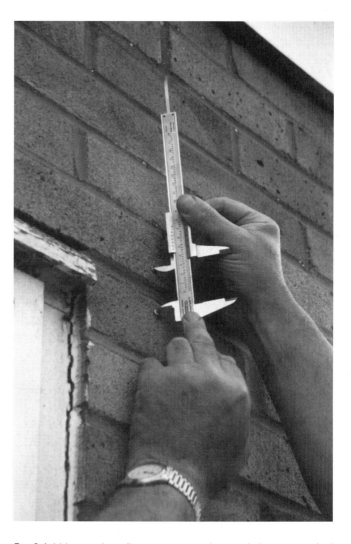

*Fig. 2.1* Using vernier calipers to measure Avongard pips over cracked masonry.

*Fig. 2.2* Using a precise automatic level to carry out a plumb and level survey.

### 2.3.4 Access equipment

A sectional ladder is essential for gaining access to lofts, carrying out plumb monitoring etc. The sectional aluminium type in short 1.0 m lengths extending to 3 m is the most useful.

### 2.3.5 Additional items

These include a good pair of small binoculars or a folding telescope for examining inaccessible roof slopes and high walls, and a small hand-held compass.

An automatic 35 mm camera with flash attachment and fast colour film is very useful for recording cracked walls etc., and with the word-processing technology now available it is easy to incorporate photographs into the report alongside the relevant text, so making the report user-friendly. The use of a digital camera will allow the photographs to be imported into the word processor when compiling the report.

A hand-held lamp and a strong torch with spare batteries are required for examining under floors, in roof spaces etc. Where a major roof examination is to be carried out it would be preferable to use a wire-guarded lamp connected to the mains using a long lead.

*Fig. 2.3* Using a borescope to check wall ties in a cavity wall.

For surveying masonry walls a borescope, as shown in Fig. 2.3, can be very useful for examining the cavity to check on mortar build-up, location of wall ties, positions of dpc and cavity trays over lintels, etc.

Another useful tool is a moisture meter. This can be used for checking on the moisture content of timber floors, checking for the presence of damp, etc. Suitable moisture meters can be purchased from Protimeter Ltd, Marlow, Bucks SL7 1LX.

## 2.4  Desk study

During an investigation of a damaged property it may not be obvious what has caused the damage, and even examination of the foundations and ground conditions may not produce a clear causation. It will therefore be necessary to extend the investigation by means of a desk study.

The results from a desk study should assist in deciding on the best type of site investigation to be carried out, and for use in the design of remedial measures. The purpose of a desk study is to collect as much factual and historical information as possible on the building being examined, and to obtain information about the site, the surrounding land and its former usage. The most useful sources of information on developed sites can be found from the following:

- local authority archives;
- Coal Authority (old coal records, mine shafts, open-cast mining);
- old Ordnance Survey plans;
- British Geological Survey and local geological memoirs;
- aerial survey photographs;
- local knowledge from the local people;
- any existing site investigations;
- various statutory bodies, e.g. water authorities;
- regional mineral valuers (the Inland Revenue hold vast records on mineral workings etc.);
- builder or contractor for the building;
- architect for the building;
- National House Building Council;
- public libraries;
- Railtrack Ltd (construction shafts for tunnels);
- Department of the Environment, Transport and the Regions.

Useful information concerning the former land usage, geology, type of existing services and the general topography can be obtained from the above sources at a relatively low cost.

### 2.4.1  The site

*Green field.*

Potential problems:

- clay heave;
- trees;

- poor natural ground conditions;
- sloping ground;
- mining, past and active;
- streams;
- activities on adjoining land;
- high water table.

*Former uses of site.*

Potential problems:

- previous industrial development;
- existing substructures/foundations;
- old quarries;
- opencast mining;
- reclaimed land;
- variable quality of filled land;
- old landfill sites;
- contaminated land.

Old Ordnance plans indicate old ponds, previous hedge lines and old watercourses. Evidence of previous developments – old industrial mill buildings for instance – can be overlain over the building being investigated.

Examination of geological maps and local memoirs provides information on mineral workings, mineshafts, coal outcrops and strata succession of the drift materials.

Aerial survey photographs give valuable information on previous developments, old railway lines or railway cuttings, and previous trees that existed prior to development. In addition old landslips, mineshafts etc. can also be detected.

Consultation with the Coal Authority is always wise in known mining areas. They are the custodians for records of coalmining, and have many detailed records of the local coal field, mineshaft positions and records and plans of opencast coal workings. There should also be records relating to any mining subsidence claims in the locality of the property being investigated.

Railtrack Ltd holds records of old underground railway tunnels, with locations of the construction shafts used during their construction.

Regional mineral valuers collect records for the Inland Revenue, and these records extend to all mineral extractions, Elland Flag workings and quarry workings.

### 2.4.2  Desk study checklist

See box on following page.

## 2.5  Geological maps and memoirs

An understanding of the geology below a building is a fundamental requirement in determining the cause of a subsidence-related defect. The British Geological Survey (BGS) at Keyworth, Nottinghamshire, is the national repository of geological records. Among the records held are the field observations used to produce the maps including approximately 250 000 records relating to mineshafts and boreholes drilled.

*Site topography*

Salient vegetation and drainage.

1  Are there any springs, ponds or water courses on or near the site?
2  Is the site steeply sloping?
3  Are there any signs of previous tree growth on site?

*Ground conditions*

1  Is the site in a known mining area for coal or other minerals?
2  What is the geological strata succession below the site?
3  Are the clay soils in the high-plasticity range?
4  Is there any evidence of slope instability?
5  What data are available on soil strengths?

*Previous development*

1  Type of development?
2  Size of development?
3  Existing services?

*Identification of ground conditions*

By consulting geological records and maps a lot of information can be obtained about the soil conditions below the surface:

1  type of drift materials, e.g. sands, clays shales, mudstones;
2  thickness of various strata bands – Usually indicated on borehole records and geological maps;
3  positions of any old mineshafts, geological faults, old backfilled opencast coal workings, whinstone dykes and buried glacial valleys.

---

The BGS will provide geological information in response to a request in writing, but they make a charge for this service. When requesting information, a copy of an Ordnance plan should be sent with the building or site location indicated.

The maps published by the BGS show the solid geology (main strata), and there are also maps showing the drift (superficial deposits).

The main scale of the maps is now 1 : 10 000 scale. Many of the maps produced since the beginning of the twentieth century were based on the old scale of 6 in to one mile (i.e. 1 : 10 560). There are other scaled old maps available for examination in the BGS archives.

These plans can be supplemented by detailed geological memoirs for the area being considered. There are 18 regional handbooks of the geology of Great Britain available. The geological maps show the distribution of geological formations printed over the topographical base maps, and it can be seen that they give a good indication of the structure and types of strata occurring in particular areas.

In recent years the BGS and the Department of the Environment have produced engineering geological maps of various major cities, such as Belfast, Leeds, and Bradford. These maps show old quarries, opencast mines, areas of infilled ground, mineshafts and coal seams.

### 2.5.1 Superficial formations (drift)

These maps generally indicate the following geological features:

- landslip areas;
- blown sand deposits;
- peat;
- alluvium, lacustrine and floodplains;
- river terrace deposits;
- marine or estuarine alluvium;
- buried glacial channels;

- morainic drift deposits;
- glacial sand and gravel;
- boulder clays and glacial drift;
- faults.

### 2.5.2 Solid formations

These maps generally indicate the following geological features:

- Permo-Triassic;
- mudstones, sandstones, Upper Permian marls, gypsum beds;
- magnesian limestones, basal Permian sands;
- Carboniferous:
  - middle coal measures – mudstones, shales, sandstones, coal seams;
  - lower coal measures – mudstones, shales, coal seams, sandstones;
  - Millstone Grit series – mudstones, shales, sandstones, limestones;
  - faults;
- intrusive igneous:
  - quartz-dolerite dykes and sills of late Carboniferous or early Permian age;
  - faults.

### 2.5.3 Ordnance Survey maps

These can be obtained from Ordnance Survey, Romsey Road, Southampton, or viewed in any public library archives division. The large-scale maps are produced at 1 : 1250 scale covering the major urban areas and 1 : 2500 covering the remainder of the UK.

Maps at the old 6 in to 1 mile are available, but these are being replaced with maps at 1 : 10 000 scale. The copying of such maps for professional purposes will be subject to copyright licensing terms.

The examination of such plans can provide a lot of information about a site and its past history.

Most large public libraries have plans for the local area extending back many years. Such features as old quarries, old watercourses, mill dams, wooded areas, old industrial buildings, railway cuttings and embankments, and previous housing are shown. By going through them in historical order it is possible to build up a picture of the previous land use for the site being examined and the surrounding area.

## 2.6 Site reconnaissance

At an early stage, a thorough visual examination of the land surrounding the buildings being examined should be carried out. This reconnaissance should where possible extend beyond the physical boundaries of the site and pick up relevant features that may have had an influence, i.e. trees, mine shafts, land slips etc. Table 2.2 shows a technical questionnaire survey sheet, which should be used as an *aide-mémoire* when carrying out the walk-over survey. The walk-over survey around the area may reveal old or existing railway cuttings in which the soil stratification is evident to see. There may be other buildings around the site in question that are showing signs of structural distress as a result of weak ground or past mining movements. Figures 2.4–2.9 indicate salient features to look out for.

There may be a vacant site in close proximity, or the site itself may be vacant and surrounded by intensive development. In such circumstances it is always wise to

*Table 2.2* Technical questionnaire survey sheet

| Site Address: Water Lane, York | Date visited: 14.3.96 |
| --- | --- |
| 1. Give description of the site | Pasture land |
| 2. Give directions and gradients of slopes. Are there any indications of hill creep? | Site fairly level, no signs of any instability |
| 3. Presence of trees, streams, marshy areas etc. | Yes. A large copse of poplar trees in north-eastern corner |
| 4. Are there any soil exposures in trenches, nearby railway cuttings etc.? | Yes. Deep sewer excavation adjacent to site shows clays become stiffer and laminated |
| 5. Is there any indication that the site has been filled? | No signs evident |
| 6. Has the site been previously developed? Give indications of likely foundations. | There were old terraced houses on the site, and foundations will be shallow |
| 7. Are any old basements present? If so give details of depth and extent. | No |
| 8. Have you taken ground levels and obtained the location of trial pits and boreholes? | Yes |
| 9. Are there nearby buildings? Any signs of cracking? | Yes. No visible signs of damage. The public house is suffering from damage due to a nearby willow tree |
| 10. Is the site in a mining area for coal or other minerals? | No |
| 11. Have you obtained all information on the ground conditions from the trial pits? | Yes |
| 12. Have you obtained shear strengths of the soils using a shear vane? | Yes |
| 13. What samples have been taken? | 6 bulk samples for PI classification |
| 14. Have groundwater samples been taken? From which trial pits? | No groundwater encountered |

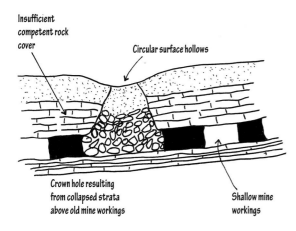

Fig. 2.4 Crown holes.

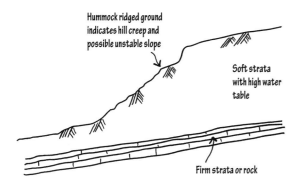

Fig. 2.5 Slope stability.

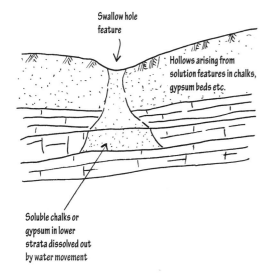

Fig. 2.6 Solution features.

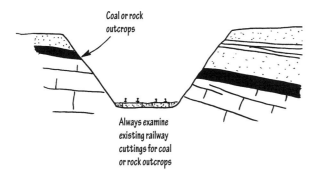

Fig. 2.7 Coal outcrops.

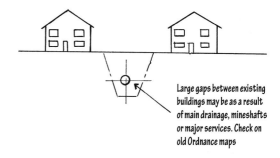

Fig. 2.8 Gaps in existing development.

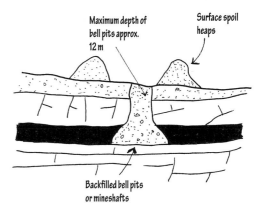

Fig. 2.9 Bell pits.

ask why that piece of land has not been developed. Often such sites have been infilled valleys that have a main drain down the centre, so making future development difficult, or there may have been a previous quarry on the site, subsequently backfilled.

When carrying out a site reconnaissance, whenever possible have some relevant district maps, geological maps and aerial photographs. Any changes will then be easily identified. Always ensure that permission to walk over adjoining land has been sought from the landowners or the occupier.

Where it is apparent that changes to the site have taken place, it is useful to refer back to the local authority engineers'/surveyors' departments, and also to talk to local inhabitants. Often longstanding residents of the area can remember various events that took place in the past, such as mass burials of animals, filling in of a mineshaft, or the location of a factory millpond. All such information should, however, be used with caution, as on occasions people's memories can be inaccurate.

To obtain general information on a site or building:

1  Walk the whole site, making careful observations and notes.
2  Observe and record any changes from the plans.
3  Carry out a cursory inspection of any surrounding buildings, and record any significant points such as cracking, or age of buildings.
4  Record any obstructions, such as trees, telephone or electric poles, gas and water mains and any main drainage and manholes.

5 If adjacent to a river or canal, check on the condition of the banks. Have they been raised to prevent flooding? Is the river tidal?
6 If underpinning works are likely to be proposed, check the buildings closest to the one to be repaired.

Points to consider when inspecting a site:

● Hill slopes which appear to be wrinkled may indicate that the land is prone to landslips. Any trees leaning over on a slope may be an indication of creep taking place in the upper strata.
● Unusual areas of green vegetation, reeds, rushes and willow trees generally mean that the underlying ground is wet or poorly drained.

● Site is sloping: will retaining walls be required?
● Site bouncy underfoot: indicative of high water table.
● Surface hollows: may be solution features in chalk or gypsum.
● Surface cracking: indicative of clay plasticity.
● Street names – e.g. Coal Pit Lane, Spring Rise, Quarry Road – indicate previous usage.

*Note*: Always bear in mind that a site examined in the summer months could be completely different in autumn and winter owing to groundwater variations and other seasonal climatic factors.

## Points to remember

● Always carry out an initial inspection of the building with the defects, and examine the surrounding area.
● If necessary do a walk-over survey.
● Talk to local inhabitants to obtain information about previous land use of the site.
● Observe and record any changes to the site that are different from the plans or maps.
● Always check the public library archives for old Ordnance plans.
● Make a note of unusual street names such as Quarry Road, Spring Lane.
● Take into account the potential for groundwater levels to change with the seasons.

● Do not allow yourself to be influenced by other reports until you have done your survey.
● Always make sure you have the right equipment to complete the survey.
● In coalmining areas always consult with the Coal Authority for information on past or active mining.
● Always check the geological maps for possible geological faults, buried glacial channels, slope instability, old opencast workings, coal outcrops etc.
● Make a note of any large mature trees and hedges.

## Useful addresses for desk study information

National House Building Council regional offices (see Chapter 1)
British Geological Survey, Knicker Hill, Keyworth, Nottinghamshire NG12 5JJ. Tel. (0115) 936 3100
Coal Authority Mining Records, Bretby Business Park, Ashby Road, Burton-upon-Trent, Staffordshire DE15 0RD. Tel. (01283) 553463
Coal Authority Mining Reports, 200 Litchfield Lane, Berry Hill, Mansfield, Notts NG18 4RG. Tel. (01623) 638364
Topographical maps: Ordnance Survey, Romsey Road, Maybush, Southampton SO16 4GU. Tel. (023) 8079 2000
Regional Mineral Valuers (contact the Inland Revenue for the local addresses)
Old mines and quarries: Department of the Environment, Transport and the Regions, 76 Marsham Street, London SW1P 4DR. Tel. (020) 7890 3856
Railway tunnels, construction shafts etc.: Railtrack Ltd, Engineering Records, Railtrack House, Euston, London NW1 2EE. Tel. (020) 7557 8000
Aerial surveys
England: Air Photo Unit, Enquiry & Research Services, NMRC Kemble Drive, Swindon, Wilts SN2 2GZ. Tel. (01793) 414716

Scotland: Air Photographs Officer, Royal Commission for Ancient & Historical Monuments of Scotland, 16 Bernard Terrace, Edinburgh EH8 9NX. Tel. (0131) 662 1456
Wales: Air Photographs Officer, Central Register of Air Photography for Wales, The National Assembly for Wales, Room G003, Crown Offices, Cathay Park, Cardiff CF10 3NQ. Tel. (029) 2082 5111
Arboriculture Advisory and Information Service, Alice Holt Lodge, Wrecclesham, Farnham, Surrey GU10 4LH. Tel. (01420) 22022
Building Research Establishment, Bucknalls Lane, Garston, Watford WD2 7JR. Tel. (01923) 664000
The Subsidence Adviser, Belgrave Centre, Talbot Street, Nottingham NG1 5GG. Tel. (0115) 971 2776
South Yorkshire coalfield records: South Yorkshire Mining Advisory Services, Barnsley Metropolitan Borough Council, Central Offices, Kendray Street, Barnsley S70 2TN. Tel. (01226) 772684
Cheshire Brine Subsidence Compensation Board, Richard House, 80 Lower Bridge Street, Chester CH1 1SW. Tel. (01244) 602576
Abandoned mines (other than coal), tips and quarries: The Health & Safety Executive hold records at the Abandoned Mines Records Office, Regina House, 259–269 Old Marylebone Road, London NW1 5RR. Tel. (020) 7717 6000

# References

*Building Research Establishment*

*Site investigation for low-rise buildings. Desk study*, BRE Digest 318 (February 1987).

*British Standards Institution*

BS 5930 : 1999 *Code of practice for site investigations.*

*Institution of Civil Engineers*

*Ground subsidence*, Thomas Telford, London (1977).

*Institution of Structural Engineers*

*Subsidence of low-rise buildings* (1994).

*National House-Building Council*

*NHBC Standards* (1999):
    Chapter 4.1, Land quality – Managing ground conditions.

M.F. Atkinson, *Structural foundations manual for low-rise buildings*, E & FN Spon (1993).
M.J. Tomlinson, *Foundation design and construction*, 5th edn, Longman Scientific and Technical (1993).

# Chapter 3
# Site investigation equipment and methods

## 3.1 Introduction

Investigations of major structural defects are usually carried out in two stages:

1 an initial examination and appraisal to confirm the structural damage, and to decide whether that damage justifies the further expense of detailed site investigation and testing;
2 a further investigation, if necessary, to determine the cause of the structural movement, and to provide sufficient information to decide on the likely remediation works.

This further investigation may involve excavating trial pits, opening up the fabric of the building, drilling deep boreholes, or setting up long-term monitoring gauges. Such works should be carried out by a qualified structural or civil engineer, or a chartered surveyor.

Once the further field work has been carried out, the information obtained, together with the desk study data, should assist in the determination of the cause of the damage. Table 3.1 lists the appropriate site investigation techniques.

## 3.2 Initial investigation

The initial survey should record any structural cracking. If the cracking evident is minor, it is possible that this is due to initial settlement having taken place as the building loaded the ground below the foundation. In such cases it is prudent to carry out a plumb and level survey and fit some Avongard tell-tales over the cracks and monitor the building for approximately 6–12 months.

If the monitoring and level survey show no appreciable changes then it is unlikely that the cracking will be progressive.

## 3.3 Trial pits

Where the cracking in the building fabric is found to be excessive and progressive it will be necessary to excavate some trial holes in various places to determine the nature

*Table 3.1* Site investigation techniques

| Technique | Applicability |
|---|---|
| Direct field observation | Observation of landslip, mining subsidence. Where coal mines are exposed in opencast operations they should be entered and surveyed only by specialist mining personnel |
| Trial pit investigation, trenching | A quick and cost effective technique. For examining the in-situ density and nature of filled ground it is probably the best method. Pits can be hand dug if access is poor. With machines the depths of pits are limited to about 6 m |
| Cable percussion borings | This is the standard method for drilling soft rocks and soils. The tripod rig is easily set up, and is very mobile. This method also allows for in-situ testing of the soils |
| Rotary boreholes | Used for investigating through rock strata and for mining investigations using compressed air or water flushing techniques |
| Rotary percussion drilling | A quick and relatively cheap open-hole drilling technique, more suited to hard rocks |
| Dynamic probing | A fast technique involving hammering a rod into the soils or soft rocks to establish the relative resistance. Can be useful when investigating filled ground overlying natural strata as it can provide good correlation |
| Ground surface geophysics | Seismic, gravimeter, resistivity meter, proton magnetometer and radar scanning. Some interference can result where metal is buried |
| Aerial surveying | This is a very useful technique for detecting old landslip features, old mineshafts, and previous tree planting |
| Direct surveying and monitoring | Used for surveying old mine workings or tunnels that are safe to enter. Often used in old limestone, ironstone and chalk workings to correlate old mining plans |

*Fig. 3.1* Hand auger tool.

of the ground supporting the building. If these trial pits show that the ground conditions and foundations are adequate for the loads carried, then other causes must be considered.

To determine the best place to dig a trial pit, stand back and examine the crack pattern and ask yourself the following questions:

1  Are the cracks wider at the top than at the bottom?
2  Do the cracks continue below the dampcourse?
3  Is there any evidence of undersailing of the masonry on the dampcourse?
4  Is there any distortion of window frames or doors?
5  Does the crack pattern have a sagging mode or a hogging mode?
6  Are there any significant trees close to the building?
7  Are the cracks mirrored through into the internal leaf of the wall?
8  Are there any drains close to the building?
9  Is there any evidence of previous repairs or repointing?
10  Do the crack widths exceed 3 mm?
11  Is the cracking consistent with a pattern of movement?
12  Is there any movement to the roof under cloaking and fascia board?
13  Are the walls measurably out of plumb?
14  Check the aspect of the wall: is it south facing?
15  Should expansion joints have been used on long terraced blocks?

These observations will assist in deciding where to put the trial pits or boreholes. Trial pits should be approximately 1.20 m square, and should extend down to the underside of the foundation base. If the foundation base is deeper than 1.20 m then consideration must be given to providing temporary support if the pit is to be entered.

*Fig. 3.2* Mini soil survey rig.

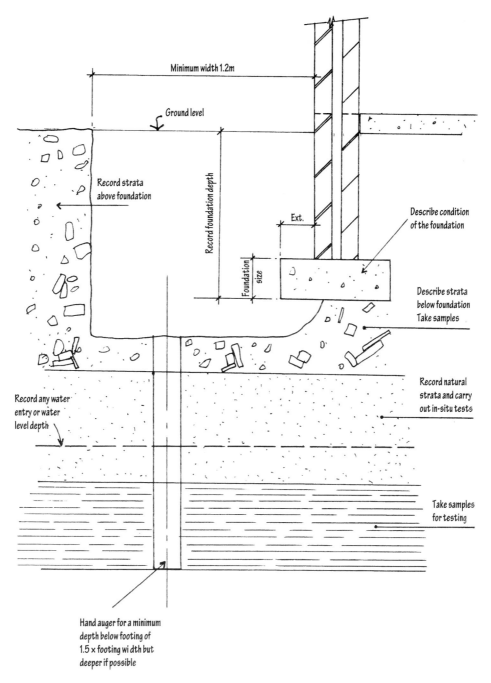

*Fig. 3.3* Typical trial pit.

From the base of the trial pit, the depth of strata can be determined by using a hand auger, as shown in Fig. 3.1. If difficulty is experienced in furthering the auger hole, it may be necessary to employ a mini-soils survey firm who have special equipment that is pushed into the ground by a hammer worked off a car battery, as shown in Fig. 3.2.

The trial pit should be accurately logged, and bulk samples taken for testing later. These samples should be at least 1.0 kg of material.

Figure 3.3 shows a typical trial pit with the necessary information added, such as:

● the depth of the foundation;
● the thickness of the foundation;
● the easement on the foundation base;
● the type of strata below the foundation;
● the condition of the foundation and substructure masonry;
● the level of any groundwater entry, and whether it was sampled;
● the total depth augered down below the foundation base;
● the nature of the strata encountered, e.g. stony clays, wet loose sands.

If the soils are cohesive, vane test result readings can be carried out at various depths to determine the undrained shear values of the soils below the foundation. Shear

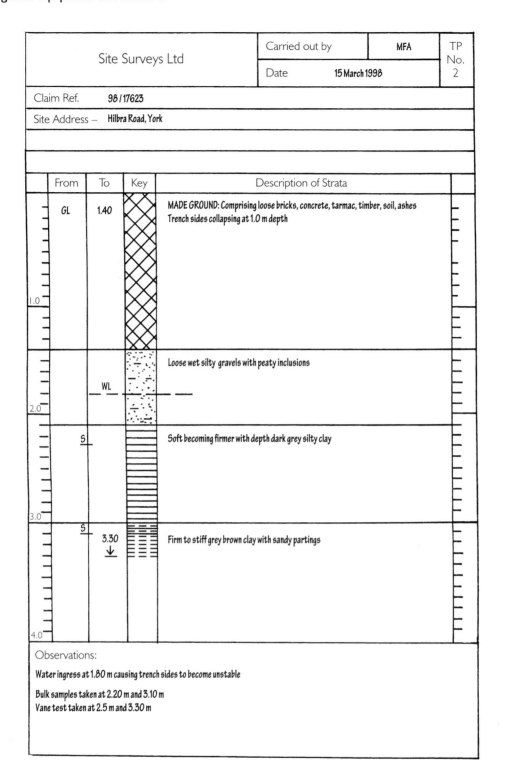

| | From | To | Key | Description of Strata | |
|---|---|---|---|---|---|
| | | | | **Site Surveys Ltd** — Carried out by MFA — TP No. 2 — Date 15 March 1998 | |

*Fig. 3.4* Typical trial pit log.

vane values, when doubled, will give an approximate allowable bearing capacity for the clays.

If the soil conditions are granular, a Mackintosh probe or simple peg test could be carried out to determine the relative density of the strata. The Mackintosh probe is driven into the ground by a standard hammer blow dropped from a set distance. The number of blows required to drive the probe 75 mm into the ground is recorded. The instrument is similar to the standard penetration test, and a summation of the total blows taken to drive 300 mm will enable a relative density to be obtained.

Figure 3.4 shows the trial pit log for the strata encountered in Fig. 3.3.

## 3.4 Boreholes

Boreholes are generally used where the depths of strata to be investigated are too deep for hand augering, or if the strata contain stones that prevent the auger from penetrating. When investigating deep backfilled quarries, mine workings etc., boreholes are the most useful and reliable method to obtain the soils data.

A borehole investigation should provide:

- information on the depth and description of the various strata encountered;
- the depths of any water entry, and its final standing level;
- the depths at which the various tests are carried out;
- the types of sample taken;
- where possible, the ground levels at the borehole location. This is very important when doing a rotary percussive borehole for possible mine workings, as the coal seam may have a steep dip, and without ground levels it is very difficult to draw accurate strike lines of the seam.

Groundwater levels can vary because of seasonal effects, and this should be taken into account when examining the data. In addition, the borehole records the ground conditions at the borehole only. The ground between boreholes can be correlated, but the accuracy cannot be guaranteed.

## 3.5 In-situ testing

### 3.5.1 Standard penetration test (SPT)

If the strata are granular or made ground containing granular materials, the relative density of the ground can be determined by taking SPT tests during the advancement of the borehole.

Table 3.2 shows the soil densities for a range of SPT results. The SPT test is carried out by dropping a spoon from a standard height and recording the number of blows to drive the spoon 300 mm. The results are read in 75 mm increments, and the first two sets are discounted as they are bedding-in blows. The remaining four sets of 75 mm penetration are added together.

SPT tests can be used in cohesive strata, and Table 3.3 shows the range of undrained shear strengths for various N values.

While SPT values are a useful guide to soil strengths, they should only be used as a guide. When used in filled ground containing bricks, stone and other mixed materials, the values could be quite high, and it would be dangerous to put too much reliance on such results alone and ignore the possibility of future collapse settlement.

**Table 3.2** Standard penetration test values for granular strata

| Consistency | N value (no. of blows/300 mm) |
|---|---|
| Very dense | > 50 |
| Dense | 30–50 |
| Medium dense or compact | 10–30 |
| Loose | 4–10 |
| Very loose | < 4 |

**Table 3.3** Standard penetration test values for cohesive soils

| Consistency | Undrained shear strength ($kN/m^2$) | N value |
|---|---|---|
| Very stiff | > 150 | > 20 |
| Stiff | 75–150 | 10–20 |
| Firm | 40–75 | 4–10 |
| Soft | 20–40 | 2–4 |
| Very soft | < 20 | < 2 |

For the determination of soil density from trial pits, the Mackintosh probe or the Perth Penetrometer are useful tools. These measure penetrations by a blow count when a weight is dropped down the shaft a specific height.

### 3.5.2 Shear vane tests

These can be taken in situ in cohesive soils. The shear vane apparatus is shown in Fig. 3.5; it can be used with extended rods if the trial hole or borehole is too deep. If the value recorded on the gauge is doubled this will give an allowable bearing capacity, based on a factor of safety of 3.0.

This instrument can be very useful when investigating clay shrinkage damage, as there is a correlation between the soil strengths and the desiccation of the soils.

### 3.5.3 Undrained triaxial test

When using boreholes in cohesive strata, the clay strength can be determined in the laboratory by subjecting 38 mm diameter clay samples to an undrained compression test. These tests define the shear strength values, and from

*Fig. 3.5* Shear vane apparatus for determining unconfined shear strengths of clay soils.

*Table 3.4* Requirements for well-compacted cast *in-situ* concrete of 140–450 mm thickness exposed on all vertical faces to a permeable sulphated soil or fill materials containing sulphates

| | Concentration of sulphate and magnesium | | | | | | | |
| | In soil or fill | | | | | | | |
| | By acid extraction (%) | By 2:1 water/soil extract | | In groundwater (g/l) | | Cement type | Minimum cement content (kg/m³) Notes 1 and 2 | Maximum free water/cement ratio Note 1 |
| Class | SO$_4$ | SO$_4$ | Mg | SO$_4$ | Mg | | | |
|---|---|---|---|---|---|---|---|---|
| 1 | <0.24 | <1.2 | | <0.4 | | A–L | Note 3 | 0.65 |
| 2 | | 1.2–2.3 | | 0.4–1.4 | | A–G<br>H<br>I–L | 330<br>280<br>300 | 0.50<br>0.55<br>0.55 |
| 3 | If >0.24 classify on basis of 2:1 extract | 2.3–3.7 | | 1.4–3.0 | | H<br>I–L | 320<br>340 | 0.50<br>0.50 |
| 4 | | 3.7–6.7 | <1.2 | 3.0–6.0 | <1.0 | H<br>I–L | 360<br>380 | 0.45<br>0.45 |
| | | 3.7–6.7 | >1.2 | 3.0–6.0 | >1.0 | H | 360 | 0.45 |
| 5 | | >6.7<br>>6.7 | >1.2<br>>1.2 | >6.0<br>>6.0 | >1.0<br>>1.0 | As for Class 4 plus surface protection | | |

Notes 1  Cement content includes pfa and slag.
2  Cement contents relate to 20 mm nominal maximum size aggregate. In order to maintain the cement content of the mortar fraction at similar values, the minimum cement contents given should be increased by 40 kg/m³ for 10 mm nominal maximum size aggregate and may be decreased by 30 kg/m³ for 40 mm nominal maximum size aggregate as described in Table 8 of BS 5328: Part 1.
3  The minimum value required in BS 8110: 1985 and BS 5328: Part 1: 1990 is 275 kg/m³ for unreinforced structural concrete in contact with non-aggressive soil. A minimum cement content of 300 kg/m³ (BS 8110) and maximum free water/cement ratio of 0.60 is required for reinforced concrete. A minimum cement content of 220 kg/m³ and maximum free water/cement ratio of 0.80 is permissible for C20 grade concrete when using unreinforced strip foundations and trench fill for low-rise buildings in Class 1.
Source: Based on *BRE Digest 363*, July 1991.

*Table 3.5* pH values

| Classification of acidity | pH value |
|---|---|
| Negligible | >6.0 |
| Moderate | 5.0–6.0 |
| High | 3.50–5.0 |
| Very high | <3.50 |

these the allowable bearing capacity can be determined using the appropriate factors of safety.

### 3.5.4 Atterberg limit tests

From the bulk samples of cohesive strata the liquid limit, plastic limit, moisture content and plasticity index can be determined. These results can then be compared with the Casagrande plasticity chart to confirm the clay classification. These results are essential when dealing with damage caused by trees.

*Table 3.6* Acids in groundwater: probable aggressiveness to ordinary Portland cement concretes (This table is intended as a broad guide only)

| Category | pH range | Comments |
|---|---|---|
| (a) | 7.0–6.5 | *Attack probably unlikely* |
| (b) | 6.5–5.0 | *Slight attack probable*<br>Where the pH and chemical analysis of the groundwater suggest that some slight attack may occur, and if this can be tolerated, then the concrete should be fully compacted, made with either ordinary Portland cement or Portland blastfurnace cement and aggregates complying with BS 882 (BSI, 1973). The maximum free water/cement ratio should be 0.50 by weight with a minimum cement content of 330 kg/m³. |
| (c) | 5.0–4.50 | *Appreciable attack probable*<br>For conditions where appreciable attack is probable, the concrete should be fully compacted, made with either ordinary Portland cement or Portland blastfurnace cement and aggregates complying with BS 882. The maximum free water/cement ratio should be 0.45 by weight with a minimum cement content of 370 kg/m³. |
| (d) | <4.50 | *Severe attack probable*<br>Where severe attack is likely, the concrete should comply with the requirements of category (b), but should have the lower water/cement ratio of 0.45. In addition the concrete should be physically protected by coatings of bitumen, asphalt or other inert materials reinforced with a glass-fibre membrane. High alumina and supersulphated cements are generally recognised to be more acid-resisting than concretes made with ordinary Portland cements. |

### 3.5.5 Chemical tests

The bulk samples or groundwater samples should be tested for soluble sulphates, i.e. soluble sulphur trioxide ($SO_3$), and for the soil's pH value.

BRE Digest 363 (January 1996) lists the various concrete mixes to resist sulphate attack. These are shown in Table 3.4.

When examining a number of samples in an investigation, greater emphasis should be placed on those tests that fall into the higher classification. For example, if seven samples out of ten are found to be within the Class 1 range and hence non-aggressive, while the other three fall into the Class 2 range, then the action taken should be based on the Class 2 category of risk.

The pH values of soils and groundwater can be critical when determining the suitability of concrete in the ground. BRE Digest 363 caters for this situation. If the soils are acidic the sulphate classification may have to be raised one or more classes; these values are shown in Tables 3.5 and 3.6.

## 3.6 Plumb and level survey

When assessing a damaged building, the temptation to jump to conclusions has to be strongly resisted. What may look like part of the building moving down may actually be some adjoining section moving up.

When dealing with problems likely to have been caused by trees or mining, it is always prudent to initiate a plumb and level survey around the building at dampcourse level. The initial survey levels will be a benchmark for further monitoring.

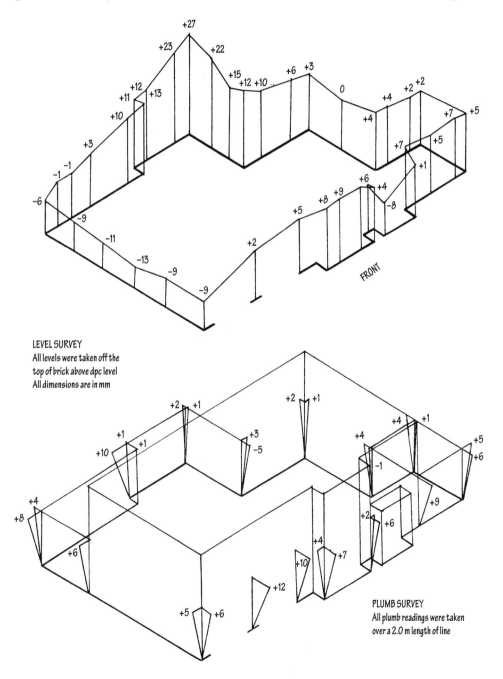

Fig. 3.6 Plumb and level survey.

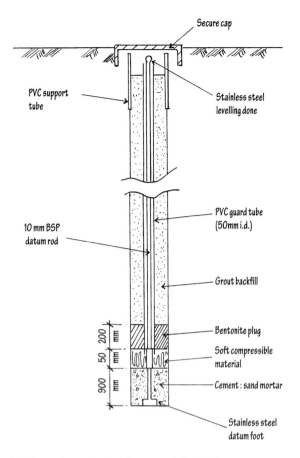

Secure cap

PVC support tube

Stainless steel levelling done

10 mm BSP datum rod

PVC guard tube (50mm i.d.)

Grout backfill

Bentonite plug

200 mm
50 mm
900 mm

Soft compressible material

Cement : sand mortar

Stainless steel datum foot

*Fig. 3.7* Deep datum bedded in mortar (after BRE).

This plumb and level survey, in conjunction with sketches of the damaged elevations, is an essential part of such investigations, and will enable the mode of damage and the cause of the movements to be better understood.

Figure 3.6 shows a typical plumb and level survey of a detached property, with the verticality of the walls shown on the isometric picture of the property. The relative horizontal levels around the building at dampcourse level are also shown. Care must be taken in assessing these levels, as some variation could have resulted during construction. Most large surveys should be carried out using a precise automatic level in conjunction with a precise invar staff, and with this equipment it is possible to obtain accurate levels to within ±0.50 mm. As well as reading accurate movements, the rates of change between levelling points need to be determined. For some jobs it may be possible to use a standard builder's dumpy level and staff if it is necessary only to record the directions of the movement over several seasons.

It is possible to purchase special stainless steel levelling plugs, which are threaded onto a screw-in socket section, and the levelling staff can then be seated onto a flat-top section. This ensures a more consistent reading at each level recording session. Between levelling sessions the plug can be removed and replaced with a perspex bung to cut out the risk of vandalism or accidental damage to the levelling station. These levelling plugs can be fixed with a good mortar mix or epoxy resin and sand filler if a quick-setting fixing is required.

For smaller buildings, especially when levelling floors, the use of a water level will provide accurate relative levels quickly, and the equipment can be operated by one person. Water levels are very useful when checking levels of floors, as they enable a grid of points to be checked quickly.

When carrying out a precise survey it is wise to use a datum that will not be affected by any climatic changes in the soils. Deep manholes are useful, or old buildings with deep cellars. If such a datum cannot be found within a close distance then a purpose-made deep datum can be

*Fig. 3.8* Avongard tell-tales.

installed at an appropriate depth. This generally consists of a vertical steel rod installed and concreted into a pre-bored hole. This system was developed by the BRE for long-term monitoring over many years' duration, and is shown in Fig. 3.7. The depth of the steel rod will be dependent on the type of strata below the building and the distance from trees.

## 3.7 Long-term monitoring

Movements in buildings can be caused in a number of ways: for example, by shrinkage or heave of clay soils, by coal extraction using longwall mining methods, by landslip settlement of deep fills, by thermal expansion, by tunnelling works, and by settlement of foundations under load. There may be situations where these movements need to be recorded to confirm the cause of damage.

These movements can be measured by:

● precise levelling techniques;
● installing crack-measuring equipment such as Avon-guard tell-tales (Fig. 3.8);
● installing Demec pips across cracks that can be measured using a vernier measuring device (Fig 3.9).

In slope stability situations a direct measurement of the slope movements can be obtained by installing simple wire extensometers or inclinometers in plastic tubes. In order to establish the displacement, a point on the landslide body is connected by a wire to a benchmark on the stable terrain. Readings can be taken mechanically or by using electromagnetic equipment.

*Fig. 3.9* Measuring Avongard pips with vernier calipers.

## Case history 3.1

### Introduction

This case history involved a block of flats in York that initially was considered to have been damaged by large trees close to the building. A major damage claim was submitted to the insurers two years after completion of the building for assumed settlement cracking. This cracking had progressively increased over the autumn and winter period.

Following notification of the damage in November 1991, an examination of the property was carried out, together with an initial water level check. This was decided because the pattern of cracking on the rear elevation suggested the movements to be similar to a ground heave mode. The water level check on the dampcourse levels also confirmed that the cracking on the rear elevation was not consistent with downward movement. Other elevations checked did indicate downward movements, and this was logical in that there were trees close by, and the crack patterns were consistent with this movement.

The two-storey terrace block of private flats was built in 1989 and was 12.50 m long and 9.75 m deep. The building was constructed using 275 mm cavity brick construction with a lightweight internal blockwork skin.

Roof construction was trussed rafters with clay pantiles. The foundations were deep trench-fill concrete.

Trial pits excavated down to the foundations at the rear elevation revealed them to be at a suitable depth for the distance from the trees at the front of the building. The foundations at the front of the building were not at an appropriate depth as defined in the *NHBC Standards* Chapter 4.2.

Examination of old Ordnance survey plans did not show any previous trees under the footprint of the building; an aerial photograph showed some vegetation, but the species and height of the trees could not be determined. The builder confirmed that some trees had been removed, but was unable to provide any precise details. It was decided therefore to carry out a precise plumb and level survey in order to determine the pattern of movement. As the trench-fill foundations had been cast directly against the firm clays it was considered that clay heave could be occurring along the rear elevation or within the building footprint.

As the building was occupied it was decided to install ten levelling stations using flat steel bars 6 mm thick set in the dpc. The joint was bedded with an epoxy mortar.

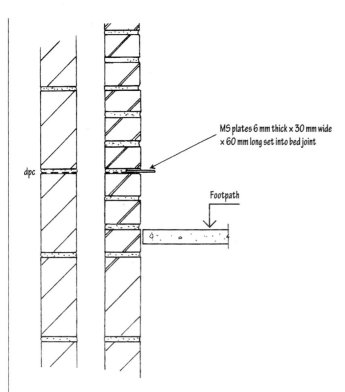

MS plates 6 mm thick × 30 mm wide
× 60 mm long set into bed joint

dpc

Footpath

*Fig. 3.10* Levelling plates.

These levelling stations are shown in Fig. 3.10. The datumn used was the cast iron cover of a manhole 5.0 m deep in a nearby main road.

Levels were taken after installation of the plates and then at 3 month intervals from January 1992 until March 1993.

### Initial survey

The levels recorded over a 15-month period are shown in Table 3.7. Each levelling station shows a two-column entry: the reduced level on the left, and the recorded movement on the right relative to the initial level reading. These levels were checked twice at each level session and the average of the two readings used. The locations of the levelling stations are shown in Fig. 3.11.

Apart from initially indicating the direction of movement, the level monitoring clearly demonstrated the long-term nature of rehydration in clay soils when trees have been removed. This building also suffered from subsidence arising from clay desiccation due to the closeness of mature trees that were retained.

The results clearly confirmed that the rear of the dwelling was moving up at a gradual rate. The section of heave appeared to be located between levelling points No. 2 and No. 3. This movement occurred below the large rear windows.

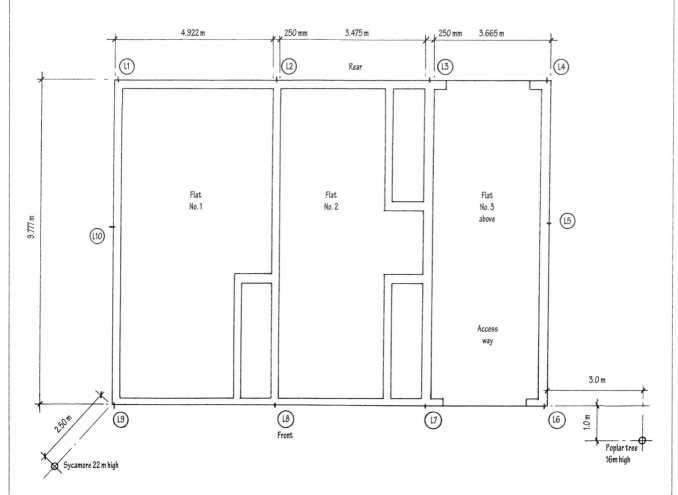

*Fig. 3.11* Level survey points.

*Table 3.7* Survey results

| Date | L1 | | L2 | | L3 | | L4 | | L5 | | L6 | | L7 | | L8 | | L9 | | L10 | |
|---|---|---|---|---|---|---|---|---|---|---|---|---|---|---|---|---|---|---|---|---|
| Jan 92 | 10.010 | – | 10.031 | – | 10.025 | – | 10.000 | – | 10.010 | – | 9.990 | – | 9.995 | – | 9.995 | – | 9.990 | – | 10.000 | – |
| Apr 92 | 10.013 | +3 | 10.038 | +7 | 10.033 | +8 | 10.005 | +5 | 10.008 | −2 | 9.985 | −5 | 9.993 | −2 | 9.995 | 0 | 9.988 | −2 | 10.002 | +2 |
| Jul 92 | 10.013 | +3 | 10.042 | +11 | 10.038 | +13 | 10.007 | +7 | 10.005 | −5 | 9.980 | −10 | 9.990 | −5 | 9.990 | −5 | 9.985 | −5 | 10.000 | 0 |
| Oct 92 | 10.010 | 0.0 | 10.036 | +5 | 10.036 | +11 | 10.003 | +3 | 10.005 | −5 | 9.985 | −5 | 9.988 | −7 | 9.985 | −10 | 9.985 | −5 | 10.002 | +2 |
| Jan 93 | 10.010 | 0.0 | 10.030 | −1 | 10.032 | +7 | 10.002 | +2 | 10.008 | −2 | 9.990 | 0 | 9.992 | −3 | 9.990 | −5 | 9.990 | 0 | 10.000 | 0 |
| Mar 93 | 10.012 | +2 | 10.030 | −1 | 10.030 | +5 | 10.002 | +2 | 10.008 | −2 | 9.992 | +2 | 9.992 | −3 | 9.990 | −5 | 9.990 | 0 | 9.995 | −5 |

By plotting the recorded movements against timescale it was evident that the heave had reached its peak in October 1993. Once the pattern of movements had been established, the terrace building was in summer 1994 underpinned down to a moisture-stable level to stabilise the foundation movements, in accordance with the *NHBC Standards* Chapter 4.2. During these underpinning works the soils at the rear of the building were checked for signs of desiccation when the foundations were underpinned.

To prevent any further upward movement a 50 mm thick layer of low-density Claymaster was inserted on the inside face of the underpinning concrete.

## Points to remember

- Always carry out an initial investigation and appraisal to confirm any damage.
- To avoid incorrect diagnosis in heave or subsidence situations always carry out an initial plumb and level survey. Such a survey should demonstrate the extent of movement that has occurred and its direction. This movement can be correlated with the estimated clay heave.
- Level monitoring provides a quick and cheap way of establishing the extent of the differential movements occurring to a building affected by trees or other types of subsidence. Do not rely on visual observation, as the eye can be easily deceived.
- Always carry out long-term levelling surveys using a deep datum. If one is not close by, install one to ensure an accurate set of results.
- Always carry out a full and proper site investigation to establish the full depth of desiccation of any clay soils affected by trees. This will enable any underpinning to be set at the right depth.
- Always ensure that the site investigation extends at least 1 m below the deepest underpinning.
- If previous tree growth is suspected, examine aerial survey photographs and old Ordnance plans.

## References

### Building Research Establishment

*Low-rise buildings on shrinkable clays, Parts 1, 2 and 3*, BRE Digests 240 (August 1980), 241 (September 1980), 242 (October 1980).
*Assessment of damage in low-rise buildings*, BRE Digest 251 (1981).
*Sulphate and acid resistance of concrete in the ground*, BRE Digest 363 (January 1996).
*Monitoring building and ground movements by precise levelling*, BRE Digest 386 (1996).

### British Standards Institution

BS 1377 : 1990 *Methods of test for civil engineering purposes.*
BS 5837 : 1991 *Guide for trees in relation to construction.*

### National House-Building Council

*NHBC Standards* (September 1999):
Chapter 4.1, Land quality – managing ground conditions.
Chapter 4.2, Building near trees.

# Chapter 4
# Structural reports

## 4.1 Introduction

Most of the time spent on a structural survey of a defective building is devoted to assessing the structure, and the damage that has resulted: finding out which walls support the roof and the floors, and seeing whether there is a pattern to the damage. Together with incidental notes, the 'expert' then needs to compile a main report. This may be a factual report, or a report that contains recommendations and a remediation strategy.

A common practice for the person doing the survey is to have copies of the report forms written up as site notes; the engineer or surveyor fills in the details during the course of inspection. Some people use a hand-held tape recorder to compile the site details, but this requires a lot of practice.

The most important part of the main report is the prognosis for the future: will all the evidence collected enable a remediation strategy to be developed? It is therefore essential that the report provides a concise and legible account of the structural damage to the building, and recommends essential remediation works:

- The main report should be well thought out.
- It must define the objectives.
- It must be logical.
- The data collected must be well presented.
- The report must be accurate in relation to technical data, and test results etc. If you are not certain of the facts, say so. Only use words such as 'maybe' or 'perhaps' when there is an element of doubt.
- The report must be written in simple English, so that the owner of the building – who more than likely is a layperson – can easily understand it.
- The report should be completed so that all the relevant data collected are recorded.
- Good grammar is essential, as it forms the basis of good communication.

## 4.2 Report format

The main report on a major structural defect claim will vary from engineer to engineer, and will be dependent on the nature of the defects. The format of the report should always follow a basic framework, and should be presented in a logical sequence as follows:

1 introduction;
2 a description of the building and its surroundings;
3 details of the investigation carried out;
4 results of any tests carried out;
5 details of any specialist reports that were necessary;
6 conclusion;
7 concise recommendations;
8 appendices, if any.

During the compilation of the report, it is essential that the test results and data collected are systematically checked for accuracy.

In the report avoid long words: keep the text simple and easy to understand.

When writing the conclusion, re-examine and check the facts to ensure that the conclusions reached are based on sound engineering judgement. Above all else they must be logical, as logic is the science of reasoned argument.

The final recommendations will rely heavily on personal experience, and the expert must apply all his or her professional skills and judgement to provide a way forward for the building owner. If only a factual report is required, then the information in the report must be sufficiently concise to enable another engineer to prepare remediation proposals. The report should highlight any items that have been based on assumptions, and which will need to be checked out on site once remediation work has commenced: for example, foundation depths may vary, as may the depth to rock.

The recommendations should take into account the nature of the building and any adjacent buildings. Taking underpinning as an example:

- When underpinning foundations, ensure that the strata below the underpinning remain competent.
- Ensure ease of excavation.
- Ensure access to carry out the works.
- Address any possible problems that could result from water ingress.
- Will partial underpinning be successful?
- If using piles, will vibrations be a problem?
- Recommend that any further foundation works are onto good bearing strata, and that any settlements will be within an acceptable magnitude.

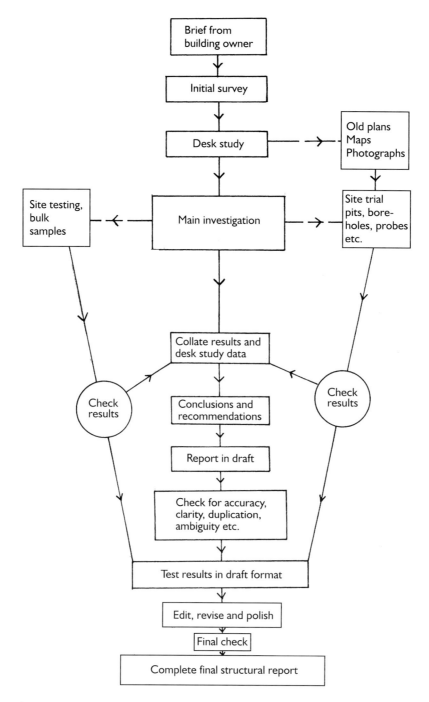

*Fig. 4.1* Stages of a structural report.

- Recommend the correct type of pile, taking into account the strata and surrounding buildings.

Figure 4.1 shows the steps leading up to the preparation of a structural report.

## 4.3 Survey report example

The following factual case history is included as an example of how to carry out and report on a major structural damage claim. This claim was investigated by Alan Wood & Partners, Consulting Engineers, on behalf of the NHBC, and the author is grateful for their permission to print the report in full.

This case history relates to a group of bungalows and shops that were suffering damage as a result of clay shrinkage caused by several large poplar trees. It could have been featured in Chapter 9 under foundation defects, but it is included here as it provides a good example of how to carry out and prepare a structural report. Such claims can take many months to resolve, especially if trees have tree preservation orders on them. In addition, many local authority tree conservation bodies put retaining trees as more important than preventing ongoing damage to people's property and it is often left to the insurers to prove that the trees should be removed. This claim was in a small village in East Yorkshire.

## 1.0    Introduction

1.01    This Report is prepared in accordance with instructions from the NHBC in April 1997.

1.02    The Report is required to establish the cause of the structural damage to the bungalows affected and is to provide recommendations as to the necessary works required to prevent progressive damage occurring in the future.

1.03    It is considered that tree roots from the adjacent Poplar trees which are in the ownership of the Local Authority are the cause of the damage.

1.04    This report therefore examines all the available information and discusses its findings and recommendations in Section 7.0.

## 2.0    Background information

2.01    The bungalows are a series of semi-detached properties, built in 1985. The builder has since gone into liquidation. They are of traditional construction with brick facings and internal blockwork cavity construction with a concrete tiled trussed rafter roof.

2.02    From examination of the building plans held in the Local Authority Archives it is understood that the dwellings have been constructed off a flat raft slab with timber floor joists on brick sleeper walls. The raft slab was intended to be placed on a cushion of imported granular fill.

2.03    In close proximity to the bungalows and adjacent shopping development are three mature Black Poplar trees.

## 3.0    Historical information relating to the claim

3.01    Cracking to the bungalows was first investigated in April 1990 by the NHBC, which had provided a ten year structural warranty. At that time long-term monitoring was recommended. This claim followed a very dry summer in 1989 and a relatively dry winter.

3.02    Further investigations were carried out by NHBC engineers in November 1990 and it was concluded that the cracking resulted from subsidence caused by the close proximity of these high water demand trees. It was recommended that the bungalows be underpinned and the offending Poplar trees be removed.

3.03    Meetings held on site with the Local Authority Tree Conservation Dept in late 1990 proved to be unsuccessful in that permission to remove the trees was refused. In view of this refusal, a concrete root barrier was introduced under the direction of the NHBC engineers between the trees and the bungalows, and following the installation of the barrier the masonry cracking was repaired.

3.04    In August 1995 further similar cracking was reported by the bungalow owners and these were measured at 6 mm to 7 mm in both internal and external walls. This followed a very dry summer.

## 4.0    Initial inspection

4.01    Following our appointment by NHBC in April 1997 an initial inspection of the bungalows was carried out to record the extent of the structural damage.

4.02    Cracking was evident in the walls of all rooms, with the exception of the bathroom.

4.03    The cracks were generally towards the tops of the walls, i.e. at ceiling level and around tops of windows.

4.04    It was noted during our visit that there were several mature Poplar trees in close proximity to the bungalows.

4.05    During our visit there was heavy rain which was seen to pond at the edge of the adjacent supermarket car park, alongside the Poplar trees, i.e. away from the rainwater gullies.

4.06    Photographs taken during our visit are appended to this report.

## 5.0    Investigations

5.01    To establish that the trees were causing the damage it was necessary to undertake various investigations of the foundations and soil strata. A specialist site investigation firm was employed to carry out some deep boreholes and a qualified arboriculturist was employed to provide advice on the trees.

5.02 An initial level survey was undertaken around the dampcourse level to establish the extent and location of any movements in the bungalow walls.

5.03 Following the drilling of boreholes at the front and rear of the nearest bungalow the specialist site investigation firm was commissioned to carry out soil tests to confirm the condition of the substrata below the raft foundation and to confirm whether these soils had been affected by the Poplar trees.

5.04 A specialist arboricultural report was sought from a qualified arboriculturist on the effects of the Poplar trees.

5.05 Hand auger boreholes were also drilled in the front garden to establish the presence and depths of any live tree roots. These root samples were sent away for identification and determination of their starch content.

5.06 Examination of old maps of the area revealed the presence of a large tree which had existed directly under the bungalow in 1890 but in 1927 the tree had been removed and replaced by buildings.

## 6.0    Examination of the specialist investigation reports

6.1.0 *Engineer's level survey*

6.1.1 The level survey showed that the front external corner of the nearest bungalow to the Poplar trees was the lowest point at 45 mm below the rear party wall corner which was the highest point. This front corner being the nearest to the trees and the rear corner being the furthest away.

6.2.0 *Engineer's hand auger investigation*

6.2.1 This hand auger investigation, carried out in the front garden, revealed extensive root growth between the concrete root barrier and the bungalow.

6.2.2 A number of large live roots were found near to the surface, with numerous other roots all the way down to a depth of 2.20 metres below ground level.

6.2.3 Many of the lower roots however appeared to be dead, but a live root sample was found at a depth of 2.20 metres down.

6.2.4 Samples of live roots at 600 mm and 2.20 metres were sent away for identification and found to be Poplar roots.

6.3.0 *Borehole investigation report*

6.3.1 The boreholes confirmed that the bungalow raft was bearing onto a stiff brown clay at a depth of 675 mm below ground level. The rear part of the raft foundation was built directly onto made ground which extended down for a depth of 1.80 metres.

6.3.2 Fine roots were encountered down to 2.40 metres below ground level at the front corner and 900 mm down at the rear corner.

6.3.3 A graph of the moisture content results shows a significant reduction at the front corner below a depth of 1.80 metres and down to 2.40 metres, although due to the made ground it is difficult to make comparisons.

6.3.4 A graph of the clay shear strengths shows a general increase in strength over the full depth of 2.60 metres at the front corner.

6.3.5 The soil suction results showed higher readings at the front corner in the upper soils, but lower readings in the lower soils.

6.3.6 The clay soils beneath the raft front corner have a medium to high shrinkage potential.

6.3.7 In the Engineering Comments, the report states that the test results appear to support the possibility of soil shrinkage by tree root action.

6.4.0 *Arboriculturist's report*

6.4.1 The report shows that the nearest Poplar tree is 20 metres high and 9.50 metres away from the bungalow.

6.4.2 This nearest tree (A) and also the adjacent Poplar tree (B) are considered to be of sufficient size, vigour and proximity to cause damage at the present time.

6.4.3 The root barrier installed in 1991 is considered to be too short and too shallow to be effective.

6.4.4 The construction of the root barrier and lack of pruning will have affected the stability of the tree.

6.5.0  *Old maps*

6.5.1  The maps of 1890 and 1927 show that there has been both a large tree and later some buildings on the site of this property.

## 7.0   Discussion of the findings

7.01  The dates of the reported cracking to the bungalows in early 1990 and late 1995 are significant, since the cracking follows two of the driest summers in recent years when many properties were damaged as a result of tree root action.

7.02  The level survey shows a significant fall to the front external corner, and yet the specialist site investigation revealed that softer made up ground is at the rear external corner.

7.03  The soil tests results indicate that more moisture has been extracted at the front external corner than the rear corner.

7.04  The arboriculturist's report confirms that the property is at risk of damage from at least two of the trees despite the presence of a root barrier.

7.05  The report is critical of the root barrier design, in that it is too short and too shallow.

7.06  The hand auger investigation in the front garden shows that the roots from the Poplar trees have bypassed the barrier, confirming the arboriculturist's comments.

7.07  You will see from the above that the investigations have shown conclusively that the tree roots from the Poplar trees are the main cause of the damage.

## 8.0   Recommendations

8.01  In view of the tree roots from the Poplar trees being the cause of the damage, the obvious solution to the problem is the removal of the offending trees.

8.02  The owners of the trees must therefore be advised of the findings and asked to remove the offending trees.

8.03  If the trees are removed, then a period of at least 12 months monitoring should be carried out to establish that the property has stabilised.

8.04  Following confirmation of stability, a cosmetic scheme of remedial work can be prepared and the work carried out.

8.05  In the event that the owners of the trees refuse to remove the trees, then a scheme of piled underpinning of the whole of the property would be required, followed by the cosmetic repairs to the superstructure.

C. Pollard, C. Eng, M.I. Struct. E
for and on behalf of Alan Wood & Partners

## Appendices

## Photographs

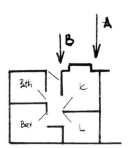

*Fig. 4.2* View A.

*Fig. 4.3* View B.

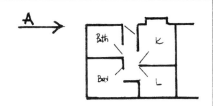

*Fig. 4.4* Standing water.

*Fig. 4.5* View A.

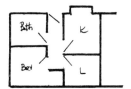

*Fig. 4.6* View A.

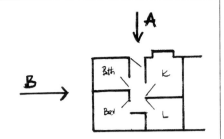

*Fig. 4.7* View A.

*Fig. 4.8* View B.

*Fig. 4.9* View A.

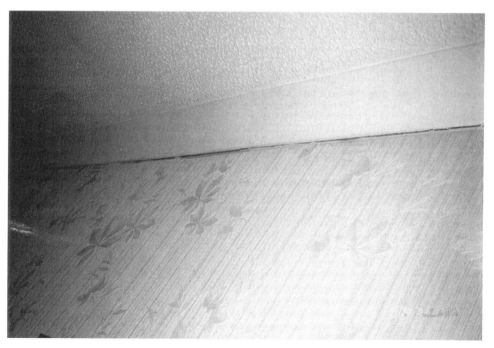

*Fig. 4.10* View B.

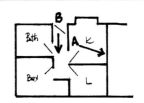

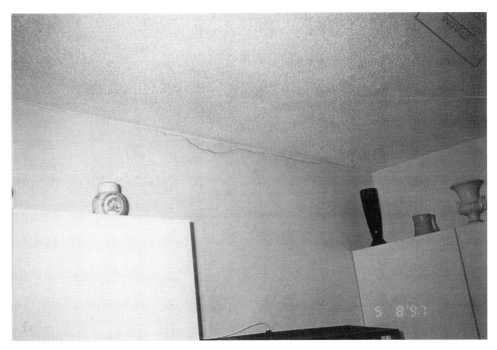

*Fig. 4.11* View A.

*Fig. 4.12* View B.

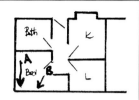

*Fig. 4.13* View A.

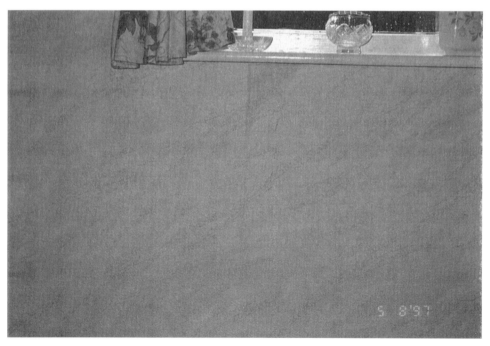

*Fig. 4.14* View B.

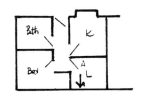

*Fig. 4.15* View A.

*TLP Ltd ground investigation report*

TLP Ground Investigations
Geotechnical Engineers and Geologists

3 Silica Crescent,
Scunthorpe,
South Humberside DN17 2XA
Tel. 01724 842520

Alan Wood & Partners,
341 Beverley Road,
Hull HU5 1LD

Date: 4 August 1997

For the attention of Mr C. Pollard

Ref – *Boreholes at Claim Reference No. 97/53478*

Further to your instructions we are pleased to enclose details of the boreholes excavated during June at the above property, together with the available in situ and laboratory test data. The excavations were located adjacent to the gable wall of the dwelling at the positions indicated on the enclosed sketch (Fig. 4.16). You will note that Borehole No. 2 was shifted slightly to the position BH2A, in order to avoid underground service pipes.

1.0 *Sampling and testing*
1.1 The borings were advanced using dynamic drilling equipment, taking undisturbed core samples reducing in size from 100 mm to 50 mm, to depths of approximately 2.70 m to 2.80 m b.g.l. In situ hand shear vane tests were performed where possible, as the borings were advanced and also on sections of the recovered core material. Borehole No.2A was extended to 7.65 m depth by driving a penetrometer and monitoring the resistance to penetration.

2.0 *Foundations*
2.1 The foundation supporting the gable wall was seen to comprise a shallow concrete footing, and although obscured by drainage pipes in TP2 it was seen in TP1 to be approximately 120 mm in thickness with a projection of approximately 150 mm and founded at approximately 0.67 m beneath the surface on natural deposits of alluvial silty clay. Figure 4.17 shows the relevant details.

3.0 *Ground conditions*
3.1 Beneath the superficial covering of topsoil, Borehole No.1 (Fig. 4.18) encountered natural soils initially represented by a firm to stiff 'crust' of brown, orange brown and grey mottled alluvial silty clay. The strength of the deposit, however, deteriorated with depth, and at around 2.10 m b.g.l. the deposits passed down into a sequence of soft and damp, brown and dark grey, slightly organic silty clay containing occasional pockets of decomposed vegetation. At 2.70 metres depth the boring penetrated a layer of loose, wet brown silty sand and gravel, and it was in this wet granular horizon after a short penetration that the boring was terminated.

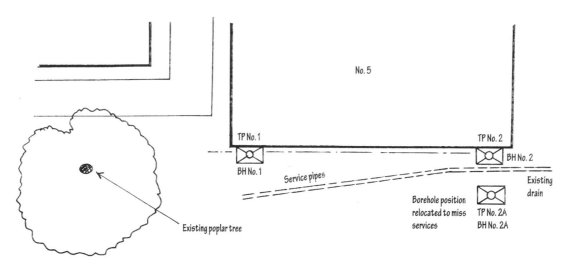

*Fig. 4.16* Location plan of trial pits and boreholes.

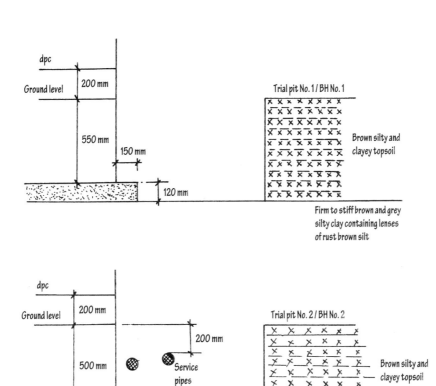

*Fig. 4.17* Existing foundation details.

3.2 In BH2A (Fig. 4.19), pockets of rotting paper were observed within the upper topsoil layer and what appeared to be a natural sequence of alluvial silty clay between 0.90 m and 1.55 m depth was actually underlain by a thin layer of soft and damp silty and sandy clay containing fragments of brick and other assorted gravel. This lower layer was clearly of unnatural origin, indicating that all the material above must be made-up ground also.

3.3 The made ground in BH2A appeared to extend to around 1.80 m depth, where it rested on deposits of brown and grey, slightly organic, silty clay, which, as in BH1, rested on a layer of loose, wet, brown, silty sand and gravel.

3.4 BH2A was extended by driving a penetrometer and monitoring its resistance to penetration. Very little resistance to penetration was noted until around 4.50 m depth. Percussive cable tool borings taken at other sites in the area have established that firm to stiff glacial boulder clay generally underlies the very soft alluvium at around 4.00 m to 5.00 m depth, and therefore it appears very likely that the increase in resistance noted at 4.50 m depth at this property represents the point where the weak alluvial sediments give way to more competent deposits of silty, sandy gravelly boulder clay. A further increase in the resistance to penetration noted at around 6.00 m to 7.00 m depth may represent the development of stiff to very stiff deposits of boulder clay or alternatively compact granular strata.

3.5 Groundwater seepages were encountered in the borings, and the details have been noted on the borehole logs. Seepages were first noted in BH2A at around 1.60 m depth, and stronger infiltrations were experienced in both borings on penetrating the wet sand and gravel layers at around 2.60 m to 2.70 m depth.

3.6 Frequent roots were noted on the upper soil profile at BH1, with the finer roots extending to around 2.40 m depth. Occasional fine roots were noted in BH2A extending to around 0.90 m depth.

4.0 *Laboratory and in situ testing*

4.1 Moisture content determinations were undertaken on soil samples recovered from the borings, and the results have been tabulated on the summary laboratory data sheet and also plotted on Fig. 4.20.

| T.L.P. Ground Investigations. | **Borehole Record** Dynamic Probe / Sampler. | Location : | | Borehole No.  1. |
|---|---|---|---|---|
| Carried out For    NHBC., C/o. Alan Wood & Partners. | | Ground Level | Co-ordinates | Date      16.6.97 |

| Description | Reduced Level | Legend | Depth & Thickness | Samples/Tests | | | | Field Records |
|---|---|---|---|---|---|---|---|---|
| | | | | Depth | samples Type | No | Test | |
| Brown silty and slightly clayey **Topsoil**. | | | (0.60) | | | | | |
| frequent roots | | | 0.60 | | | | | |
| Firm to stiff, brown, orange brown and grey mottled,  silty **Clay** containing occasional lenses of rust brown silt. | | | | 0.75 | | | Vane | 94kN/m² |
| | | | | 1.00 - 2.10 1.10 | U | 1. | Vane. | 75kN/m² |
| **Alluvium** | | | (1.50) | 1.40 | | | Vane. | 59kN/m² |
| Becoming more silty with depth. | | | | 1.70 | | | Vane. | 64kN/m² |
| finer roots extending to 2.40m. depth. | | | | 2.00 | | | Vane. | 55kN/m² |
| | | | 2.10 | 2.10 - 2.80 | U | 2. | | |
| Soft and damp, brown and dark grey, slightly organic,  silty **Clay** containing occasional pockets of decomposing organic material. | | | (0.60) | 2.30 | | | Vane. | 40kN/m² |
| | | | | 2.60 | | | Vane. | 25kN/m² |
| Loose, brown,  silty **Sand** and **Gravel**. | | | 2.70 2.80 | | | | | |
| **Observations.** Groundwater seepages encountered in granular strata at 2.70m. b.g.l. | | End| of b|rehole. | | | | |

| S.P.T. : | Where full penetration has not been achieved the number of blows for the quoted penetration is given (Not 'N' value) | Samples/Test Key. | | Remarks | Logged by R. B. |
|---|---|---|---|---|---|
| | | D   Disturbed Sample B   Bulk Sample | | | Scale 1:25 |
| Depths: | All depths and reduce levels in metres. Thickness given in brackets in depth column. | W   Water Sample U   Undisturbed Core sample S   Standard Penetration Test V   Vane Test | | | Fig. |

*Fig. 4.18* Log of borehole no. 1.

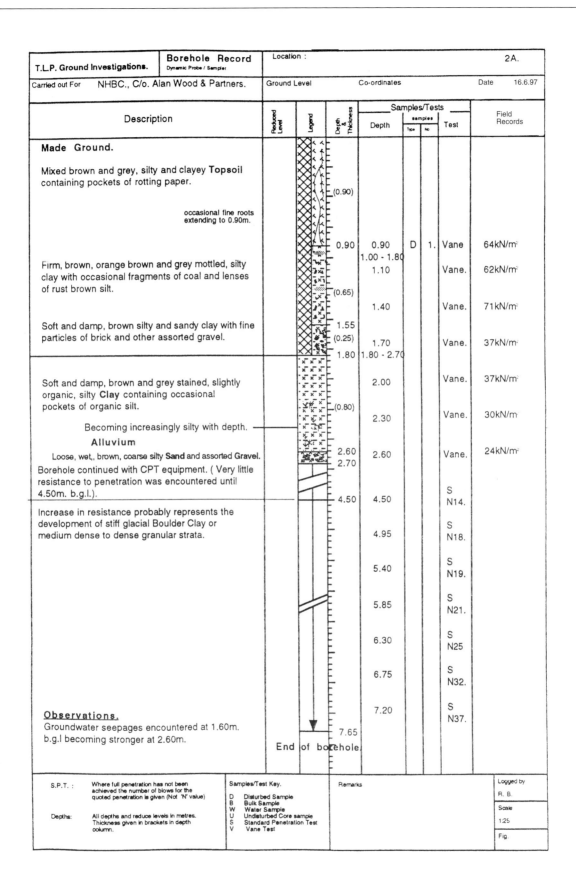

| T.L.P. Ground Investigations. | Borehole Record Dynamic Probe / Sampler | Location : | | | | | | | 2A. |

Carried out For    NHBC., C/o. Alan Wood & Partners.    Ground Level    Co-ordinates    Date    16.6.97

| Description | Reduced Level | Legend | Depth & Thickness | Samples/Tests | | | | Field Records |
| | | | | Depth | Type | No | Test | |
|---|---|---|---|---|---|---|---|---|
| **Made Ground.** | | | | | | | | |
| Mixed brown and grey, silty and clayey **Topsoil** containing pockets of rotting paper. | | | (0.90) | | | | | |
| occasional fine roots extending to 0.90m. | | | | | | | | |
| | | | 0.90 | 0.90 | D | 1. | Vane | 64kN/m² |
| Firm, brown, orange brown and grey mottled, silty clay with occasional fragments of coal and lenses of rust brown silt. | | | (0.65) | 1.00 - 1.80 1.10 | | | Vane. | 62kN/m² |
| | | | | 1.40 | | | Vane. | 71kN/m² |
| Soft and damp, brown silty and sandy clay with fine particles of brick and other assorted gravel. | | | 1.55 (0.25) | 1.70 | | | Vane. | 37kN/m² |
| | | | 1.80 | 1.80 - 2.70 | | | | |
| Soft and damp, brown and grey stained, slightly organic, silty **Clay** containing occasional pockets of organic silt. | | | | 2.00 | | | Vane. | 37kN/m² |
| | | | (0.80) | 2.30 | | | Vane. | 30kN/m² |
| Becoming increasingly silty with depth. | | | | | | | | |
| **Alluvium** | | | | | | | | |
| Loose, wet,, brown, coarse silty **Sand** and assorted **Gravel.** | | | 2.60 | 2.60 | | | Vane. | 24kN/m² |
| Borehole continued with CPT equipment. ( Very little resistance to penetration was encountered until 4.50m. b.g.l.). | | | 2.70 | | | | | |
| | | | 4.50 | 4.50 | | | S N14. | |
| Increase in resistance probably represents the development of stiff glacial Boulder Clay or medium dense to dense granular strata. | | | | 4.95 | | | S N18. | |
| | | | | 5.40 | | | S N19. | |
| | | | | 5.85 | | | S N21. | |
| | | | | 6.30 | | | S N25 | |
| | | | | 6.75 | | | S N32. | |
| | | | | 7.20 | | | S N37. | |
| **Observations.** Groundwater seepages encountered at 1.60m. b.g.l becoming stronger at 2.60m. | | | 7.65 | | | | | |
| | | End of borehole | | | | | | |

| S.P.T. : | Where full penetration has not been achieved the number of blows for the quoted penetration is given (Not 'N' value) | Samples/Test Key. | Remarks | Logged by R. B. |
| Depths: | All depths and reduce levels in metres. Thickness given in brackets in depth column. | D  Disturbed Sample B  Bulk Sample W  Water Sample U  Undisturbed Core sample S  Standard Penetration Test V  Vane Test | | Scale 1:25 Fig. |

*Fig. 4.19* Log of borehole no. 2A.

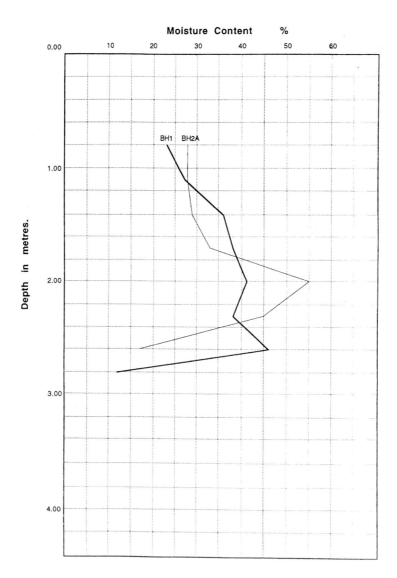

Fig. 4.20 Moisture content profiles.

4.2 Atterberg limit tests performed on selected samples of the alluvium recorded plastic index values between 28% and 47%, suggesting that the material is of medium to high shrinkage/swell potential as shown in Fig. 4.21.

4.3 Shear vane tests were taken in situ and on sections of the recovered core samples. The results have been noted on the borehole logs and are also plotted on Fig. 4.22.

4.4 Selected undisturbed samples from BHs 1 and 2A were subjected to soil suction analysis using the filter paper method of testing (BRE Information Paper IP/4), and the results are tabulated on Fig. 4.23.

4.5 A chemical test performed on a sample of groundwater obtained from BH2A recorded a water-soluble concentration of 0.29 g/l with a pH of 7.9 (Class 1, BRE Digest 363).

5.0 *Engineering comments*

5.1 The excavation at TP1/BH1 has revealed that the dwelling is constructed on a shallow foundation supported on initially firm to stiff silty clay, which progressively decreases in strength until more competent soils develop around 4.50 metres depth. Although the consistency of the alluvium does deteriorate, the strength profile obtained with the shear vane tester indicates that it is of adequate bearing capacity to sustain the nominal loads of the dwelling without excessive settlement. BH2A, however, penetrated deposits of made ground to approximately 1.80 metres depth. The consolidation of this layer could be suspected of causing foundation settlement particularly during

# Summary of Laboratory Test Data

Client :   NHBC., C/o. Alan Wood & Partners.

Location :

| Sample Details | | | Classification | | | |
|---|---|---|---|---|---|---|
| No.<br>Type | Depth<br>m. | Description | w<br>% | LL<br>% | PL<br>% | PI<br>% |
| BH1 | 0.80 | Silty Clay. | 23.0 | | | |
| " | 1.10 | " | 27.0 | 46.0 | 18.0 | 28.0 |
| " | 1.40 | " | 36.0 | 52.0 | 19.0 | 33.0 |
| " | 1.70 | " | 38.0 | | | |
| " | 2.00 | " | 41.0 | 68.0 | 21.0 | 47.0 |
| " | 2.30 | Organic silty clay. | 38.0 | | | |
| " | 2.60 | " | 46.0 | | | |
| " | 2.80 | Sand & Gravel. | 12.0 | | | |
| BH2A | 0.80 | Made Ground. | 28.0 | | | |
| " | 1.10 | " | 28.0 | 44.0 | 16.0 | 28.0 |
| " | 1.40 | " | 29.0 | | | |
| " | 1.70 | Made ground | 33.0 | | | |
| " | 2.00 | Organic silty clay. | 55.0 | 59.0 | 24.0 | 35.0 |
| " | 2.30 | " | 45.0 | | | |
| " | 2.60 | Sand & Gravel. | 17.0 | | | |
| | | | | | | |
| BH2A | 1.60 | Groundwater pH 7.9, SO4 0.29g/l | | | | |

Notes    U   Undisturbed          NP   Non Plastic

         B   Bulk                 *  Coarser particles (i.e. > 425µm. ) were removed for
                                  Atterberg limit tests.
         D   Disturbed

*Fig. 4.21* Atterberg limit results on clay samples.

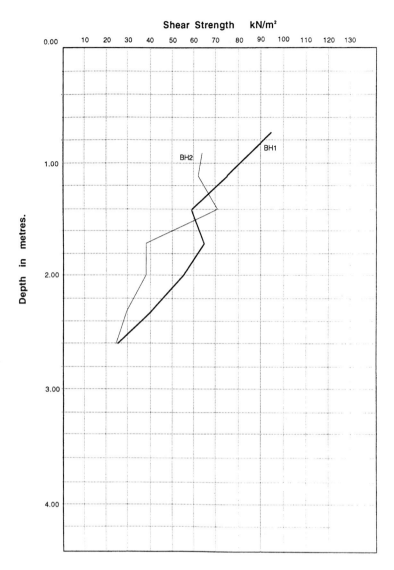

*Fig. 4.22* Shear strength profiles.

the early history of the dwelling but if more recent instability has been experienced then other factors may be involved.

5.2  Atterberg limit tests have indicated that the alluvial silty clay beneath the foundations has a medium to high shrinkage potential (PI 28% to 47%), and as such it could experience significant volume change with variations in moisture content. Since the foundations are quite shallow they will be vulnerable not only to soil movements associated with seasonal changes in moisture content but also to the additional effects of soil moisture depletion by root action.

5.3  Roots were encountered in the soil cores from both borings but were more prolific in BH1 where the finer roots extended to around 2.40 metres depth. Depletion in the soil moisture by these roots particularly during the summertime could therefore be contributing to additional soil shrinkage beneath the dwelling at this time.

5.4  Since made ground was present to 1.80 m depth in BH2A, it is difficult to make a direct comparison of the moisture content profiles at the respective locations. However, from a close inspection of the results, there does appear to be a zone of drier alluvial soil between 1.80 m and 2.30 m depth in BH1, and from a comparison of the soil suction values the higher values obtained on samples from BH1 within this zone may be further evidence of soil desiccation, albeit very slight.

5.5  The shear strength values obtained on the soil cores from BH1 were generally higher than those values obtained on cores from equivalent depth in BH2. This fact, taken in context with the moisture content,

# Results of Soil Suction Determination.
# Filter Paper Method.

SITE:

DATE: 16.6.97.
TEST PERIOD: 10 Days.

| BH No. | Depth m. | Position. | Paper dry mass $Md$ / g. | Paper moisture content $Wp$ / % | Sample moisture content $W$ / % | Suction kN/m² | Mean Suction kN/m |
|---|---|---|---|---|---|---|---|
| BH1 | 1.10 | 1 | 0.3513 | 103.45 | 27 | 11 | 11 |
| " | " | 2 | 0.3472 | 101.18 | | 11 | |
| " | " | 3 | 0.3488 | 105.07 | | 10 | |
| " | 1.40 | 1 | 0.3457 | 96.42 | 36 | 13 | 12 |
| " | " | 2 | 0.3404 | 98.70 | | 13 | |
| " | " | 3 | 0.3518 | 101.79 | | 11 | |
| " | 1.70 | 1 | 0.3426 | 72.63 | 38 | 25 | 27 |
| " | " | 2 | 0.3430 | 70.31 | | 29 | |
| " | " | 3 | 0.3469 | 70.09 | | 28 | |
| " | 2.00 | 1 | 0.3455 | 77.60 | 41 | 22 | 25 |
| " | " | 2 | 0.3521 | 74.02 | | 26 | |
| " | " | 3 | 0.3483 | 71.39 | | 26 | |
| " | 2.30 | 1 | 0.3025 | 108.89 | 38 | 10 | 10 |
| " | " | 2 | 0.3053 | 106.99 | | 10 | |
| " | " | 3 | 0.3124 | 110.27 | | 9 | |
| " | 2.60 | 1 | 0.3067 | 115.75 | 46 | 8 | 8 |
| " | " | 2 | 0.3093 | 118.29 | | 8 | |
| " | " | 3 | 0.3084 | 117.31 | | 8 | |
| BH2A | 1.10 | 1 | 0.3459 | 72.35 | 28 | 27 | 26 |
| " | " | 2 | 0.3498 | 70.69 | | 21 | |
| " | " | 3 | 0.3514 | 69.19 | | 31 | |
| " | 1.40 | 1 | 0.3473 | 64.36 | 29 | 37 | 36 |
| " | " | 2 | 0.3429 | 65.12 | | 36 | |
| " | " | 3 | 0.3496 | 66.75 | | 34 | |
| " | 2.00 | 1 | 0.3021 | 134.25 | 55 | 6 | 6 |
| " | " | 2 | 0.2994 | 128.87 | | 7 | |
| " | " | 3 | 0.3016 | 130.98 | | 6 | |
| " | 2.30 | 1 | 0.3056 | 129.21 | 45 | 7 | 7 |
| " | " | 2 | 0.3067 | 132.67 | | 6 | |
| " | " | 3 | 0.3035 | 127.59 | | 7 | |

*Fig. 4.23* Soil suction test results.

soil suction and Atterberg limit test results, appears to support the possibility that soil shrinkage resulting from depletion in soil moisture by root action is a likely contributory factor to more recent instability which may have been experienced by the dwelling.

5.6 Guidance on alternative remedial action to stabilise the dwelling can be obtained from Table 2 of BRE Digest 298; however, it is suggested that advice is also taken from an arboriculturist to establish the most appropriate action to take regarding the hedge and the nearest of the poplars growing adjacent to the front of the dwelling from which the roots observed in the bore holes probably emanate. Root samples have been enclosed with this report should you wish to commission a more positive identification.

5.7 If underpinning is undertaken, any underpinning foundations should be extended down onto a moisture-stable zone in competent strata where the bearing capacity is capable of sustaining the imposed foundation loadings. BH2A established that soft alluvium deposits may be present to around 4.50 m depth, and therefore it would appear that a system of underpinning piles will provide the most appropriate bearing solution to stabilize the dwelling.

5.8 The increase in resistance to penetration of the dynamic probe noted at around 4.50 m depth is likely to indicate the development of more competent glacial strata, and from the results of the penetration tests these deposits clearly become quite strong at around 5.0 m to 6.0 m depth. This would seem to represent an approximate depth of penetration for underpinning piles. Assuming that the material at this depth is a stiff boulder clay, then the allowable end bearing capacity for piles is estimated to be around 400 $kN/m^2$. Unit shaft adhesion in the stiff clay would be approximately 25 $kN/m^2$.

We trust that this preliminary information is sufficient for your present requirements. If, however, you require any further assistance, please do not hesitate to contact us.

For TLP Ground Investigations

R.L. Trattles

*Northern Resource Consultants report*

# Tree Report

# Claim Ref. 97/53478

## Section A – Introduction

A1.0    THE AUTHOR

A1.1    My name is David C. Houldershaw. I am a Chartered Forester, Fellow of the Arboricultural Association, Member of the Academy of Experts and Member of the British Institute of Agricultural Consultants.

A1.2    A past Northern England Regional Chairman of the Institute of Chartered Foresters, author and consultant with over sixteen years' professional experience, I have a specialist interest in the relationship between trees and buildings.

A2.0    THE PROPERTY

A2.1    A modern semi-detached bungalow with various trees and shrubs present.

A3.0    THE DAMAGE

A3.1    Damage has been noted to the front of the bungalow.

A4.0    THE REPORT

A4.1    The author surveyed the property in order to report on any trees or shrubs which may be associated with the current identified damage.

A4.2    The preparation of this report has been assisted by reference to a range of material published by various authors including BRE, Cutler and Richardson, Kostler, Matteck and Shigo etc. This in addition to my knowledge of tree physiology, tree root morphology and extensive experience of trees and their interaction with buildings.

David C. Houldershaw BSc (For) MICFor, F Arbor A, MAE, MBIAC
NORTHERN RESOURCE CONSULTANTS
22 Ethel Crescent
Knaresborough
North Yorkshire HG5 0DJ
Tel. (01423) 868952

Consultants in forestry, landscape and the environment

October 1997

## Section B – Site plan

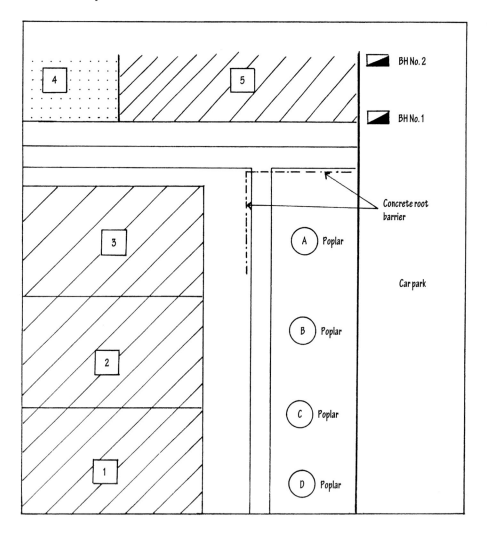

*Fig. 4.24* Site plan showing trees.

## Section C – Trees

C1.0   GENERAL

C1.1   In this section each tree is described briefly. Distance to buildings is to the closest point. Remaining life is the period during which further expansion growth can be expected; physical life may be somewhat longer.

C2.0   TREE A

C2.1   Height: 20 m; Spread: 13 m; Diameter: (breast height): 75 cm; Age: c. 60+ years; Remaining life: 30+ years; Species: Hybrid black poplar (Poplar).

C2.2   Relatively vigorous maturing deciduous tree of a high water demand species.

C2.3   c. 9.50 metres from No. 5 and well within the distance where damage might reasonably be anticipated. Significantly closer to owner's bungalow.

C2.4   Planted long before the bungalow was built.

C3.0   TREE B

C3.1   Height: 21 m; Spread: 13 m; Diameter (breast height): 55 cm; Age: c. 60+ years; Remaining life: 30+ years; Species: Hybrid black poplar (Poplar).

C3.2 Relatively vigorous maturing deciduous tree of a high water demand species.

C3.3 c. 14 m from No. 5 and within the distance where damage might reasonably be anticipated. Significantly close to owner's bungalow.

C3.4 Planted long before the bungalow was built.

C4.0 TREE C

C4.1 Height: 21 m; Spread: 12 m; Diameter (breast height): 50 cm; Age: c. 60+ years; Remaining life: under 30 years; Species: Hybrid black poplar (Poplar).

C4.2 Relatively vigorous deciduous tree of a high water demand species. Decaying lower stem cavity developing. Decaying root buttress. Possibility of significantly short remaining safe life.

C4.3 c. 20 m from No. 5 and well within the distance where damage may occur. Significantly close to owner's bungalow.

C4.4 Planted long before the bungalow was built.

C5.0 TREE D

C5.1 Height: 20 m; Spread: 12 m; Diameter (breast height): 55 cm; Age: c. 60+ years; Remaining life: 30+ years; Species: Hybrid black poplar (Poplar).

C5.2 Relatively vigorous deciduous tree of a high water demand species. Large branch at one-third height over path/car park in danger of early breakage.

C5.3 c. 25 m from No. 5 and within the distance where damage may occur. Significantly close to owner's bungalow.

C5.4 Planted long before the bungalow was built.

C6.0 OTHER VEGETATION

C6.1 The gardens of No. 5 contain a number of shrubs which are considered too small to cause damage at the present time.

C7.0 ROOT BARRIER

C7.1 In 1991 a 2 metre deep concrete root barrier was installed between tree A and Nos. 3 and 5.

C7.2 The specification for this barrier was inadequate, being too shallow and too short to be effective. There will have been a short-term benefit in slicing through some of the tree roots which would otherwise have been developing in the area of No. 5. There will however have been some reduction in stability, possibly significant, and the apparent lack of pruning to offset this is a concern from a safety standpoint.

C7.3 The nature of the site is such that I do not consider a root barrier to be appropriate here.

C8.0 OTHER INFORMATION

C8.1 The site investigation report prepared by TLP Ground Investigations indicates the bungalow to have relatively shallow footings and to be built at least in part on a medium to high shrinkability clay with roots present to a depth of c. 2.40 metres.

C8.2 Root samples from this investigation were sent for analysis by Tree Roots and Wood Technical Services in August 1991, who identified the species to be either willow or poplar. The samples lacked starch content, suggesting that these particular roots were dead though not sufficiently decayed to prevent identification.

C8.3 Further to this report, a further trial hole was augered at the front of the bungalow. It is reported that significant rooting was evident to a depth of 2.20 metres. Samples from this hole were identified by Tree Root and Wood Technical Services in September 1997 as either willow or poplar with a moderate starch content, i.e. living roots.

## Section D – Conclusions

D1.0 TREES CAPABLE OF CAUSING DAMAGE AT THE PRESENT TIME

D1.1 The following trees are considered to be of sufficient size, vigour and proximity to cause damage at the present time:
Poplar (A). A relatively vigorous maturing high water demand tree with over 30 years' remaining life during which further expansion growth is expected. 20 m height, 13 m overall spread with a dense crown. 9.50 metres from the bungalow with a probable zone of influence at the present time of around 15 m or so.
Poplar (B). A relatively vigorous maturing high water demand tree with over 30 years' remaining life during which further expansion growth is expected. 21 m height, 13 m overall spread with a dense crown. 14 m from the bungalow with a probable zone of influence at the present time of around 15 m or so.

D1.2   In both instances the presence of the root barrier will have interrupted some of the root development in the direction of No. 5. Given the short length and inadequate depth however it is likely that there will be sufficient living root concentration to cause damage at the present time. This is confirmed by the two site investigation visits, which indicate the presence of a significant quantity of poplar roots throughout the soil profile to a depth of 2.0 metres.

D2.0   TREES CAPABLE OF CAUSING FUTURE DAMAGE

D2.1   The following trees are capable of causing damage to No. 5:

| Tree | Species | Description | Life | Damage potential | Comments |
|------|---------|-------------|------|------------------|----------|
| A | Poplar | Relatively vigorous high water demand tree; 20 m height, 9.50 m from No. 5 | 30+ | Very high | Of sufficient size to cause damage at the present time |
| B | Poplar | Relatively vigorous high water demand tree; 21 m height, 14 m from No. 5 | 30+ | High | Of sufficient size to cause damage at the present time |
| C | Poplar | Relatively vigorous high water demand tree; 21 m height, 20 m from No. 5 | <30 | Moderate to low | Close to the size where damage may occur |
| D | Poplar | Relatively vigorous high water demand tree; 20 m height, 25 m from No. 5 | 30+ | Low | |

D3.0   FUTURE GROWTH

D3.1   All four trees are relatively vigorous and have the potential for further growth.

**Old maps**

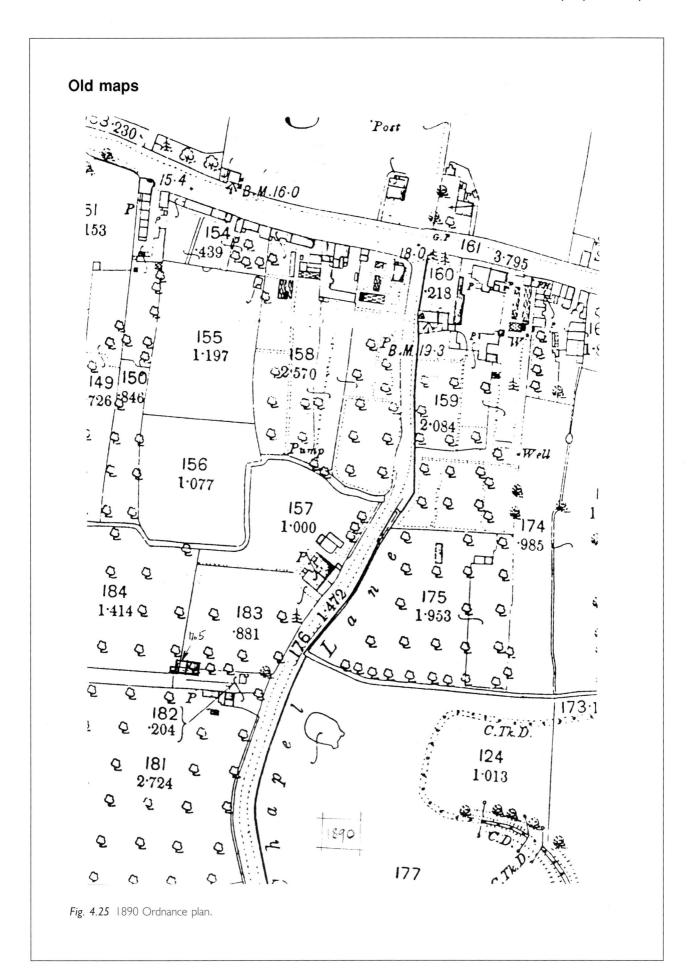

*Fig. 4.25* 1890 Ordnance plan.

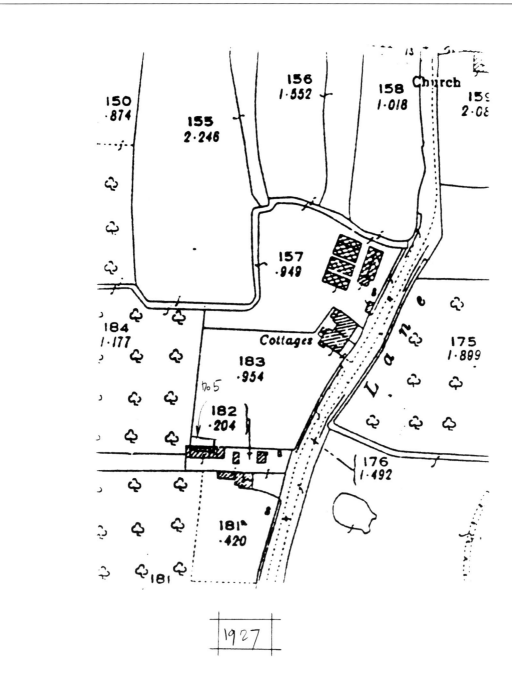

Fig. 4.26  1927 Ordnance plan.

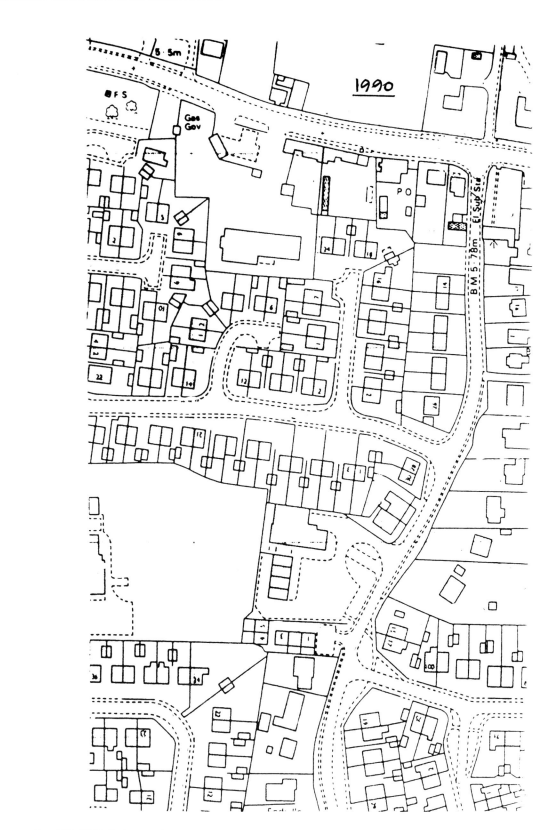

*Fig. 4.27* 1990 Ordnance plan.

## Points to remember

- Ensure that the final report has been checked for accuracy in terms of test results, technical data and presentation.
- The information in the report should be presented in a logical order. Logic is the science of reasoning, and is vital to understanding.
- Make sure that the report is written in simple English.
- Make sure that the evidence contained in the report supports any recommendations made.

## References

*Institution of Structural Engineers*

*Subsidence of low-rise buildings* (1994).
*Guide to surveys and inspections of buildings and similar structures* (1991).

*Royal Institution of Chartered Surveyors*

*Guidance notes on structural surveys of commercial and industrial property*, Building Surveyors Division (1989).

# Chapter 5
# Roofs

## 5.1 Introduction

In the UK, trussed rafters are used in 95% of pitched roof constructions. This shift away from the traditional cut roof using spars and purlins has resulted in a knowledge gap in the construction of traditional roofing.

It is not surprising therefore that the majority of serious structural defects occur on traditional roofs, especially those where there are large or complicated hip features and other complex geometrical shapes. Usually the main defects result from a lack of triangulation of the roof spars and ceiling ties, which results in lateral movements at the wall plate level.

The majority of roofs are constructed using trussed rafters; these are fully triangulated roofing elements, and roof spread of such construction is rare. However, roof spread can occur where raised-tie trusses are used, and the truss designer should ensure that the horizontal displacements at the wall plate do not exceed 6 mm at each support from dead and imposed loads.

While the trussed rafter industry has a reputation for producing good-quality rafters, it must be recognised that the creation of a sound roof is reliant on good design, sound materials, high quality fabrication, and good workmanship on site. It is a fact that, no matter how good the proprietary structural element is, bad workmanship on site can reduce its effectiveness and, regretfully, in some cases result in a roof structure that is structurally inadequate, containing major defects.

## 5.2 Trussed rafter construction

Trussed rafters can now be designed to accommodate a variety of geometrical shapes using stress-graded soft-woods, as indicated in Fig. 5.1. Most trussed rafters are fabricated using metal plates with punched-out teeth. For the light industrial market, timber trusses fabricated using plywood gussets at the joints with nailed connections are an alternative. Factory-produced trusses are the stronger, as the toothed plates are mechanically pressed together on both sides of the timber joint to form an intrinsic structural element. These types of truss are generally used at 600 mm centres when supporting concrete interlocking tiles or thin slate. Trusses are normally manufactured from 35 mm thick timber, but for 'attic' trusses it is general to use 47 mm thick timbers, owing to the higher floor loading on the bottom chord.

**Major structural defects** can occur in a trussed rafter roof as a result of:

- inadequate design and construction;
- the use of poor-quality timber;
- faults in fabrication;
- bad workmanship and poor storage on site;
- inadequate bracing.

## 5.3 Inadequate design and construction

Trussed rafters are precisely designed structural elements, which when constructed together should result in a load-sharing roof structure. Major structural problems can develop in a building if the design of the roof is incorrectly carried out. The type of design faults that can occur are:

- failure to identify the correct dead loading, i.e. some roof coverings are heavier than others;
- failure to identify the correct snow loading, i.e. altitude factor and snow drift;
- inadequate allowance for water tank loads, i.e. non-standard tanks;
- excessive lateral movement on raised tie trusses at the wall plate bearing – it is good practice to limit the height of the raised tie to less than half that of the roof rise;
- inadequate anchorage for uplift due to wind forces, more common on shallow pitch roofs, mono-pitch roofs, and roofs that are subject to wind from below the ceiling, i.e. car ports, garage links, inset balconies etc.;
- failure to provide for torsional movements behind the girder trusses that support long-span trusses – this torsional effect is more likely when the span of the trusses being supported by the girder truss exceeds 8 m, and should be allowed for by the roof truss designer.

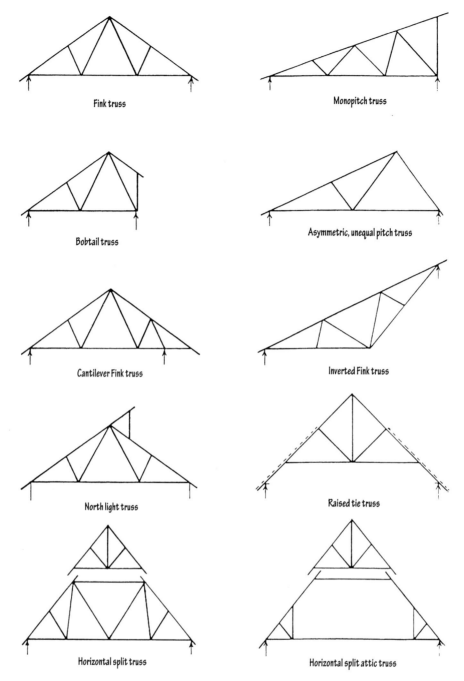

*Fig. 5.1* Typical trussed rafter profiles.

In some situations, the trussed rafter designer prepares a design that requires lateral bracing to be provided to strut members. The provision and positioning of this bracing are the responsibility of the builder or the building designer. The building designer must ensure that the design of the roof as a whole, and its connection to and compatibility with the supporting structural elements, are satisfactory in regard to the overall stability of the building. The trussed rafter designer is usually employed by the truss fabricator.

The building designer may be an architect, a chartered structural engineer or a builder. The building designer must liaise with the trussed rafter designer to ensure that all the structural aspects of the roof design as a whole are

considered, and that all wind loadings, precise site location and deflection criteria are available and acceptable to all parties.

### 5.3.1 Design loadings

● **Dead loads:** These are the loads that make up the permanent structure, and include finishes, ceilings, services, insulation and water tanks. In certain regions in the UK local planning conditions require roofs to be covered in traditional stone flag tiles or thick slate, and this must be allowed for in the design. In such situations the natural stone tiles or slates require a larger overlap than the normal interlocking roof tiles, and can weigh up to 2.0 kN/m². It is considered prudent to reduce the truss spacings from 600 mm to 450 mm, and design the trusses for the actual loading.

● **Imposed loads**: These are the loads due to people, light storage, snow load and wind loads.

### Water tanks

BS 5268 : Part 3 requires the designer to allow for a water tank in a domestic property. Tanks are generally 230 L to 300 L capacity supported on a properly designed platform over at least three trusses. In the truss design an allowance for two point loads of 450 N per truss applied at the supporting node points must be made. Figure 5.2 shows a typical tank support in accordance with Trussed Rafter Association guidelines.

### Snow loads

The correct snow load to be applied is detailed in BS 6399 : Part 3. Its value depends on the geographical location of the site and its altitude above mean sea level. If the roof geometry is complicated, with valleys and varied storey heights that could cause snow to drift, then the designer of the roof trusses must be made aware of this. Too many trussed roofs are designed as if they are forming part of separate buildings, with no regard given to the effects of linked developments.

### Light storage loading

For domestic roofs a light storage imposed load of 250 N/m² is allowed for to be carried by the bottom tie over the whole roof area. In addition a 900 N allowance for a human load must be considered, placed to give the maximum stresses in the bottom tie.

For trusses placed at maximum centres of 600 mm with a plasterboard ceiling this human load can be reduced by sharing 25% of the point load into the adjoining trusses, and the resulting load becomes 675 N per truss.

### Wind loads

The major consideration with wind loading is uplift. This is particularly true when the roof pitch is shallow, as the uplift forces increase, and also in cases where the roof coverings are lightweight, i.e. composition tiles.

### Raised-tie trusses

These are used in roofs where high level windows form part of the design, and by virtue of their geometry horizontal forces are imposed at the wall plate level. To ensure that the maximum allowable displacement of 6 mm is not exceeded the truss designer must ensure that the top chord of the truss is of adequate size. It may be necessary to add a scab timber as shown in Fig. 5.3, and this would generally be plated up in the factory.

### 5.3.2 Remediation of inadequately designed roofs

If a trussed rafter roof is showing signs of excessive deflection then the first things to check are the location, type of roof coverings and the loading conditions used in the calculations. Table 5.1 details the loading criteria to be applied to domestic roofs.

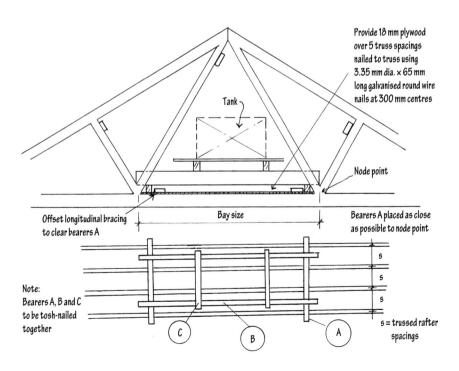

Provide 18 mm plywood over 5 truss spacings nailed to truss using 3.35 mm dia. × 65 mm long galvanised round wire nails at 300 mm centres

Tank

Node point

Offset longitudinal bracing to clear bearers A

Bay size

Bearers A placed as close as possible to node point

Note: Bearers A, B and C to be tosh-nailed together

s = trussed rafter spacings

*Fig. 5.2* Water tank supports.

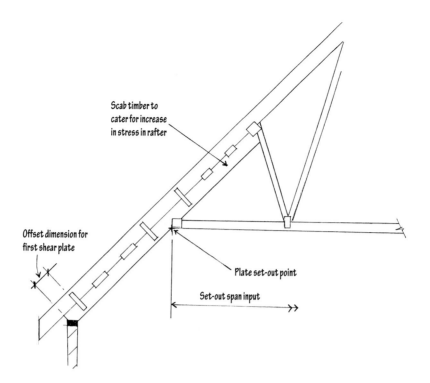

*Fig. 5.3* Raised tie truss.

*Table 5.1* Summary of roof loading criteria

| Rafter loads | Location | Duration |
|---|---|---|
| *Dead loads*: Concrete tiles 685 N/m²udl (measured along the slope) or as specified | Full length | Long term |
| *Imposed loads*: Snow load as BS 6399 Part 3 (except drift loads) | Full length | Medium term |
| *Drift loads or* | Full length | |
| 900 N human load | Centre of bay | Short term |
| *Wind*: Calculated in accordance with CP3 Chap. 5 for Class B structures | Full length | Very short term |
| *Ceiling tie loads*: | | |
| *Dead*: 250 N/m² udl *plus* | Full length | Long term |
| 2 × 450 N concentrated loads for water tank or actual load if greater | At two nodes nearest to water tank | Long term |
| *Imposed loads*: 250 N/m² udl | Full length | Long term |
| 900 N human load reduced where appropriate to 675 N per truss | Centre or either end of any bay | Short term |

In situations where the design of a trussed rafter roof has proved to be inadequate, and structural damage has occurred or there is a potential for damage to occur, then remediation techniques that can be applied without taking the roof off are limited. Often the only practical solution is to remove the roof trusses and reconstruct, hopefully utilising the existing trusses in the remediation scheme designed for the correct loading conditions. Where buildings are in the form of terraces with various storey heights, such that a bungalow lies next to a two- or three-storey house, then snow can build up on the lower roofs. In such cases the actual snowdrift loading condition must be determined when the roof is re-designed. In such cases it is often necessary to remove the roof coverings and tiling battens and provide additional trusses at closer centres, or provide new trusses throughout.

In some situations it may be possible to insert transverse beams or purlins at specific node points to reduce the rafter spans, but this should be done only with the agreement of the truss fabricator or designer.

For localised overstressing it is often possible to add scab timbers bolted to the existing trusses. These scab timbers must be properly designed, and the bolted connections must be detailed for compliance with BS 5268 Part 3.

## Case history 5.1

### Introduction

This case history highlights a combination of inadequate design, poor workmanship and the use of substandard materials to support the water tank in a bungalow roof. The bungalow had been surveyed for a valuation, and it was reported that there was an excessive deflection over the central section of the ceiling together with cracking and distortion of the plasterboard.

An examination of the roof void revealed that the water tank was supported on a sheet of chipboard that was resting across two trusses in the central section of the duo-pitch roof. The tank was larger than the standard 230 L capacity, and was assessed at 320 L capacity. When the tank was emptied the deflection of the ceiling chords recovered by an appreciable amount.

The use of chipboard alone to support a water tank is a contravention of the NHBC requirements; water tanks must be supported in accordance with the Trussed Rafter Association recommendations. These require the tank to be supported by four trusses.

Owing to the long-term residual deflection that had resulted, it was considered necessary to jack the trusses back level and provide an 18 mm plywood composite decking between the node points, placed over a length of six truss spacings. The tank itself was then supported on properly designed bearers as shown in Fig. 5.2, with the loads being applied at the node points each side of the tank spread over four trusses.

### Calculations of tank supports

The total weight of the tank plus platform was approximately 3.500 N. Shared over four trusses the eight truss node point loadings came to 437.50 N.

Using SC 3 timbers:

Basic stress in bending = 5.30 N/mm$^2$

Depth modification factor $K_7 = 1.13$

### Trimmer C

Try 63 mm × 100 mm

$$\text{Span 1.20 m} - \text{udl} = \frac{3.20}{1.2 \times 2} = 1.33 \text{ kN/m}$$

$$\text{Maximum moment} = \frac{1.33 \times 1.20^2}{8} = 0.239 \text{ kN m}$$

$$\text{Section modulus, } Z_{reqd} = \frac{0.239 \times 10^6}{1.13 \times 5.30} = 39.90 \times 10^3 \text{ mm}^3$$

$$Z_{prov} = \frac{63 \times 100^2}{6} = 105 \times 10^3 \text{ mm}^3$$

Use 63 mm × 100 mm for trimmer C

### Trimmer B

Reaction onto trimmer B = 0.50 × 1.33 = 0.666 kN

Span of trimmer B = 2.50 m max.

$$\text{Max. BM} = 0.666 \times \frac{(2.50 - 0.75)}{2.0} = 0.582 \text{ kNm}$$

Try 63 mm × 150 mm

$K_7 = 1.08$

$$Z_{reqd} = \frac{0.582 \times 10^6}{5.30 \times 1.08} = 101.67 \times 10^3 \text{ mm}^3$$

$$Z_{provided} = \frac{63 \times 150^2}{6} = 236 \times 10^3 \text{ mm}^3$$

Use 63 mm × 150 mm for trimmer B.

*Table 5.2* Timber sizes for tank support timbers

| Tank capacity to marked waterline | Minimum member size (mm) | | Max. trussed rafter span for fink (m) | Max. bay size for other configurations (m) |
|---|---|---|---|---|
| | SC 3 timber a and c | SC 3 timber b | | |
| Water tank not more than 300 L on four trussed rafters | 47 × 72 | Two 35 × 97 or One 47 × 120 | 6.50 | 2.20 |
| | 47 × 72 | Two 35 × 120 or One 47 × 145 | 9.00 | 2.80 |
| | 47 × 72 | Two 35 × 145 | 12.00 | 3.80 |
| Water tank not more than 230 L on three trussed rafters | 47 × 72 | 1/47 × 97 | 6.50 | 2.20 |
| | 47 × 72 | 2/35 × 97 or 1/47 × 120 | 9.00 | 2.80 |
| | 47 × 72 | 2/35 × 120 or 1/47 × 145 | 12.00 | 3.80 |

*Trimmer A*

Span between trusses $= 0.60$ m

Maximum bending moment $= \dfrac{0.666 \times 0.60}{4} = 0.099$ kNm

Section modulus, $Z_{reqd} = \dfrac{0.099 \times 10^6}{1.13 \times 5.30}$

$= 16.680 \times 10^3$ mm$^3$

$$Z_{prov} = \frac{50 \times 75^2}{6} = 46 \times 10^3 \text{ mm}^3$$

Use 50 mm × 75 mm deep SC3 for trimmer A. The *ITPA Technical Handbook* provides standard timber sections for standard-size tanks, as shown in Table 5.2.

### 5.3.3 Specification for proprietary timber roof trusses

#### 1.0 General requirements

1.1   The 'Engineer' referred to in this Specification shall mean the Consulting Structural or Civil Engineer who has responsibility for the overall design of the building.

1.2   The terms 'Approved' and 'Approval' shall mean the written approval of the Consulting Engineer.

1.3   All British Standard Codes of Practice, Specifications or any other Building Standards referred to in this Specification shall be the latest version including all amendments published before the final tender date, unless otherwise stated.

1.4   The design, fabrication and erection works described in this Specification shall be carried out to the entire satisfaction of the Consulting Engineer. Clause 1.3 refers to the minimum standards of acceptance which may be supplemented or modified by this Specification.

1.5   No variance to this Specification or to the Clause 1.3 provisions will be allowed without the written agreement of the Consulting Engineer.

#### 2.0 Design requirements to relevant British Standards

2.1   The design, fabrication, storage, handling and erection of timber trussed rafters shall be in accordance with BS 5268 : Part 3 *Code of practice for trussed rafter roofs*, which shall apply as a minimum standard of acceptance in matters not covered in this Specification.

2.2   The trussed rafters shall be designed and fabricated by an approved truss fabricator specialist.

2.3   The trussed rafter fabricator specialist shall prepare drawings and calculations for submission to the approved regulatory bodies for approval, which shall contain information in accordance with Clause 46.3 of BS 5268 : Part 3.

#### 3.0 Materials

3.1   All timber and steel plating used in the manufacture of trussed rafters shall comply with Section 2 of BS 5268 : Part 3.

3.2   All timber used in the manufacture of trussed rafters shall be treated with an approved organic solvent type of preservative, strictly in accordance with the recommendations of BS 5268 : Part 5. (Not an NHBC requirement.) Such preservative treatment shall not increase the risk of corrosion of the metal plate fasteners nor adversely affect any glued joints or nails.

3.3   All mechanical fasteners shall be in accordance with BS 5268 : Part 3, Clause 6.0.

3.4   The moisture content of the timber at the time of fabrication shall not exceed 20%. BS 5268 : Part 3, Clause 5.1.3.

#### 4.0 Design requirements

4.1   The trussed rafters shall safely support, without excessive deflection, all combinations of dead and imposed loads as specified in BS 6399 : Part 3 and the wind loads specified in BS 6399: Part 2 1997.

4.2   The trussed rafters shall be capable of resisting without damage the forces and deflections imposed during handling, transportation and erection.

4.3   Under the uniformly distributed dead and imposed loads on the rafters and ceiling ties the long-term deflection in the bottom ceiling tie shall not exceed 0.003 of the effective span unless it is pre-cambered to take account of dead loads.
      Notwithstanding this limitation, the deflection at any node point under long-term loading only shall be in accordance with BS 5268 : Part 3, Clause 10.

4.4   A summary of design loads is listed in Clause 15 of BS 5268 : Part 3 and in Table 5.1 in this chapter.

#### 5.0 Fabrication

5.1   All trussed rafters shall be fabricated strictly in accordance with the recommendations of Section 6.0 of BS 5268 : Part 3.

#### 6.0 Stability bracing

6.1   Stability bracing, wind bracing, lateral restraint strapping, holding downs traps etc. to ensure overall stability shall be the responsibility of the Consulting Engineer and shall be shown on the Architect's or the Consulting Engineer's drawings.

6.2    Any bracing required to be fitted on site in order to prevent buckling of compression members shall be specified by the trussed rafter fabrication specialist, and the size, position and method of fixing of this bracing shall be clearly indicated on the truss detail drawings. The supply and fixing of this truss bracing shall not be charged as an extra to the Main Contract.

### 7.0 Site storage, handling and erection

7.1    Handling, storage and erection of timber trussed rafters shall be in accordance with Section 7.0 of BS 5268 : Part 3.

7.2    Trusses shall be handled and transported in such a manner that the trusses are not damaged.

7.3    Where trussed rafters are banded together for transport, care must be taken to ensure that the timbers on the outside are not damaged by excessively tight bands. Timber packs should be used to prevent this.

7.4    Trussed rafters shall be stored on site on raised bearers to keep the timbers clear of the ground to prevent distortion and deterioration. The trusses shall be stacked vertically and covered with waterproof sheeting which is well ventilated.

7.5    Trusses should be lifted onto the wall plates in accordance with the Trussed Rafter Association guidance notes. Trusses should never be lifted over scaffolding on their weak axis.

   Prior to nailing on the tiling battens, all trussed rafters should be erected vertically and parallel, at the specified centres with no bowing evident in the length of the truss.

7.6    Adequate provision shall be made for the introduction of any temporary bracing during the erection process to resist wind forces and to prevent trusses moving out of plumb. Such temporary bracing is to remain in position until replaced by permanent wind bracing.

### 8.0 On-site modifications

8.1    Under no circumstances should timber trusses be cut, notched or drilled without the written approval of the Consulting Engineer or the Truss Fabricator Specialist.

8.2    Any repairs needed to trusses after erection shall be in accordance with the provisions of BS: 5268 : Part 3 and shall be approved in writing by the job Architect or the Consulting Engineer or truss fabricator.

### 9.0 Approvals prior to fabrication

9.1    Design calculations and detailed truss drawings are to be submitted to the Architect for dimensional approval and to the Consulting Engineer and Approved Regulatory Bodies for structural approval prior to fabrication. Sufficient time must be allowed for in the manufacturing programme for such approvals.

## 5.4 Poor-quality timber and faults in fabrication

In 1995 new rules were introduced for the specification of structural timber. This required that all structural softwoods of less than 100 mm thickness to be used for internal applications such as floors, walls and roofs should be graded at a moisture content of 20% or less. The timber must have a stress grade stamp indicating that it is DRY or KD (kiln dried) timber.

Structural softwoods for use in high-moisture environments, i.e. fully exposed outdoors, will be marked WET to indicate that the timbers are not suitable for internal applications.

In addition to improving the strength and stability of timber, correct drying eliminates a lot of shrinkage and distortion after fixing. It also improves the penetration of preservative treatments. This will result in less distortion in floors and less springing in floor deckings, and as floor strutting is less likely to loosen a stiffer floor will result.

Trussed rafters should be fabricated in accordance with BS 5268 : Part 3, and must have a minimum thickness of 35 mm for spans up to 11 m and 47 mm for a 16 m span. BS 5268 : Part 3, Clause 6.2.

When delivered to site they should be stacked vertically, off the ground, and covered with waterproof sheeting to ensure that their moisture content is not increased by wet weather.

Where preservative treatment is specified, it should be based on the requirements of BS 5268 : Part 5 (1989). This Code of Practice lists various preservation methods available. For trussed rafters the preferred method is the double vacuum organic solvent treatment, which allows truss plating to be carried out without having to wait for the timber to dry out, as is the case with CCA-treatment, which actually increases the moisture content of the timber.

Joint connector plates must be manufactured from hot-dipped galvanised mild steel. For special situations and a longer life, special plates can be used: for example, stainless steel and thicker plates can also be specified where large forces have to be accommodated at the joints.

While most truss manufacturers produce good-quality trusses, especially those firms who are members of the TRADA quality assurance scheme, defects can occur that may get past the factory quality control. Typical faults that can result are:

● excessive gaps between two adjoining members (the average gap should not exceed 1.5 mm);
● timber fabricated with wane in the plated area;
● top chords and bottom chords with excessive wane on the bearing surfaces;
● different timber thicknesses at a node point (the difference in thickness should not exceed 1 mm);
● metal plates not fully bedded;
● excessive dead knotting in the plate areas;

- plates badly aligned such that the maximum 5 mm tolerance between the opposing plates is exceeded;
- failure to provide an adequate camber in long-span trusses. This would normally be the product of the span × 0.0005. For example, a 10 m span truss would have a pre-camber of $0.0005 \times 10 \times 1000 = 5$ mm.

Defects as described rarely result in major structural damage to the whole roof, as there is an element of load sharing. However, fabrication faults that have resulted in localised damage should be attended to by repairing the damaged areas.

## 5.5 Defects due to bad workmanship on site

Trussed rafters are very slender structural elements, and must be handled with care by site operatives. The eight most common defects found on sites are:

- trusses that have become distorted and damaged by being badly stored on site. Trusses can also be damaged by incorrect handling procedures when lifting them onto the roof: pulling them over the scaffold on their weak axis can lead to plates being loosened and timber being damaged;
- trusses that have been damaged by site plant, vehicles etc.;
- trusses with excessive moisture content due to lack of protection or delays in battening and felting. This can cause corrosion of the metal plates and weakening of the timber;
- trusses that have been cut to provide access openings, service access etc. It is a general rule that timber trusses should never be modified on site without the

Fig. 5.4 Inadequate water tank platform.

prior agreement of the truss fabricator, in writing and with the appropriate drawings;
- inadequate diagonal and longitudinal bracing. This can result in trusses being erected out of plumb with little resistance to wind forces;
- trusses spaced out in excess of 600 mm;
- poor load distribution of water tank supports, leading to local distortion of ceiling ties and plaster cracking. Figure 5.4 shows a site pallet used as a tank support;
- trusses erected out of plumb with inadequate bracing.

## Case history 5.2

A newly built bungalow was surveyed and found to have roof trusses severely out of plumb. The initial survey showed that many of the trusses were 75 mm out of plumb. Most of the remaining trusses were built such that the plumb line passed outside the bottom tie, and the trusses were also severely bowed on plan. It was considered that the bowing of the trusses initially resulted from the manner in which the trusses had been erected. It was established later that a single joiner erected the roof, and the trusses had been pulled up over the scaffold on their weak axis. In addition the joiner had not provided sufficient temporary bracing during the erection, and in the completed roof there was a serious omission of the essential permanent diagonal and longitudinal bracing. On several trusses web members had split, and some attempts had been made to brace these damaged areas locally by adding short longitudinal lengths of timber.

Following the initial survey report, a major damage claim was made to the warranty insurance providers.

The roof was subsequently inspected by a chartered structural engineer, and during the plumb and level

survey it became evident that the bricklayers had followed the line of the out-of-plumb trusses when constructing the gable peaks. The engineer recommended that the roof be rebuilt and the gable walls be taken down to wall plate level and be rebuilt plumb.

There was a common misconception, held by the builder and roof truss fabricators, that such bad workmanship could be put right by the introduction of extra lateral bracing, or by infilling between the trusses with plywood sheets in the plane of the webs.

Such remediation, had it been carried out, would have made the problem worse. Additional dead load would have been introduced onto trusses that were clearly unable to support the existing vertical loads without further distress.

The only proper and long-term solution was to strip the roof tiles and battens and replace the defective trusses, fitting the required longitudinal and diagonal bracings and replacing the roof tiles. The diagonal bracings were securely fixed as shown in Fig. 5.5 to ensure that the lateral forces due to wind on the gables were transferred into the masonry.

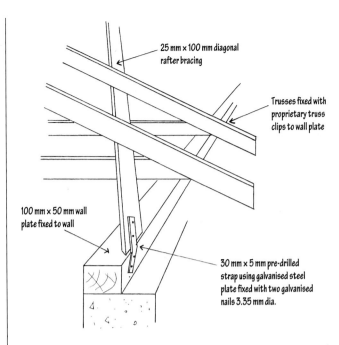

Fig. 5.5 Diagonal bracing fixing to wall plate.

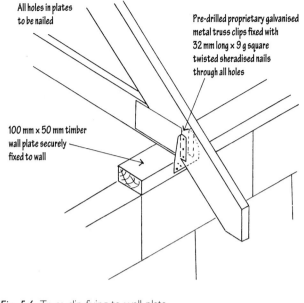

Fig. 5.6 Truss clip fixing to wall plate.

As most of the trusses had been damaged or distorted during erection, it was deemed necessary to have new trusses fabricated throughout. The gable walls were taken down to wall plate level and rebuilt. During these works the lateral restraint straps were built in at roof and eaves levels, diagonal and longitudinal bracing were installed (Figs 5.6 and 5.7), and following felting and tiling the interior plasterboard ceilings and internal walls were made good.

Table 5.3 TRA maximum deviations of trusses from vertical

| Rise of trussed rafter (m) | Deviation from vertical (mm) |
| --- | --- |
| 1 | 10 |
| 2 | 15 |
| 3 | 20 |
| 4 or more | 25 |

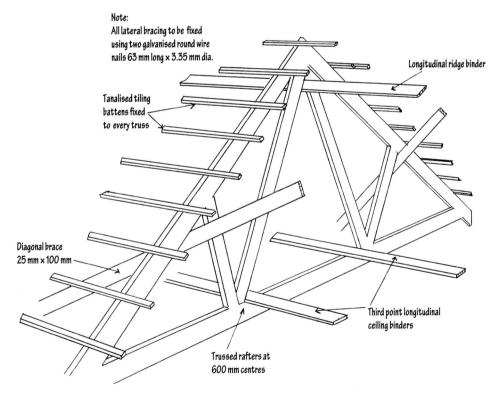

Fig. 5.7 Lateral roof bracing.

## Case history 5.3

### Introduction

A large bungalow had been constructed using traditional brick and block walls and a trussed rafter roof covered with concrete interlocking tiles. The interior walls were lightweight timber stud and plasterboard partitions, which had been erected following the construction of the main shell.

Shortly after moving in the owner, concerned about severe cracking that was occurring in ceilings and internal walls, lodged a major structural damage claim with the warranty provider. This roof was inspected by a chartered engineer, and a report recommending remedial works was produced.

Figure 5.8 shows the plan shape of the bungalow and the internal non-loadbearing partitions. The 9.50 m span trussed rafters spanned from front to rear, and a triple girder truss was used to support the trusses spanning at right angles.

### Survey results

A survey of the roof space revealed the following structural defects:

● The roof trusses spanning onto the triple girder truss had been inadequately supported at the girder truss position. The galvanised metal shoes had not been fully nailed, and in addition the girder truss bottom chords were being subjected to torsional forces because the large span, approximately 9.50 m, was being supported from one side. It was also evident that the triple girder truss had not been nailed together in accordance with the truss fabricator's

recommendations: this had caused local separation of the girder truss members to develop, and the resulting downward movement over the three trusses had damaged the ceiling (Fig. 5.9).
● The stud partitions had been built up tight to the underside of the plasterboard ceilings. The ceilings had been placed prior to partitions going in as the builder had given the prospective purchasers an option of installing the internal partitions to suit their own layout. The rotational movements of the roof trusses and the lack of free vertical movement at the head of the partitions were the causes of most of the ceiling damage.

### Remedial works

The main remediation work involved correcting the defective supports on the girder truss and nailing up the triple girder truss in accordance with the fabricator's instructions. The girder truss was supported by bearers and Acrow props positioned at the node points. The supported trusses were also supported by props close to the girder truss to relieve some of the load, and these were jacked up until finger tight. The few nails that were joining the triple trusses were sawn, allowing the trusses to be jacked level. Owing to the combined thickness of 105 mm it was considered that renailing the trusses could result in the timbers splitting. It was decided therefore to bolt the trusses together on the centre lines of the sections using 8 mm bolts at 600 mm centres.

During the refixing of the supporting shoes (Fig. 5.10) it was decided to take out the torsional effect on the

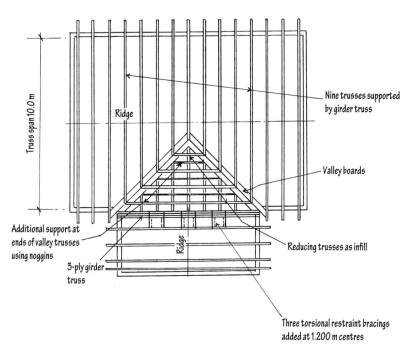

*Fig. 5.8* Plan of bungalow.

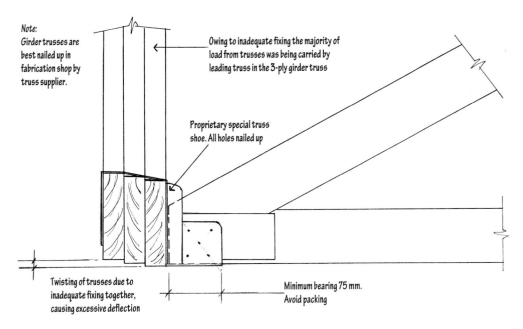

*Note:*
*Girder trusses are best nailed up in fabrication shop by truss supplier.*

*Owing to inadequate fixing the majority of load from trusses was being carried by leading truss in the 3-ply girder truss*

*Proprietary special truss shoe. All holes nailed up*

*Twisting of trusses due to inadequate fixing together, causing excessive deflection*

*Minimum bearing 75 mm. Avoid packing*

*Fig. 5.9* Distortion of triple girder truss.

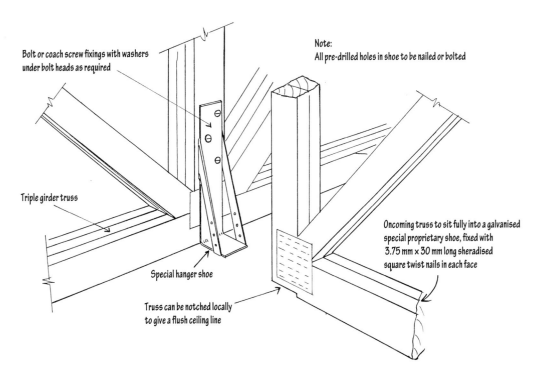

*Bolt or coach screw fixings with washers under bolt heads as required*

*Note: All pre-drilled holes in shoe to be nailed or bolted*

*Triple girder truss*

*Special hanger shoe*

*Truss can be notched locally to give a flush ceiling line*

*Oncoming truss to sit fully into a galvanised special proprietary shoe, fixed with 3.75 mm × 30 mm long sheradised square twist nails in each face*

*Fig. 5.10* New shoe fixing onto girder truss.

girder truss by providing stiffeners and a plywood diaphragm, as shown in Fig. 5.11. This plywood diaphragm extended back to the next truss, with the vertical noggins inserted between each supported truss at 600 mm centres. Once all the trusses had been fixed in place and adequately supported the Acrow props were removed and the ceiling was made good.

The diagonal roof bracing was found to be inadequate and was replaced as shown in Fig. 5.12, with two fixings per truss using 3.35 mm × 65 mm galvanised round wire nails. The diagonal bracing was also overlapped a minimum of two truss spacings.

To prevent deflection of the trusses loading the non-loadbearing partitions it was necessary to modify the partition head details, and these are shown in Fig. 5.13. On completion of these remedial works the ceilings were repaired, and plaster Gyproc covings were installed prior to redecoration.

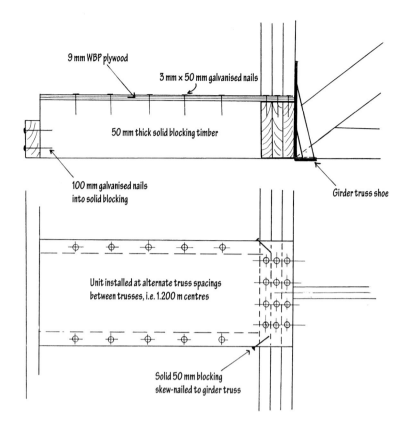

9 mm WBP plywood

3 mm × 50 mm galvanised nails

50 mm thick solid blocking timber

100 mm galvanised nails
into solid blocking

Girder truss shoe

Unit installed at alternate truss spacings
between trusses, i.e. 1.200 m centres

Solid 50 mm blocking
skew-nailed to girder truss

Fig. 5.11 Torsional restraints using plywood and timber diaphragms.

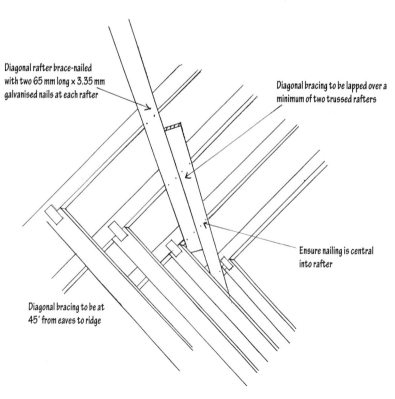

Diagonal rafter brace-nailed
with two 65 mm long × 3.35 mm
galvanised nails at each rafter

Diagonal bracing to be lapped over a
minimum of two trussed rafters

Ensure nailing is central
into rafter

Diagonal bracing to be at
45° from eaves to ridge

Fig. 5.12 Diagonal rafter bracing details.

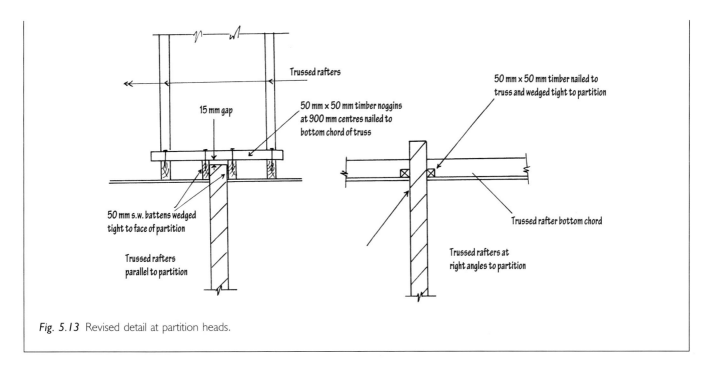

Fig. 5.13 Revised detail at partition heads.

## 5.6 Traditional cut roofs

This type of roof construction is essentially a roof, cut and assembled on site from individual sections of timber. For the simple duo-pitch roof with brick gables it consists of a series of roof spars spanning onto transverse purlins. Purlins can be timber, steel beams or – on larger spans – special glulam timber beams or fabricated timber girder trusses. Ceilings are supported from the purlins on hangers and ceiling binders.

When using solid section timber for purlins, standard tables are available that give the sizes required based on the span and area of roof being supported, and Table 5.4 (extracted from the Approved Documents Section A1/2) illustrates these. For situations where the purlins fall outside the standard tables they must be calculated using the appropriate allowable timber stresses, generally C16 or C24 grade.

One of the most common defects encountered in cut roofs occurs when the purlins are built in perpendicular to the roof spars. The sloping purlin has a tendency to sag down the roof slope as shown in Fig. 5.14, owing to

bending about its weaker axis. To prevent this occurring would require a much thicker timber section, which would be uneconomic.

In addition, the purlin cannot be seated easily into hangers at party wall positions, but the main problem is the tendency for the rafters to move down the slope, causing a horizontal thrust at the wall plate, which could lead to cracking in the masonry lower down. This horizontal thrust also increases proportionally to the vertical deflection in the purlin. For this reason purlins should never be designed with large deflections.

Despite what traditionalists say, the *golden rule* should be to provide ample purlins that are built in vertically. This allows the roof rafters to be birdsmouthed over the purlin, thus ensuring that the rafter loading is applied vertically and therefore preventing the tendency to 'sag' down the slope, with subsequent roof spread.

### 5.6.1 Ceiling ties

Owing to the large spans on domestic roofs it would be uneconomic to provide large-section ceiling joists to span between walls. Figure 5.15 shows a typical purlin and common rafter roof with the ceiling ties supported off vertical hangers via binders and purlin connections. Because of the long lengths required it is often necessary to form a joint in ceiling ties, and these joints should be designed to take the tension load in the ties resulting from the roof coupling action at the rafter feet.

Another defect that can occur results when ceiling or floor joists are built at right angles to the main rafters, as shown in Fig. 5.16. This is generally a result of bad design, but cannot be avoided on buildings with hipped roofs. The lack of a horizontal tie on two opposing sides can results in roof spread, and if the purlins deflect excessively, then major damage to the roof and masonry below can develop.

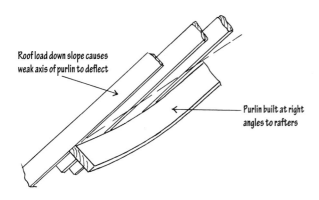

Fig. 5.14 Purlins sagging sideways.

*Table 5.4* Purlins supporting rafters
(imposed load 0.75 kN/m²). Maximum clear span of rafter (m). Timber of strength class SC3 and SC4

| | Dead load (kN/m²) excluding the self-weight of the rafter | | | | | | | | |
|---|---|---|---|---|---|---|---|---|---|
| | Not more than 0.50 | | | More than 0.50 but not more than 0.75 | | | More than 0.75 but not more than 1.25 | | |
| | Spacing of purlin (mm) | | | Spacing of purlin (mm) | | | Spacing of purlin (mm) | | |
| Size of purlin (mm × mm) | 400 | 450 | 600 | 400 | 450 | 600 | 400 | 450 | 600 |
| 38 × 100 | 2.28 | 2.23 | 2.10 | 2.10 | 2.05 | 1.91 | 1.96 | 1.91 | 1.76 |
| 38 × 125 | 3.07 | 2.95 | 2.69 | 2.87 | 2.77 | 2.52 | 2.65 | 2.56 | 2.35 |
| 38 × 150 | 3.67 | 3.53 | 3.22 | 3.44 | 3.31 | 3.01 | 3.26 | 3.14 | 2.85 |
| 47 × 100 | 2.64 | 2.54 | 2.31 | 2.45 | 2.38 | 2.17 | 2.28 | 2.21 | 2.04 |
| 47 × 125 | 3.29 | 3.17 | 2.88 | 3.09 | 2.97 | 2.70 | 2.92 | 2.81 | 2.56 |
| 47 × 150 | 3.93 | 3.78 | 3.45 | 3.69 | 3.55 | 3.23 | 3.50 | 3.37 | 3.06 |
| 50 × 100 | 2.69 | 2.59 | 2.36 | 2.53 | 2.43 | 2.21 | 2.38 | 2.30 | 2.09 |
| 50 × 125 | 3.35 | 3.23 | 2.94 | 3.15 | 3.03 | 2.76 | 2.98 | 2.87 | 2.61 |
| 50 × 150 | 4.00 | 3.86 | 3.52 | 3.76 | 3.62 | 3.30 | 3.57 | 3.44 | 3.13 |
| 38 × 89 | 1.91 | 1.87 | 1.77 | 1.77 | 1.73 | 1.62 | 1.67 | 1.62 | 1.50 |
| 38 × 140 | 3.43 | 3.30 | 3.01 | 3.22 | 3.10 | 2.82 | 3.05 | 2.93 | 2.66 |
| 38 × 100 | 2.56 | 2.47 | 2.24 | 2.40 | 2.31 | 2.10 | 2.28 | 2.19 | 1.99 |
| 38 × 125 | 3.19 | 3.07 | 2.80 | 2.99 | 2.88 | 2.62 | 2.84 | 2.73 | 2.48 |
| 38 × 150 | 3.81 | 3.67 | 3.35 | 3.58 | 3.45 | 3.14 | 3.39 | 3.27 | 2.97 |
| 47 × 100 | 2.74 | 2.64 | 2.41 | 2.58 | 2.48 | 2.25 | 2.44 | 2.35 | 2.13 |
| 47 × 125 | 3.41 | 3.29 | 3.00 | 3.21 | 3.09 | 2.81 | 3.04 | 2.93 | 2.66 |
| 47 × 150 | 4.08 | 3.93 | 3.59 | 3.83 | 3.69 | 3.36 | 3.64 | 3.50 | 3.19 |
| 50 × 100 | 2.80 | 2.70 | 2.45 | 2.63 | 2.53 | 2.30 | 2.49 | 2.40 | 2.18 |
| 50 × 125 | 3.48 | 3.35 | 3.06 | 3.27 | 3.15 | 2.87 | 3.10 | 2.99 | 2.72 |
| 50 × 150 | 4.16 | 4.01 | 3.66 | 3.91 | 3.77 | 3.43 | 3.71 | 3.57 | 3.25 |
| 38 × 89 | 2.28 | 2.20 | 2.00 | 2.14 | 2.06 | 1.87 | 2.03 | 1.95 | 1.77 |
| 38 × 140 | 3.56 | 3.43 | 3.13 | 3.35 | 3.22 | 2.93 | 3.17 | 3.05 | 2.77 |

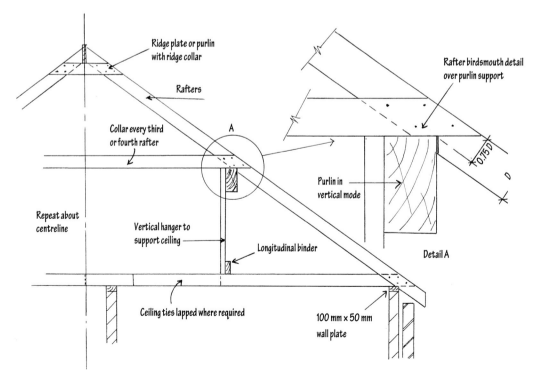

*Fig. 5.15* Cut roof with ceiling ties. Purlin in vertical mode.

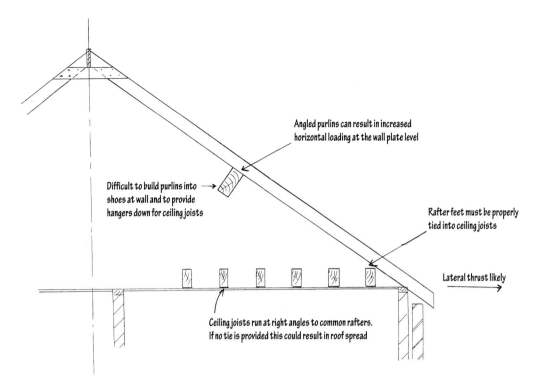

*Fig. 5.16* Lack of triangulation to roof spar feet.

It is important, therefore, when surveying such a roof, to note whether the roof rafters are triangulated, and that the roof geometry is arranged so as to form a coherent structure.

### 5.6.2 Hipped roofs

Hipped roof constructions tend to resolve wind and lateral forces through their overall geometry. On domestic roofs, with purlins properly supported and with rafter feet and hip rafters tied at the wall plate and corners, lateral forces are transferred into the ceiling plate, which transfers the loads into the supporting shear walls.

Figure 5.17 shows a typical dragon tie detail at the base of the hip rafter.

Defects can result in hipped roof constructions as a result of:

- purlins not built in vertically;
- purlins not supported properly off loadbearing walls built up from ceiling level. This may not always be possible, and props may be required supported off steel beams spanning transverse in the roof space onto the loadbearing walls;
- purlins not properly strapped together at corners to form a continuous collar, or purlins poorly scarfed over supporting walls;

- rafter feet not tied into ceiling where ceiling or floor joists spans are at right angles;
- purlins that are undersized;
- roof not having symmetrical geometry. This can result in out-of-balance forces in the hip rafters;
- defective timbers, large knots or shakes (more likely in large purlins).

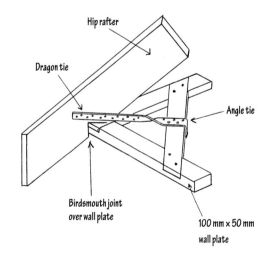

*Fig. 5.17* Dragon tie arrangement to hip girder.

## Case history 5.4

### Introduction

This dwelling had suffered from severe cracking to main internal loadbearing walls and at the junctions with the external walls. In addition the ceilings to the ground and first floor were badly cracked. The cracking is shown on the ground floor plan in Fig. 5.18.

### Survey results

In the first survey carried out, the chartered surveyor put the vertical cracking down to differential vertical loading. The internal masonry was built in lightweight autoclaved blocks, and these types of block have compressive strengths as low as 2.80 N/mm². They also suffer from abnormal contraction shrinkage movements. The owner of the house lodged a major structural damage claim with the warranty provider, and the roof was then inspected by a chartered engineer.

A check was carried out on the wall loadings, and these were found to be within the allowable compressive strength of the blocks. There was also no large differential loading occurring on the walls.

A detailed survey of the roof was carried out, and it became evident the problems resulted from a badly designed and constructed roof, which was spreading out on two sides. The main roof was a multi-hip and valley roof with unsymmetrical ridge lines, and it had been

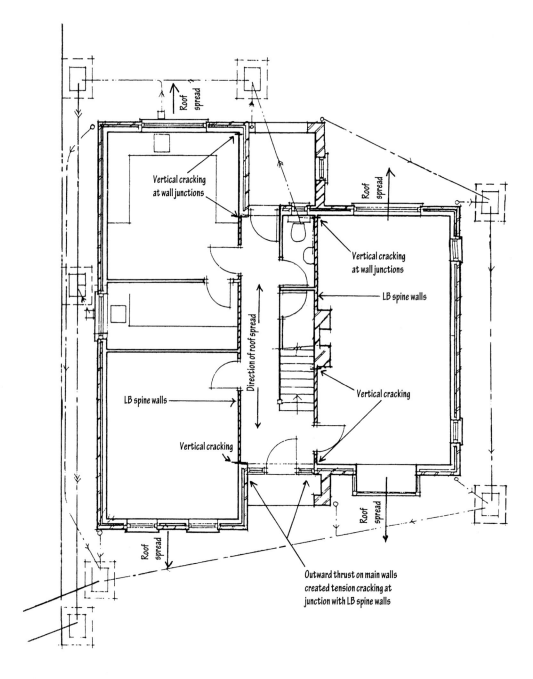

*Fig. 5.18* Ground floor plan.

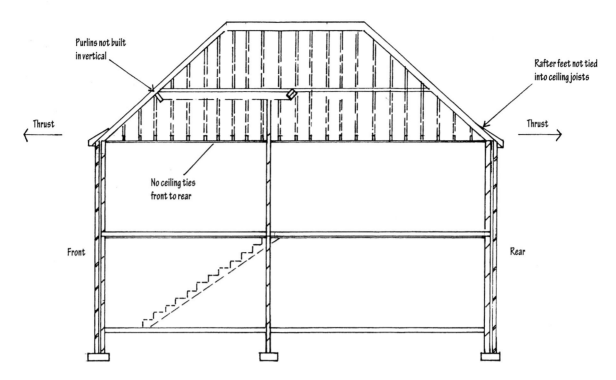

*Fig. 5.19* Cross-section through house showing roof spread.

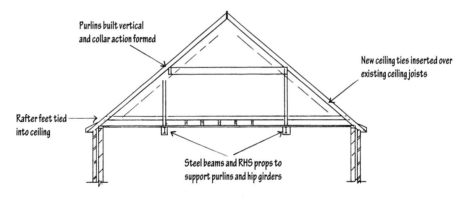

*Fig. 5.20* Steel beams supporting purlin props.

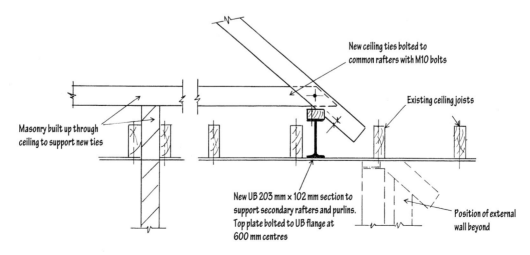

*Fig. 5.21* New ceiling ties to tie in spar feet.

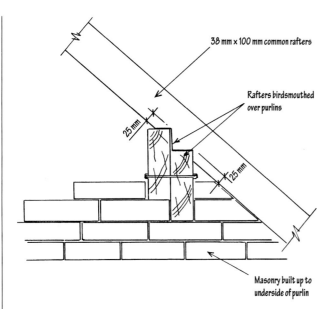

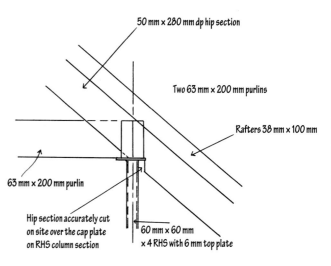

Fig. 5.23 Steel box support to purlins and hip girder.

Fig. 5.22 Rafters birdsmouthed over new vertical purlins on internal walls.

constructed using large timber purlins perpendicular to the roof spars, spanning onto blockwork piers built up from below.

On one side of the roof two steel beams had been inserted to provide support for the purlins, but these had been omitted on the other side, the purlins having been cantilevered an excessive span. In addition the end of the cantilevered purlin supported the return purlin as a point load, and this had caused a large deflection to occur. The roof spread was pushing out the front and rear walls.

This bowing out was initially resisted by the buttressing action of the two main internal loadbearing walls to the downstairs hallway and the upstairs landing. The lateral forces eventually separated the main walls from the

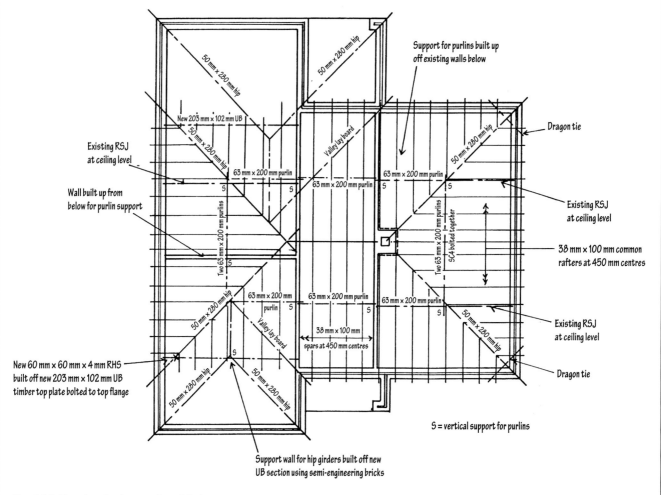

Fig. 5.24 Plan showing layout of roof timbers.

corridor walls, resulting in the internal cracking to walls and ceilings.

The following defects were noted:

- Purlins were not built vertical.
- Several purlins were cantilevered an excessive distance, and had deflections well above the allowable.
- The rafters at the front and rear roof slopes were not triangulated at the ceiling level.
- The hip girders and purlins on one side of the roof had no vertical support at their junction.
- Steel beams shown on the construction drawings to support piers for the purlins had been omitted.
- The asymmetry of the roof geometry was resulting in localised distortions in the roof.

## Remedial works

Figure 5.19 shows the original roof 'as built' with the direction of roof spread indicated. The remediation works required the roof to be totally removed and to provide protection from the elements a temporary roof was

installed off the scaffold using proprietary beams and lightweight roof sheeting.

The works carried out included:

- installing steel beams in the upper ceiling void to enable props to be built off to support the purlins and hip girders (Fig. 5.20);
- providing timber ties between the front and rear rafter feet (Fig. 5.21);
- removing badly deflected purlins, and forming a collar of purlins to the main body of the roof;
- removing all angled purlins and refixing them in a vertical mode, as shown in Figs 5.22 and 5.23.

Figures 5.24 and 5.25 show plans of the completed main roof timbers and details of the secondary hips and ceiling tie arrangement.

During the roof construction the lightweight block pillars supporting the purlins were replaced with second-class engineering bricks.

On completion of the roof repairs, all cracked walls and ceilings were repaired or replaced.

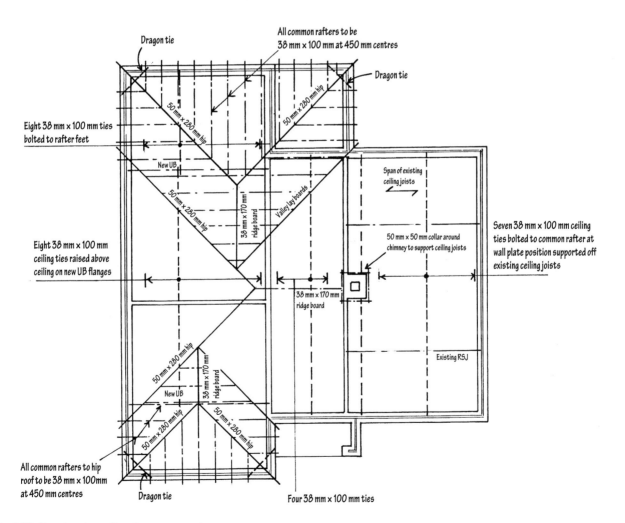

*Fig. 5.25* Plan showing ceiling tie arrangement.

## Case history 5.5

### Introduction

Following a major damage claim lodged with the NHBC by the owners of a detached bungalow, a structural survey of the roof was carried out. Using 25 mm wood blocks and a string line it was clear that the deflection of the ceiling was excessive at 24 mm, and clearly this was the cause of severe cracking to the ceiling finishes.

The roof construction was a traditional dual pitch with two 100 mm × 275 mm deep purlins spanning between gable walls over three spans. The purlins had been installed in a vertical mode, and the roof spars were well triangulated at ceiling level with an additional collar tie fixing the spars close to the ridge. Figure 5.26 shows the roof cross-section, and Fig. 5.27 shows the plan of the existing timber purlins and new steel purlins and supporting walls.

The purlin dimensions were measured, and check calculations confirmed that they were only adequate for the shorter end span of 3.20 m. However, it was decided to supplement all three spans with a steel channel bolted to the existing timber purlins. This ensured the minimum disturbance to the original roof timbers, and made it easier to link all the purlins together at the wall positions.

### Calculations

Total unit load on roof taken as 1.85 kN/m$^2$ with ceiling imposed load taken as 0.25 kN/m$^2$.
Central span 4.90 m
Purlin supports 2.675 m of roof, therefore:

Load/metre $= 2.675 \times 2.10 = 5.61$ kN/m

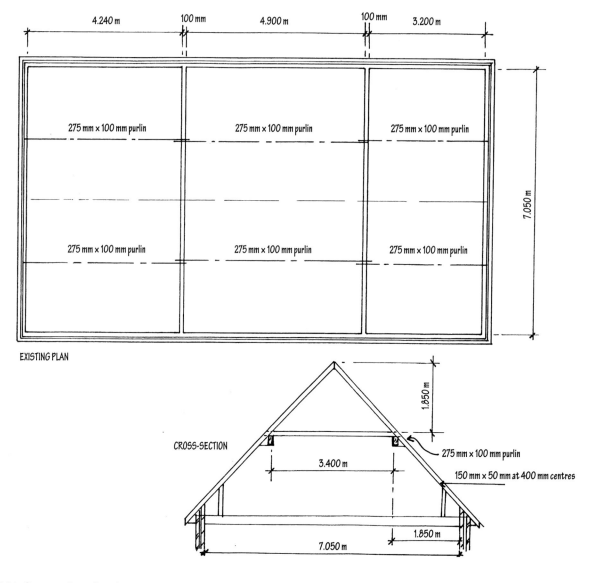

**Fig. 5.26** Cross-section of roof.

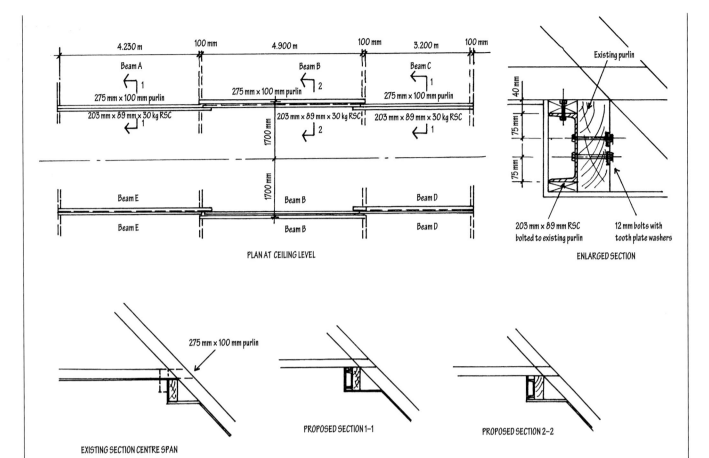

*Fig. 5.27* Plan at ceiling level showing new steel purlins.

Using SC3 grade timber:

$$\text{Maximum bending moment} = \frac{5.61 \times 4.90^2}{8} = 16.83 \text{ kNm}$$

Purlin 275 mm × 100 mm:

$$\text{Section modulus} = \frac{100 \times 275^2}{6} = 1260 \times 10^3 \text{ mm}^3$$

$$\text{Actual bending stress} = \frac{16.83 \times 10^6}{1260 \times 10^3} = 13.35 \text{ N/mm}^2$$

Permissible bending stress $= 5.30 \times 1.25 = 6.625 \text{ N/mm}^2$: therefore the timber purlins were grossly overstressed.

For the deflection check the minimum $E$ value of 5800 N/mm$^2$ must be used for a primary structural member without load sharing:

$$\text{Maximum deflection} = \frac{5 \times WL^3}{384 \ E \times I}$$

$$= \frac{5 \times 5.61 \times 10^3 \times 4.9 \times 4900^3 \times 12}{384 \times 5800 \times 100 \times 275^3}$$

$$= 41 \text{ mm}$$

where $W =$ total load on span (kN), $L =$ span, $E =$ moment of elasticity, and $I =$ second moment of area about $x$–$x$ axis.

This deflection was considered to be unacceptably high, and the claim was upheld.

## Remediation works

The two longest bays would therefore require the purlins to be changed or supplemented. To avoid major disturbance it was decided to install universal channel sections bolted to the purlins after they had been jacked up level. As the purlins had not been scarf-jointed over the internal walls it was decided to provide the new steel channel sections with a cantilevered plate section to support the new steel purlins coming from the central span.

At the supporting walls new concrete padstones were installed under the purlins prior to placing the new steel purlins.

## Calculations for new steel channel purlins (in accordance with BS 5950)

Try 203 × 89 × 30 kg/m RSC:

Span 4.90 m

$$\text{Imposed load/m}^2 = 1.0 \times 1.60$$
$$= 1.60 \text{ kN/m}^2 \times 2.675 = 4.28 \text{ kN/m}$$

$$\text{Dead load/m}^2 = 1.49 \times 1.40$$
$$= 2.08 \text{ kN/m}^2 \times 2.675 = 5.56 \text{ kN/m}$$
$$\text{Total} = 9.84 \text{ kN/m}$$

$$\text{Design bending moment} = \frac{9.84 \times 4.90^2}{8} = 29.53 \text{ kNm}$$

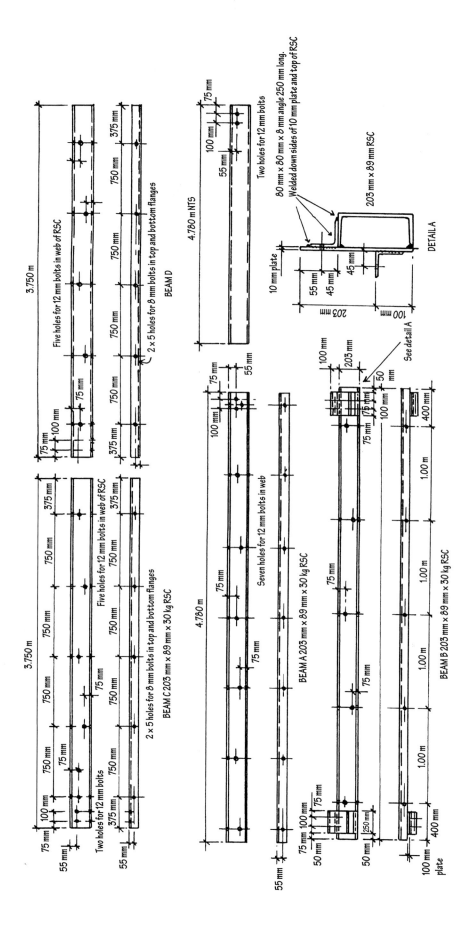

*Fig. 5.28* Beam fabrication details.

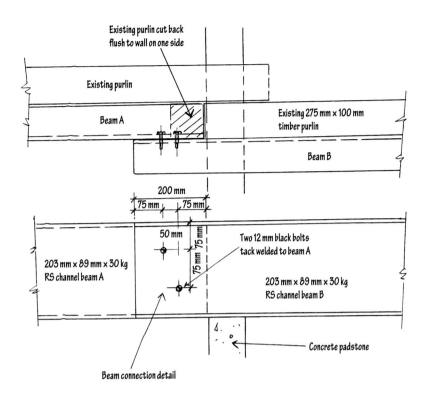

*Fig. 5.29* Steel channel fixing details.

## Effective length

Purlin is laterally unrestrained, since channel is pinned at both ends, therefore from BS 5950 Table 9:

$$L_e = 1.0L = 4.9 \times 10^3 = 4900 \text{ mm}$$

## Buckling resistance

Using a conservative approach of Clause 4.3.7.7:

$$\frac{L_e}{R_y} = \frac{4900}{26.40} = 185.60$$

where $R_y$ = minimum radius of gyration, and $L_e$ = effective length. From BS 5950 Table 20, the slenderness correction factor, $n = 0.94$
Therefore: slenderness ratio $= 0.94 \times 185.60 = 174.46$
Then with $D/t = 15.80$, $P_b = 131$ kN/mm$^2$ from BS 5950 Table 19
$$M_b = P_b \times S_x = 131 \times 286.6 \times 10^3 = 37.54 \text{ kNm}$$
where $P_b$ is the buckling strength, $M_b$ is the buckling resistance moment, and $S_x$ is the plastic section modulus about the $x–x$ axis.

As the buckling resistance moment is greater than the actual imposed moment this beam section will be adequate.

## Method statement for remediation

1 Remove plasterboard to ceiling.
2 Provide temporary propping under existing purlins and jack up as level as possible. Ensure that the props are adequately spread on floor below. If in doubt consult Engineer.
3 Cut out blockwork and provide as detailed concrete padstones on 1 : 1 : 6 mortar.
4 Props to be left in for at least 7 days until mortar has achieved its required strength.
5 Supply six 203 mm × 89 mm × 30 kg rolled steel channel sections, to be fabricated as shown in Fig. 5.28, to required lengths and prime them using a zinc phosphate primer.
6 Erect beams A, C, D and E first, bolting through predrilled holes in channel web into existing timber purlins using coach bolts and tooth plate washers at timber face. Channels to have 40 mm thick wall plate bolted to top flange prior to erecting.
7 Erect beams B and bolt onto cantilevered sections (Fig. 5.29). Bolt beam B to existing purlin.
   Skew nail existing ceiling ties to top wallplate using two 65 mm long × 3.35 mm diameter galvanised wire nails.
8 Bolt bottom timber packing to bottom flange of channel. Re-board ceiling, and make good ceiling finishes and repair plaster around padstones.

## Case history 5.6

### Introduction

This major structural damage claim related to a large double garage in Grimsby, which had developed a sagging roof with bowing fascia boards. An inspection of the roof revealed that the roof had been built using traditional spars onto timber purlins, which were built at right angles to the roof slope (Fig. 5.30). The purlins spanned onto two 200 mm × 35 mm mid-collar ties with a birdsmouthed joint in the top of the collar. This form of construction is generally referred to as the 'Grimsby roof'; it is structurally indeterminate, and relies on good craftsmanship. The roof spars sat onto a site-manufactured timber infill panel made up from 100 mm × 50 mm timbers nailed together into the end grain of the vertical sections (Figs 5.31 and 5.32).

This timber infill beam was constructed on top of the steel beam over the garage doors, and was a very weak element in the overall construction.

Only one ceiling tie and one vertical hanger had been installed, and this had resulted in lateral spreading of the spars at the wall plate. This movement had been exacerbated because the raised timber infill above the front door beam was very weak; the largest movement had occurred over the garage front.

Following a detailed survey of the roof it was decided that the most economic repair would be to remove the existing roof structure and replace it with a trussed rafter construction that would have no lateral forces at the wall plate. To support the trusses a single skin of brickwork was used to replace the timber infill panels, and this required the existing 178 mm × 102 mm rolled steel joist to be replaced with a 203 mm × 102 mm × 23 kg universal beam (Fig. 5.33).

Figures 5.34 and 5.35 show the repair solution using the trussed rafter roof built off the existing walls, reusing the spars and tiles.

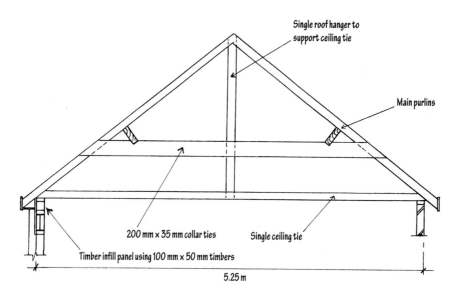

Fig. 5.30 Existing Grimsby roof to garage.

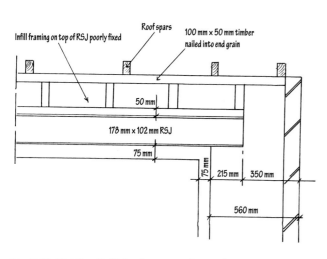

Fig. 5.31 Details of infill framing supporting roof spars.

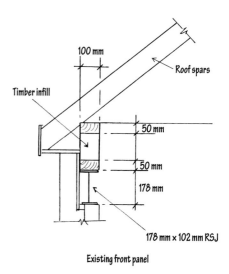

Fig. 5.32 Section through front infill beam.

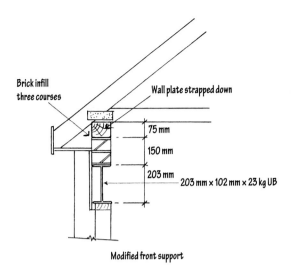

*Fig. 5.33* New front infill beam detail.

## Check on existing steel beam over front door

Designed in accordance with BS 449 (including amendment No. 8, 1990) using Grade 43 steel.

Existing section 178 mm × 102 mm × 19 kg/m RSJ.

The top flange of the beam had no effective lateral restraint: therefore permissible stresses will be reduced in accordance with Clause 19(a) and BS 449 Table 3:

|  | kN/m$^2$ |
|---|---|
| Effective span = 4.61 m |  |
| Loading to roof: tiles | 0.55 |
| battens/felt | 0.05 |
| trusses | 0.20 |
| storage | 0.25 |
| snow load | 0.75 |
| Total = | 1.80 |

$$\text{Load on beam from trusses} = \frac{1.80 \times 5.90}{2.0} = 5.31 \text{ kN/m}$$

160 mm brickwork over beam $2.20 \times 0.160 = 0.35$

Total = 5.66 kN/m

$$\text{Maximum bending moment} = \frac{5.66 \times 4.61^2}{8} = 15.03 \text{ kNm}$$

Section modulus, $Z = 153$ cm$^3$:

$$\text{bending stress} = \frac{15.03 \times 10^6}{153} = 98.23 \text{ N/mm}^2$$

$$\frac{D}{T} = 22.50 \qquad \frac{L}{r_{yy}} = \frac{4.61 \times 10^3}{23.90} = 192$$

where $D$ = depth of steel section, $L$ = span of beam, $T$ = flange thickness, and $r_{yy}$ = radius of gyration about $y$–$y$ axis.

Permissible stress from BS 449 Table 3 = 79.63 N/mm$^2$

This beam was therefore not suitable for dead load and imposed loads from snow and storage, and a new section was essential.

Try 203 mm × 102 mm × 23 kg/m:

Section modulus, $Z = 206$ cm$^3$

$$\frac{D}{T} = 21.80 \qquad \frac{L}{r_{yy}} = \frac{4.610 \times 10^3}{23.70} = 194$$

Permissible stress = 80.41 N/mm$^2$

$$\text{Actual stress} = \frac{15.03 \times 10^6}{206} = 72.96 \text{ N/mm}^2$$

Deflection (dead load + imposed)

$$= \frac{5 \times (5.66 \times 4.61) \times 10^3 \times 4610^3}{384 \times 210\,000 \times 2090 \times 10^4} = 7.58 \text{ mm}$$

$$\text{Permissible deflection} = \frac{\text{span}}{360} = \frac{4610}{360} = 12.80 \text{ mm}$$

Use a 203 mm × 102 mm × 23 kg universal beam with a 100 mm × 50 mm timber wall plate bolted on to the top flange using 10 mm bolts at 900 mm centres on staggered pitch.

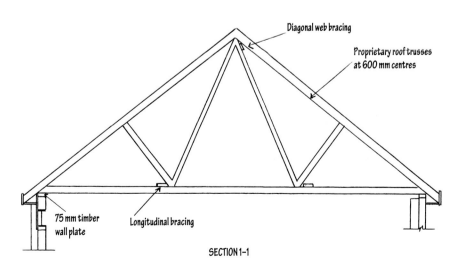

SECTION 1–1

*Fig. 5.34* New trussed rafter roof.

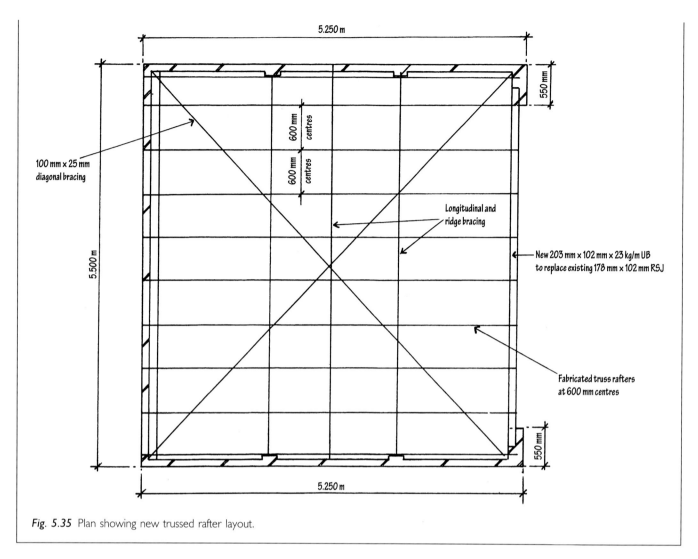

**Fig. 5.35** Plan showing new trussed rafter layout.

Tables 5.5–5.7 are span tables modified from those given in BS 5268 : Part 3 *Code of practice for trussed rafter roofs*. They have been modified to accommodate the additional loads created by water storage tanks, and are based on fink and fan truss configurations.

Reduced limiting spans are given to accommodate tanks shown in Table 5.2. However, as an alternative,

the limiting spans given in BS 5268 Part 3 (1985) may be used, but the spacing should be closed up beneath the tank support area so that S (spacing) equals 480 mm maximum, where the tank support area for this purpose should be taken as a total of five spaces for 450 L tanks and four spaces for 315 L tanks.

**Table 5.5** Rafter spans for fink trusses (mm)

|  | Roof slope (degrees) | | | | | | | | |
|---|---|---|---|---|---|---|---|---|---|
| Actual size (mm) | 15 | 17.5 | 20 | 22.5 | 25 | 27.5 | 30 | 32.5 | 35 |
| 35 × 72 | 5 717 | 5 886 | 6 047 | 6 193 | 6 314 | 6 422 | 6 537 | 6 651 | 6 763 |
| 35 × 97 | 7 150 | 7 380 | 7 573 | 7 739 | 7 892 | 8 030 | 8 167 | 8 311 | 8 464 |
| 35 × 120 | 8 459 | 8 700 | 8 933 | 9 130 | 9 324 | 9 483 | 9 640 | 9 815 | 10 008 |

**Table 5.6** Rafter spans for fan trusses (mm)

|  | Roof slope (degrees) | | | | | | | | |
|---|---|---|---|---|---|---|---|---|---|
| Actual size (mm) | 15 | 17.5 | 20 | 22.5 | 25 | 27.5 | 30 | 32.5 | 35 |
| 35 × 72 | 7 607 | 8 005 | 8 304 | 8 566 | 8 804 | 9 018 | 9 229 | 9 437 | 9 654 |
| 35 × 97 | 9 446 | 9 976 | 10 315 | 10 639 | 10 930 | 11 000 | 11 000 | 11 000 | 11 000 |
| 35 × 120 | 10 530 | 11 000 | 11 000 | 11 000 | 11 000 | 11 000 | 11 000 | 11 000 | 11 000 |

*Table 5.7* Ceiling joist spans for fink and fan trusses (mm)

| Actual size (mm) | Roof slope (degrees) | | | | | | | | |
|---|---|---|---|---|---|---|---|---|---|
| | 15 | 17.5 | 20 | 22.5 | 25 | 27.5 | 30 | 32.5 | 35 |
| 35 × 72 | 4 894 | 5 158 | 5 397 | 5 622 | 5 835 | 6 026 | 6 225 | 6 423 | 6 600 |
| 35 × 97 | 6 820 | 7 178 | 7 521 | 7 849 | 8 144 | 8 427 | 8 708 | 8 967 | 9 246 |
| 35 × 120 | 8 424 | 8 895 | 9 310 | 9 721 | 10 098 | 10 462 | 10 815 | 11 000 | 11 000 |
| 35 × 145 | 9 908 | 10 483 | 11 000 | 11 000 | 11 000 | 11 000 | 11 000 | 11 000 | 11 000 |

## Points to remember

- Ensure that water tanks in trussed roofs are adequately supported and spread over three or four trusses.
- Make sure the trusses are designed for the correct snow loading as defined in BS 6399 Part 3.
- Check that raised tie trusses have no more than 6 mm lateral deflection at each support.
- Ensure that all trusses are plumb and adequately braced.
- Always specify kiln-dried timber for structural use.
- Protect new roof trusses from the weather, and ensure that they are adequately stacked on site, using timber bearers.
- Where girder trusses are supporting secondary trusses with spans greater than 8 m provide torsional restraint behind the girder truss.
- Always construct purlins in a vertical mode, and ensure that rafters are properly birdsmouthed over purlins and wall plates.
- Always check to ensure that the rafter feet and ceiling ties are fixed together to ensure the roof is properly triangulated.

## References

*British Standards Institution*

BS 447 : 1987 *Specification for sizes of sawn and processed softwood.*
BS 449 : 1969 *Specification for the use of structural steel in building*
BS 648 : 1964 *Schedule of weights of building materials.*
BS 1282 : 1999 *Wood preservatives. Guidance on choice, use and application.*
BS 1579 : 1960 *Specification for connectors for timber.*
BS 4072 : 1992 *Copper/chromium/arsenic preparations for wood preservation.*
BS 4169 : 1988 *Specification for manufacture of glued-laminated timber structural members.*
BS 4978 : 1996, Amendment 1 *Specification for visual strength grading of softwood.*
BS 5268 *Structural use of timber*
  Part 2 : 1996 *Code of practice for permissible stress design, materials and workmanship*
  Part 3 : 1998 *Code of practice for trussed rafter roofs*
  Part 5 : 1989 *Code of practice for the preservative treatment of structural timber.*
BS 5589 : 1989 *Code of practice for the preservation of timber.*
BS 5950 *Structural use of steelwork in building*
  Part 1 : 1990 *Code of practice for design in simple and continuous construction: hot rolled sections.*
BS 6399 *Loading for buildings*
  Part 1 : 1996 *Code of practice for dead and imposed loads*
  Part 2 : 1997 *Code of practice for wind loads*
  Part 3 : 1988 *Code of practice for imposed roof loads.*

*Building Research Establishment*

*Loads on roofs from snow drifting against vertical obstructions,* BRE Digest 332 (1988).

*National House-Building Council*

*NHBC Standards,* Vol. 2, Chapters 7.1 and 7.2 (1999).

The Building Regulations, Approved Documents, Part A1 (1991)

J.A. Baird and E.C. Ozelton, *Timber designer's manual,* 2nd edn, Collins (1984).
*The Gang-Nail trussed rafter manual,* Gang Nail Systems Ltd.

*Trussed Rafter Association (TRA)*

*Technical handbook – site installation guide* (1999).

# Chapter 6
# Upper floors

## 6.1 Introduction

In low-rise dwellings upper floors are generally constructed using timber joists and timber boarding. This form of construction generally meets the Building Regulation requirements for fire and sound.

Where flats are concerned, upper floors need to have more mass to reduce the sounds that can be transmitted between floors, and also to provide better fire resistance. These floors are generally constructed using in-situ reinforced concrete, precast concrete beams with infill blocks, or prestressed planks with structural screeds.

## 6.2 Timber floors

The main structural duties of a timber floor are performed by individual joists. In a plated floor these joists should act together in carrying imposed loads and point loads without excessive deflections. There is a degree of load sharing from weaker joists onto stronger neighbouring joists, and where more than four joists are involved a factor of 1.10 can be used for bending stresses. For joists with spacings not exceeding 600 mm the permissible bending stresses can be increased by 10% to take account of this load sharing effect.

With long-span joists, this load sharing is best achieved by the provision of herringbone strutting or solid blockings between joists. These struttings need to be well fixed and anchored at the main walls.

The stiffness of a timber floor is increased even further once the plasterboard ceiling and top decking are fixed in place. Most floors are decked out in chipboard but for special situations where good composite action is needed the floor decking can be plywood sheeting. When fully nailed or screwed this results in a stronger stressed skin floor panel but it is a more expensive form of construction.

### Fire resistance

The first floor of a two-storey house is required by fire regulations to have a 30 min resistance in regard to load-carrying capacity, i.e. stability, but only 15 min resistance to flame penetration, i.e. integrity. This is referred to as the *modified half-hour* requirement.

First floors in two-storey flats, intermediate floors in maisonettes, and all upper floors in three-storey houses require the full 30 min fire resistance in all aspects.

Floors between flats or maisonettes in three- or four-storey domestic buildings are required to have 60 min fire resistance.

Where floors are built without ceilings, the flooring and joists need to be checked for adequacy using the charred sections.

**Major structural defects** can occur in a timber floor as a result of:

- inadequate design and specification;
- the use of poor-quality timber and inadequate components;
- bad workmanship on site.

### 6.2.1 Inadequate design and specification

Design defects that occur with timber floors can result from:

- failure to accurately assess the unit *dead loads*;
- failure to accurately assess the unit *imposed loads* for the building class;
- no proper allowance for partition loading, those parallel to the floor joists or those transverse to the joists;
- the use of an incorrect modulus of elasticity, $E$, when working out the floor deflections.

### Dead loads

For dead loads BS 648 *Schedules of weights of building materials* can be used to determine the unit floor loads and partition loadings. Table 6.1 lists the most commonly used materials for floors and partitions.

### Imposed loads

These depend on the class of building, e.g. domestic, commercial offices, or stores, and reference should be

**Table 6.1** Dead loads of flooring materials and partitions

| Element of structure | Unit weight (kN/m²) |
|---|---|
| Chipboard | |
| 18 mm thick | 0.150 |
| 22 mm thick | 0.170 |
| T & G boarding | |
| 19 mm thick | 0.100 |
| 25 mm thick | 0.170 |
| Plasterboard | |
| 9.50 mm thick | 0.083 |
| 12.7 mm | 0.112 |
| 19.1 mm thick | 0.171 |
| Wet plaster | |
| 3 mm skim coat | 0.068 |
| 2 coats to walls | 0.220 |
| Stud partitions (excluding plaster and skim) | |
| 75 mm × 50 mm studding at 400 centres | 0.120 |
| 75 mm × 50 mm studding at 350 centres | 0.137 |
| 75 mm × 50 mm studding at 300 centres | 0.159 |
| 75 mm × 75 mm studding at 400 centres | 0.179 |
| 75 mm × 75 mm studding at 350 centres | 0.205 |
| 75 mm × 75 mm studding at 300 centres | 0.239 |
| 100 mm × 50 mm studding at 400 centres | 0.159 |
| 100 mm × 50 mm studding at 350 centres | 0.182 |
| 100 mm × 50 mm studding at 300 centres | 0.212 |

made to BS 6399 Part 1: 1996 *Loadings for buildings. Code of practice for dead and imposed loads.*

Timber floors should be designed in accordance with the following British Standards:

- BS 8103 Part 3: 1996 *Structural design of low-rise buildings. Code of practice for timber floors and roofs for housing*
- BS 6399 Part 1: 1996 *Loadings for buildings. Code of practice for dead and imposed loads.*
- BS 5268 Part 2: 1996 *Structural use of timber. Code of practice for permissible stress design, materials and workmanship.*

### Spacing of joists

The former LCC (London County Council) Bye-Laws gave clear guidance on the spacing of joists in relation to the type of decking materials used and the ceiling finishes, and these are detailed in Table 6.2.

**Table 6.2** Suggested joist spacings

| Material | Maximum joist spacing (mm) | NHBC requirements[a] (mm) |
|---|---|---|
| 20 mm tongue and groove boarding | 450 | 16 |
| 25 mm tongue and groove boarding | 600 | 19 |
| 19 mm tongue and groove plywood | 600 | |
| 10 mm plasterboard | 400 | |
| 12.50 mm plasterboard | 450 | |

[a] For comparison

### Strutting and stiffening

Floor joists 50 mm and thinner tend to bow, and can cause cracking in ceilings. It is a general rule to provide strutting for floor spans in excess of 3.0 m, but this is also worth doing on smaller spans as it creates a much stiffer floor and minimises ceiling movement. Table 6.3 indicates the strutting requirements.

All strutting should be placed before the floor decking is placed. The strutting should be 38 mm × 38 mm timber herringbone strutting, proprietary steel herringbone strapping, or 38 mm thick solid timber strutting, as shown in Fig. 6.1.

The depth of the solid strutting should be not less than three-quarters of the floor joist depth. At the ends of the rows of strutting solid timber blocking pieces should be provided between the end joist and the masonry, as shown in Fig. 6.1.

**Table 6.3** Strutting to floor joists

| Joist span (m) | Rows of strutting |
|---|---|
| Up to 2.50 | None required |
| 2.50–4.50 | One row at mid-span |
| Over 4.50 | Two rows at equal spacings |

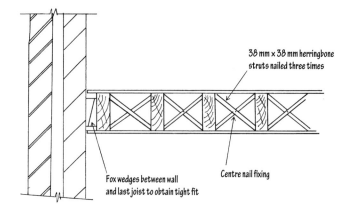

**Fig. 6.1** Herringbone strutting to timber floor joists.

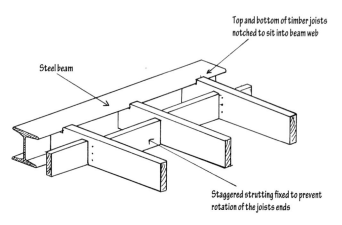

**Fig. 6.2** Joists built into steelwork.

Table 6.4 Minimum imposed floor loads

| | Uniformly distributed load (kN/m²) | Concentrated load (kN) |
|---|---|---|
| *Domestic and residential activities* | | |
| Self-contained dwelling units; communal areas in blocks of flats | 1.50 | 1.40 |
| Bedrooms and dormitories | 1.50 | 1.80 |
| Bedrooms in hotels and motels; hospital wards; toilet areas | 2.0 | 1.80 |
| Billiard rooms | 2.0 | 2.70 |
| Communal kitchens | 3.0 | 4.50 |
| Balconies | | |
| Single dwelling units and communal areas in flats | 1.50 | 1.40 |
| Guest houses; residential clubs | Same as the rooms to which they give access | 1.50/m run, concentrated at the outer edge |
| Hotels and motels | Same as the rooms to which they give access, but with a minimum of 4.0 | 1.50/m run, concentrated at the outer edge |
| *Offices and work areas not covered elsewhere* | | |
| Operating theatres, X-ray rooms and utility rooms | 2.0 | 4.50 |
| Banking halls | 3.0 | 2.70 |
| Kitchens, launderies, labs | 3.0 | 4.50 |
| Computer rooms | 3.50 | 4.50 |
| Projection rooms | 5.0 | |
| Factories, workshops and similar buildings | 5.0 | 4.50 |
| *Areas where people may congregate* | | |
| Public, institutional dining rooms, cafes and restaurants | 2.0 | 2.70 |
| Reading rooms (no storage) | 2.50 | 4.50 |
| Classrooms | 3.0 | 2.70 |
| Churches | 3.0 | 2.70 |
| Assembly areas with fixed seating | 4.0 | 3.60 |

Where joists are supported on steel beams or hangers, solid blocking should be provided at the ends between the joists to prevent rotation of the joists, as shown in Fig. 6.2.

Table 6.4 provides information on some of the various loading classes for the most common types of building in the low-rise category.

*Modulus of elasticity, E*

When working out deflections in floor joists the value of E used is the mean value for load-sharing joists.

When working out deflections for primary members such as trimmer joists or timber beams then the minimum E value must be used.

For most general purposes, including domestic floor loads, the maximum permitted deflection is 0.003 of the span, up to a maximum of 14 mm.

For certain loading conditions it is sometimes necessary to calculate the deflection due to shear as it may be significant and this deflection is added to the bending deflection.

Where timber beams comprise several members bolted together, BS 5268 Table 18 allows a modification factor $k_9$ to be applied to the E value to take account of the varying strengths of the individual timbers making up the beam.

For two timbers $k_9$ is 1.14, for three timbers 1.21, and for four timbers or more 1.24 when using softwoods. Table 6.5 lists the E values for the various strength classes.

*Partitions*

When permanent partitions are indicated, their weight must be included in the dead load, acting at the particular partition location.

In buildings where the use of other partitions is envisaged, an additional imposed load should be specified for the whole floor area. This may be taken as

Table 6.5 Dry grade stresses and moduli of elasticity

| Strength class | Bending stress (N/mm²) | Modulus of elasticity, mean (N/mm²) | Modulus of elasticity, minimum (N/mm²) |
|---|---|---|---|
| SC1 | 2.80 | 6800 | 4500 |
| SC2 | 4.10 | 8000 | 5000 |
| SC3 | 5.30 | 8800 | 5800 |
| SC4 | 7.50 | 9900 | 6600 |
| SC5 | 10.00 | 10700 | 7100 |
| SC6 | 12.50 | 14100 | 11800 |
| SC7 | 15.00 | 16200 | 13600 |
| SC8 | 17.50 | 18700 | 15600 |
| SC9 | 20.50 | 21600 | 18000 |

a UDL of not less than one-third of the load per metre run of the finished partition loads. For floors of offices, this additional uniformly distributed partition loading should not be less than 1.0 kN/m².

The majority of partitions used are plasterboard on timber studding or paramount partitions with a plaster skim coat. However, on some occasions masonry partitions may be supported off timber joist floors, and this method of construction generally leads to structural defects in the walls and floors. The NHBC standards strongly recommend that this form of construction be avoided, but if it is used the timber sections used must be designed to minimise deflections in the supporting timbers. The main problems are caused by the timber shrinking, which causes the masonry to crack, added to the timber deflecting, which causes the brittle masonry over to crack. In such situations a steel beam is the best technical solution for blockwork partitions, or consideration should be given to using timber stud partition, with the floor joists designed accordingly.

## Case history 6.1

### Introduction

The structural defects in this case history resulted from a combination of inadequate design and poor workmanship on site. The owners of the property, which was only 4 years old, had submitted a major structural damage claim to the warranty provider. Figure 6.3 shows the existing floor joist layout.

The two-storey detached house had developed cracking in the ground-floor ceilings and the upstairs partition walls. During investigation of the cause of the damage it was established that the partition walls were clinker blockwork, and these were built off the timber joist floor. Some of the partitions ran parallel to the floor joists, and when the flooring was opened up it was found that there had been a non-compliance with the NHBC requirements for mid-floor strutting and beam infill at supports.

Using a string line and timber blocks it was established that there was a deflection in the ceiling below the partitions of 22 mm. In other areas the block partition was supported by the flooring only. Such a practice could have resulted in serious problems during the investigation, as sections of the flooring required to be moved. Fortunately the decision to remove small sections of flooring turned out to be a wise one. Owing to the form of construction and the potential for ongoing movement the claim was accepted, and a remediation scheme was prepared as described.

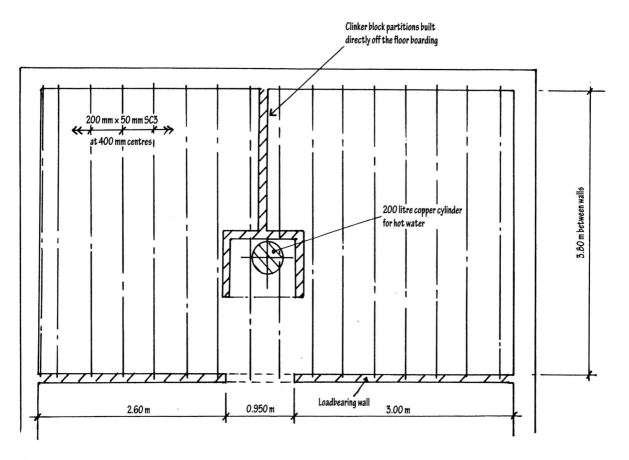

*Fig. 6.3* Existing floor joist layout.

Figure 6.4 shows the modified floor joist arrangement, with the block walls replaced by timber stud partitions.

Where timber stud partitions run parallel to the joists it is an NHBC requirement that joists have to be doubled up and fastened together under the partition.

Where partitions are of masonry construction, NHBC Chapter 6.4 – D4 (e) states that steel or concrete beams should be used to avoid the effects of shrinkage and long-term deflection, which can occur with timber beams. It also states that it is not acceptable to support masonry partitions on timber joists unless the joists are designed by an engineer in accordance with NHBC Technical Requirement R5.

Figure 6.3 shows the floor layout supporting the partitions. It was decided that the best way to resolve the problem was to take the blockwork partitions down and rebuild timber stud partitions off strengthened joists. The wall returns were supported on timber noggins. The modified layout is shown in Fig. 6.4, and the floor sections are shown in Figs 6.5 and 6.6.

Had the owners insisted on retaining the upper blockwork walls the floor would have required a steel beam system inserted to carry the blockwork partitions, and substantial propping would have been required to support the walls. As the stud walls were to be packed with glass fibre insulation the owners agreed to this remedy, which made the repairs a simpler job.

## Checking the existing floor design

Existing joists were 200 mm × 50 mm SC3 grade timber at 450 mm centres.

Span = 3.80 m clear. Total floor loading = 2.0 kN/m²

Loading on joist = 0.45 × 2.0 = 0.90 kN/m

$$\text{Bending movement} = \frac{0.90 \times 3.90^2}{8} = 1.71 \text{ kN m}$$

$$\text{Section modulus} = \frac{50 \times 200^2}{6} = 333.33 \times 10 \text{ mm}^3$$

With load sharing, $K_8 = 1.10$

$$\text{Depth factor, } K_7 = \left(\frac{300}{200}\right)^{0.11} = 1.046 \quad \text{(see Table 6.6)}$$

Basic stress for SC3 timber grade = 5.30 N/mm²

Therefore allowable design stress = 5.30 × 1.10 × 1.046

$$= 6.09 \text{ N/mm}^2$$

$$\text{Required section modulus} = \frac{1.71 \times 10}{6.09}$$

$$= 280 \times 10^3 \text{ mm}^3 < 333.333 \times 10^3$$

Deflection = Bending deflection + Shear deflection

$$= \frac{5 \times WL^3}{384EI} + \frac{12 \times 3.42 \times 10^3 \times 3.8 \times 10^3}{5 \times EA}$$

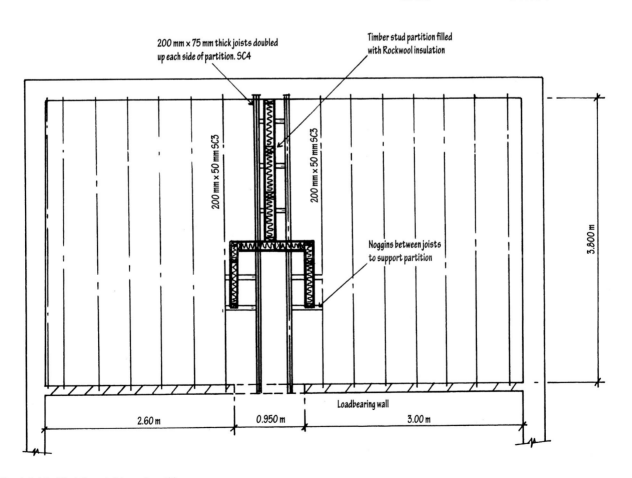

200 mm x 75 mm thick joists doubled up each side of partition. SC4

Timber stud partition filled with Rockwool insulation

200 mm x 50 mm SC3

200 mm x 50 mm SC3

Noggins between joists to support partition

Loadbearing wall

3.800 m

2.60 m

0.950 m

3.00 m

*Fig. 6.4* Modified floor joists and partitions.

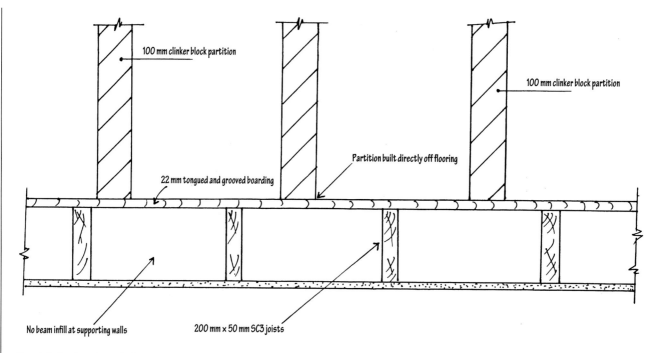

*Fig. 6.5* Section through floor showing existing construction.

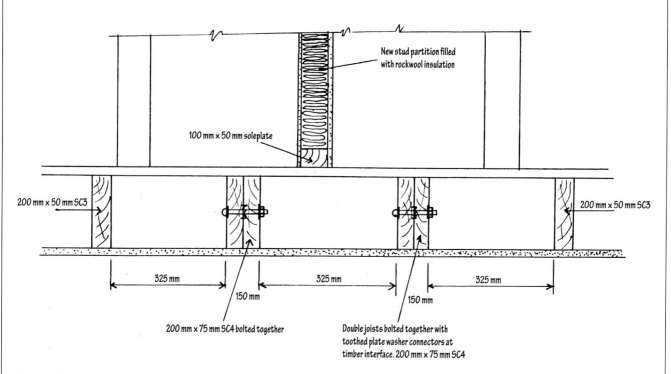

*Fig. 6.6* Section showing new joists and partitions.

$E_{\text{mean}} = 8800 \text{ N/mm}^2$ and

Area, $A = 50 \times 200 = 10.0 \times 10^3 \text{ mm}^2$

The second movement of area, $I$, is given by

$$I = \frac{50 \times 200^3}{12} = 33.333 \times 10^6 \text{ mm}$$

Deflection = Bending deflection + Shear deflection

$$= \frac{5 \times 3.42 \times 10^3 \times (3900)^3}{384 \times 8800 \times 33.33 \times 10^6}$$

$$+ \frac{12 \times 3.42 \times 10^3 \times 3.8 \times 10^3}{5 \times 8800 \times 10.0 \times 10^3}$$

$$= 9 + 0.35 = 9.35 \text{ mm} < 3800 \times 0.003$$

$$= 11.40 \text{ mm}$$

**Table 6.6** Depth modification factor $K_7$ and width modification factor $K_{14}$

| $h$ (mm) | $K_7$ | $h$ (mm) | $K_7$ | $h$ (mm) | $K_{14}$ | $h$ (mm) | $K_{14}$ |
|---|---|---|---|---|---|---|---|
| 72 | 1.17 | 270 | 1.012 | 72 | 1.17 | 270 | 1.012 |
| 75 | 1.165 | 285 | 1.006 | 75 | 1.165 | 285 | 1.006 |
| 89 | 1.143 | 300 | 1.000 | 89 | 1.143 | 300 | 1.000 |
| 97 | 1.132 | 315 | 0.994 | 97 | 1.132 | 315 | 0.983 |
| 100 | 1.128 | 350 | 0.970 | 100 | 1.128 | 350 | 0.980 |
| 122 | 1.104 | 360 | 0.964 | 122 | 1.104 | 360 | 0.969 |
| 125 | 1.101 | 400 | 0.943 | 125 | 1.101 | 400 | 0.968 |
| 140 | 1.087 | 405 | 0.940 | 140 | 1.087 | 405 | 0.956 |
| 145 | 1.083 | 450 | 0.921 | 145 | 1.083 | 450 | 0.946 |
| 147 | 1.082 | 495 | 0.905 | 147 | 1.082 | 495 | 0.945 |
| 150 | 1.079 | 500 | 0.904 | 150 | 1.079 | 500 | 0.937 |
| 169 | 1.065 | 540 | 0.893 | 169 | 1.065 | 540 | 0.936 |
| 170 | 1.064 | 550 | 0.890 | 170 | 1.064 | 550 | 0.929 |
| 175 | 1.061 | 585 | 0.882 | 175 | 1.061 | 585 | 0.927 |
| 180 | 1.058 | 600 | 0.879 | 180 | 1.058 | 600 | 0.922 |
| 184 | 1.055 | 630 | 0.873 | 184 | 1.055 | 630 | 0.915 |
| 194 | 1.049 | 675 | 0.866 | 194 | 1.049 | 675 | 0.911 |
| 195 | 1.049 | 700 | 0.863 | 195 | 1.049 | 700 | 0.908 |
| 200 | 1.046 | 720 | 0.860 | 200 | 1.046 | 720 | 0.902 |
| 219 | 1.035 | 765 | 0.855 | 219 | 1.035 | 765 | 0.898 |
| 220 | 1.035 | 800 | 0.851 | 220 | 1.035 | 800 | 0.896 |
| 225 | 1.032 | 810 | 0.850 | 225 | 1.032 | 810 | 0.891 |
| 235 | 1.027 | 855 | 0.846 | 235 | 1.027 | 855 | 0.886 |
| 250 | 1.020 | 900 | 0.843 | 250 | 1.020 | 900 | |

$k_7 = 1.17$ for solid timber beams having a depth of 72 mm or less

$k_7 = \left(\dfrac{300}{h}\right)^{0.11}$ for solid and glued laminated members having a depth between 72 mm and 300 mm, where $h$ is the depth of the joist section

$k_7 = 0.81 \times \left(\dfrac{h^2 + 92300}{h^2 + 56800}\right)$ for solid and glued laminated members having a depth greater than 300 mm

Width modification factor

$k_{14} = 1.17$ for solid timber members having a width of 72 mm or less

$k_{14} = \left(\dfrac{300}{h}\right)^{0.11}$ for solid and glued laminated members having a width greater than 72 mm

Therefore the 50 mm × 200 mm SC3 joists are adequate in bending and deflection in those areas where only flooring is being supported.

### Consider joists supporting clinker blockwork and flooring

Weight of clinker block plastered both sides is $1.25 + 0.50 = 1.75$ kN/m$^2$

Height of partition = 2.30 m, which gives a load/metre of $1.75 \times 2.30 = 4.025$ kN/m

Point load from partition returns = $0.45 \times 2.30 \times 1.75$

$$= 1.80 \text{ kN}$$

Hot water tank in cylinder cupboard = 200 L = 2 kN shared over two joists.

Joists B and C carry the block wall flooring and water tank

$R_l = 5.80$ kN   $R_r = 9.10$ kN

Maximum bending movement = 8.40 kN m

Equivalent UDL = 4.64 kN/m run and shared by two joists = 2.32 kN/m

For dead + imposed loading use mean $E$ value of 8800 N/mm$^2$ from BS 5268 Part 2, Table 8:

Maximum deflection = Bending deflection

$\qquad\qquad$ + Shear deflection

$$= \frac{5 \times (2.32 \times 3.80) \times 10^3 \times 3800)^3}{384 \times 8800 \times 33.33 \times 10^6}$$

$$+ \frac{12 \times (2.32 \times 3.80) \times 10^3 \times 3800}{5 \times 8800 \times 10 \times 10^3}$$

$$= 21.48 \text{ mm} + 0.91 \text{ mm} = 22.39 \text{ mm}$$

This deflection exceeds the 11.40 mm permissible. In addition, the blockwork will be subjected to increased deflection stresses as the timber joists shrink and creep stresses develop. In view of this it was decided the best remedial scheme would be to remove the upper blockwork and rebuild using timber stud partitions on strengthened joists.

Stud partition loading = 0.80 kN/m run

Using SC4 timber and double joists bolted together, try 200 mm × 75 mm joists.

Load/metre = $0.80 + 0.90 = 1.70$ kN/m

Point load = $0.45 \times 2.30 \times 0.80 = 0.828$ kN

Tank load = 2 kN

$R_l = 4.0$ kN   $R_r = 3.80$ kN

Maximum bending moment = 4.50 kN m

Equivalent UDL = 2.51 kN/m. For deflection, $E = E_{min} \times K_9 = 6600 \times 1.14 = 7524$ N/mm$^2$

For two joists bolted together, $k_9 = 1.14$; $k_7 = 1.046$; $k_8 = 1.10$

Basic stress for SC4 timber = 7.50 N/mm$^2$ from BS 5268. Part 2, Table 8

Allowable stress = $7.50 \times 1.10 \times 1.046 = 8.63$ N/mm$^2$

Section modulus = $\dfrac{2 \times 75 \times 200^2}{6} = 1000 \times 10^3$ mm$^3$

Therefore bending stress = $\dfrac{4.50 \times 10^6}{1000 \times 10^3}$

$$= 4.50 \text{ N/mm}^2 < 8.63$$

Deflection

$$= \frac{5 \times 2.51 \times 3.80 \times 10^3 \times 3800 \times 3800 \times 3800 \times 12}{384 \times 6600 \times 1.14 \times 150 \times 200^3}$$

$$= 9.05 \text{ mm}$$

$$\text{Shear deflection} = \frac{12 \times (2.51 \times 3.80) \times 10^3 \times 3800}{5 \times 6600 \times 1.14 \times 200 \times 150}$$

$$= 0.38 \text{ mm}$$
$$\text{Total deflection} = 9.43 \text{ mm}$$

Therefore use two SC4 joists 200 mm deep × 75 mm thick, bolted together using 10 mm black bolts and toothed plate shear connectors at the interface of the joists.

## Method statement for remediation

1   Remove skirting boards on both sides of partition and store for reuse.
2   Drain central heating and hot water system and remove water tank and radiators.
3   Take down block partitions and cart off site.
4   Lift floor boards where indicated and store for reuse.
5   Remove lounge ceiling where indicated and cart off site.
6   Remove and replace existing SC3 200 mm × 50 mm joists where shown on plan.
7   Jack up floor joists with Acrow props to original level, providing spreader plates at ground-floor level.
8   Provide twin SC4 joists below new partition, and build into end walls.
9   Provide noggins where partition will cross joists using cuts from existing 200 mm × 50 mm joists doubled up and nailed together.
10  Re-fix floor boarding and fill nail holes.
11  Repair lounge ceiling with 12 mm plasterboard and 3 mm skim coat.
12  Construct timber stud partitions to original partition layout.
13  Partitions to be 100 mm × 50 mm timber studs at 600 mm centres with mid-height noggins. Cavity to be filled with Rockwool insulation prior to fixing plasterboard.
14  Re-fix skirting boards, water tank and radiator and redecorate walls.

During the remediation works to this floor it was found that beam filling had been omitted between the joist ends at the internal loadbearing wall, and this had caused the original joists to twist at their bearings, as shown in Fig. 6.7. This was modified prior to replacing the ceiling boards.

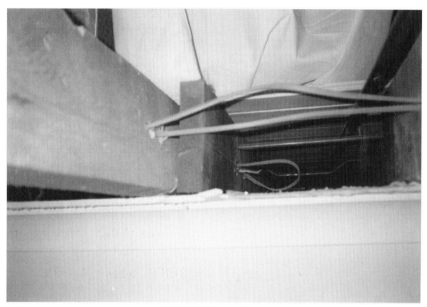

*Fig. 6.7* Beam filling omitted.

## Case history 6.2

### Introduction

The structural defects in this case history resulted from poor workmanship on site by the plumbers installing the hot water services and bath waste outlets. Because of the severe cracking to ceilings and upstairs partitions a major structural damage claim had been submitted by the occupiers to the warranty provider.

A detailed inspection of the dwelling was carried out together with a level survey of the first floor using a water level, and this confirmed that there were deflections in excess of the 14 mm allowed in BS 5268.

The layout of the floor was as shown in Fig. 6.8. Following the removal of sections of floorboarding it became evident that oversize notches close to the mid-span point of the joists had been cut in the top of the joists to accommodate water services. It was considered that this was the likely cause of the excessive deflections, and design checks confirmed this. The existing joists were 220 mm deep × 50 mm wide SC3 grade timber.

In addition the main trimmer joists trimming the staircase opening had also been notched. These trimmer joists had not deflected, thanks to the support given by the non-loadbearing stud partitions under the stairs. Following the design checks it was decided to strengthen the notched joists in situ by bolting on an additional joist, and to reroute the plumbing such that notches

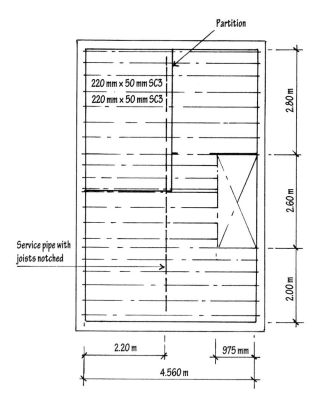

Fig. 6.8 Existing floor plan and new trimmers.

occurred in the permitted end zones. In addition two rows of herringbone strutting were added to bring the floor in line with NHBC Standards.

The trimmers at each end of the staircase were found to be inadequate because of the notching, and it was decided to replace them with either a flitched beam or a rolled steel channel section. Designs for each option were prepared.

### Design check of existing joists and trimmers

*Calculations*

|  | kN/m² |
|---|---|
| Floor loading: Imposed load | 1.50 |
| Dead load: | |
| 50 mm × 220 mm joists at 400 mm centres | 0.15 |
| 25 mm thick softwood boarding | 0.23 |
| 12.7 mm plasterboard ceiling | 0.12 |
| Dead load | = 0.50 |
| Total loading | = 2.0 kN/m² |

Paramount partitions: 85 kg/m run = 0.83 kN/m

Span of floor joists 4.56 m clear:

Effective span = 4.56 + 0.100 = 4.66 m

Loading per metre:

UDL floor loading = 2.0 × 0.40 = 0.800 kN/m

Point load from partition = 0.83 × 0.40 = 0.33 kN

$$\text{Bending moment} = \frac{0.80 \times 4.66^2}{8} + \frac{0.33 \times 4.66}{4}$$

$$= 2.17 + 0.38 = 2.55 \text{ kN m}$$

Using SC3 grade timber, basic stress 5.3 N/mm² (BS 5268 Part 2, Table 8)

Duration of loading factor $K_3 = 1.0$

As more than four joists are loaded apply load-sharing factor $K_8 = 1.10$

Depth factor, $K_7 = \left(\dfrac{300}{h}\right)^{0.11} = 1.035$ (see Table 6.6)

Therefore design stress = 5.30 × 1.10 × 1.035

$$= 6.03 \text{ N/mm}^2$$

With 20 mm notching at mid-span, joist depth is taken as 200 mm.

$$\text{Actual bending stress} = \frac{2.55 \times 10^6 \times 6}{50 \times 200^2} = 7.65 \text{ N/mm}^2$$

Therefore joists were overstressed

$$\text{Deflection} = \frac{5WL^3}{384EI}$$

$$= \frac{5 \times 0.80 \times 4.66 \times 10^3 \times 4660 \times 4660 \times 4660 \times 12}{384 \times 8800 \times 50 \times 200 \times 200 \times 200}$$

$$= 16.76 \text{ mm}$$

This exceeds the allowable deflection of 14 mm for floors.

Check deflection with partition loading of 0.33 kN at mid-span:

Deflection due to partition

$$= \frac{PL^3}{48EI} \quad \frac{0.33 \times 1000 \times 4660 \times 4660 \times 4660 \times 12}{48 \times 8800 \times 50 \times 200 \times 200 \times 200}$$

$$= 2.37 \text{ mm}$$

where $P$ = point load (kN) and $L$ = span of beam.
Total deflection = 16.76 + 2.37 = 19.13 mm. These calculated deflections were higher than the measured deflections of 18 mm, and therefore the joists required to be strengthened such that deflections were less than span × 0.003 or 14 mm maximum. The shear deflection for the UDL and point load were calculated at an additional 0.55 mm. Normally the shear deflection is so small in comparison with the bending deflection that it is ignored. However, where a beam section has been notched it could become significant.

### Remediation to floor

It was decided that the ceilings would be removed over the area of the affected joists and the joists jacked up

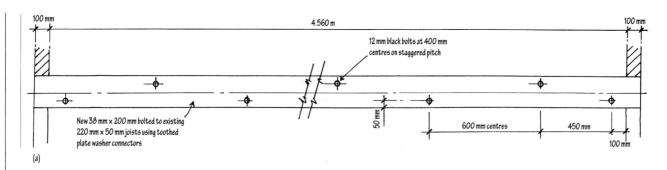

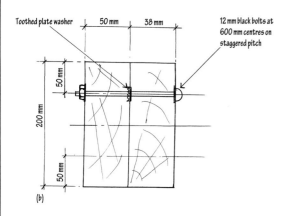

**Fig. 6.9** Fabrication details of new joists: (a) joist stiffening details; (b) connection detail.

level using Acrow props and bearers, leaving the flooring in place. Once the joists were level an additional 38 mm × 200 mm joist would be bolted to the existing joists using toothed plate washers at the timber interfaces, as shown in Fig. 6.9. It was essential that the new joists were let into the wall by 75 mm at each end in order that all the shear loading could be transferred. Owing to the tight spacing of the joists a special chuck arrangement was required when drilling the pilot holes for the bolts, to ensure that they were horizontal.

Once the floor was bolted up, the props were removed and the ceiling was re-boarded.

$$I \text{ value of } 50 \times 200 = \frac{50 \times 200^3}{12} = 33.333 \times 10^3 \text{ mm}^4$$

$E_{min} = 5800 \text{ N/ mm}^2$    For two joists $K_9 = 1.14$

$$\text{Additional } I \text{ value of new } 38 \times 200 = \frac{38 \times 220^3}{12}$$
$$= 33.718 \times 10^3 \text{ mm}^4$$

$$\text{Deflection} = 19.13 \times \frac{33.33}{33.33 + 33.71} = 9.51 \text{ mm}$$

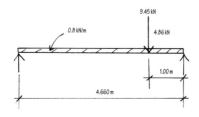

**Fig. 6.10** Trimmer loading.

This was considered a suitable deflection for the span, i.e. $< 0.003 \times 4660 = 13.98$ mm.

### Main staircase trimmers

$$\text{Load/metre on trimming joist} = \frac{2.0 \times 3.50}{2.0} = 3.50 \text{ kN/m}$$

(see Fig. 6.10)

Point load onto trimmer joist $= 3.50 \times 2.70 = 9.45$ kN

UDL on trimmer joist $= 2.0 \times 0.40 = 0.80$ kN/m

$R_a = 3.90$ kN      $R_b = 9.30$ kN

Bending moment $= 8.70$ kN m

$$\text{Equivalent UDL} = \frac{8.70 \times 8}{4.66^2} = 3.20 \text{ kN/m run}$$

$$\text{Bending stress} = \frac{8.70 \times 10^6 \times 6}{2 \times 50 \times 200^2} = 13.10 \text{ N/mm}^2 > 6.03$$

Deflection

$$= \frac{5 \times (3.20 \times 4.66) \times 10^3 \times 4660^3 \times 12}{384 \times 2 \times 50 \times 200^3 \times 5800 \times 1.14} = 44.50 \text{ mm}$$

This theoretical deflection had not materialised owing to the support being given by the staircase partition. It was considered that as this partition had no foundation the trimmer joists would be strengthened.

### Design using a Flitch beam

This type of composite beam is arranged as shown in Fig. 6.11 with the timber and steel components bolted

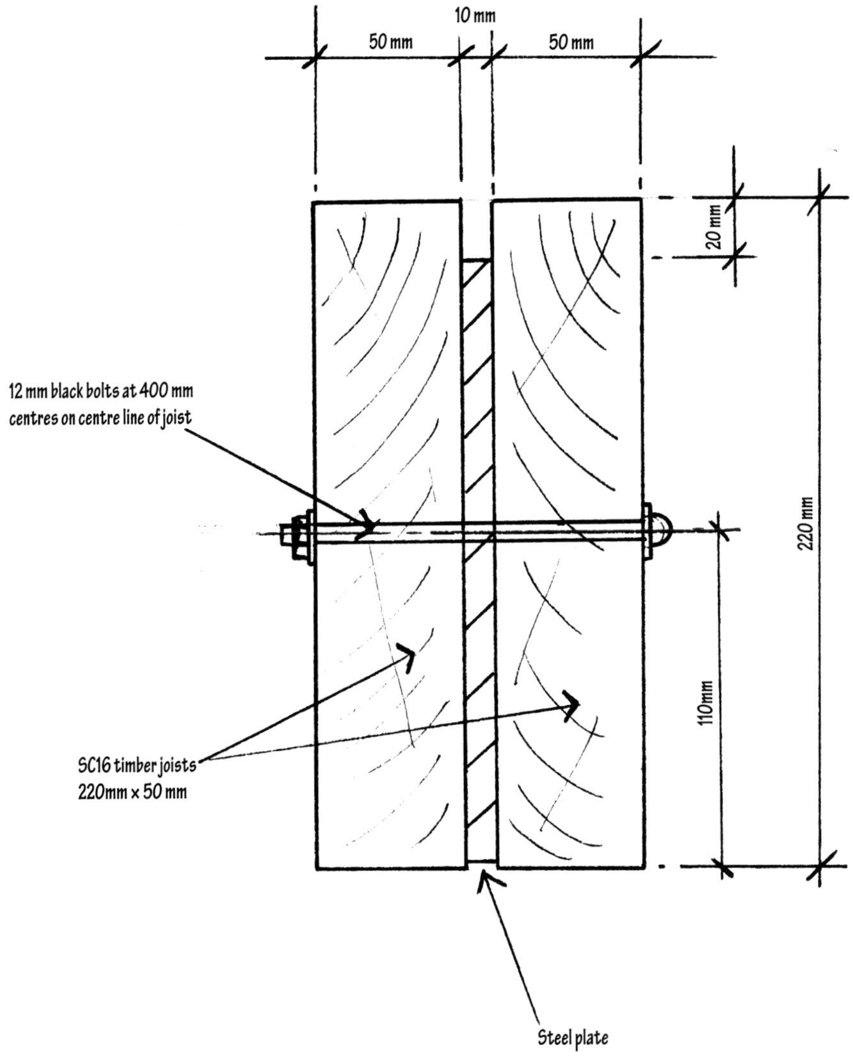

**Fig. 6.11** Section of flitch beam.

rigidly together so that both the timber and steel bend to the same radius of curvature. Therefore

$$\frac{M}{I} = \frac{f}{m} = \frac{E}{p} \qquad (6.1)$$

Because the moduli of elasticity for timber and steel are different we cannot merely substitute in the expression, but as the radius of curvature are the same, we have

$$p = \frac{E_t I_t}{M_t} \qquad (6.2)$$

$$= \frac{E_s I_s}{M_s} \qquad (6.3)$$

where $E_s$ = modulus of elasticity for steel; $E_t$ = modulus of elasticity for timber; $I_s$ = moment of inertia of steel plate; $I_t$ = moment of inertia of timber joists; $M_s$ = bending moment carried by steel plate; $M_t$ = bending moment carried by timber joists; $p$ = radius of curvature.
Now

$$M_t + M_s = \text{total bending moment, } M \qquad (6.4)$$

Therefore

$$M = M_t + M_s$$

$$= \frac{E_t I_t}{p} + \frac{E_s I_s}{p} \qquad (6.5)$$

$$= \frac{1}{p}(E_t I_t + E_s I_s) \qquad (6.6)$$

Therefore

$$\frac{1}{p} = \frac{M}{E_t I_t + E_s I_s} \qquad (6.7)$$

But from equation (6.1):

$$\frac{E}{p} = \frac{f}{m}$$

so that

$$\frac{E_t}{p} = \frac{f_t}{m_t} \qquad (6.8)$$

where $f_t$ = intensity of stress in timber and $m_t$ = farthest distance from neutral axis. Therefore

$$f_t = \frac{E_t m_t}{p}$$

$$= E_t m_t \frac{M}{E_t I_t + E_s I_s}$$

from equation (6.7), similarly:

$$f_s = E_s m_s \frac{M}{E_t I_t + E_s I_s}$$

and

$$M_t = \frac{M}{1 + E_s I_s / E_t I_t}$$

Clear span = 4.56 m.

Equivalent UDL = 3.20 kN/m + self-weight

Try 220 mm × 50 mm SC16 timber joists with 200 mm × 10 mm mild steel plate (Fig. 6.11):

Designed in accordance with BS 5268: Part 2: 1996

Depth of timber section, $d$ = 220 mm
Width of timber section, $b$ = 50 mm
Depth of steel, $d_s$ = 200 mm
Width of steel section = 10 mm
Length of bearing, $L_b$ = 100 mm
Timber $E_{mean}$ = 8800 N/mm$^2$
Timber $E_{min}$ = 5800 N/mm$^2$
Modulus of elasticity for steel = 205 000 N/mm$^2$
Load sharing factor, $K_8$ = 1.10
Depth factor, $K_7$ = 1.035

For deflection, modification factor, $K_9$ = 1.14

$E = E_{min} \times 1.14 = 5800 \times 1.14 = 6612$ N/mm$^2$

Modular ratio, $MR = \dfrac{E_{steel}}{E_{timber\,mod}} = \dfrac{205\,000}{6612} = 31.0$

Equivalent UDL = 3.20 kN/m
Self-weight = (15.70 + 12.98) kg/m
Design load = 3.46 kN/m

Max. bending moment = $\dfrac{3.46 \times 4.66^2}{8.0} = 9.39$ kN m

Basic grade stress for SC16 = 5.30 N/mm$^2$

Permissible stress = $5.30 \times 1.10 K_4 \times 1.035 K_7$

$$= 6.034 \text{ N/mm}^2$$

Permissible steel stress for grade 43 steel plate < 40 mm thick = $P_{bc}$ = 180 N/mm$^2$

### Bending stresses

$$I_s = \frac{10 \times 200^3}{12} = 6\,666\,666.66 \text{ mm}^4$$

$$I_t = \frac{2 \times 50 \times 220^3}{12} = 88\,733\,333 \text{ mm}^4$$

From

$$f_t = E_t m_t \frac{M}{E_t I_t + E_s I_s}$$

$$f_t = 6612 \times 110$$

$$\times \frac{9.39 \times 10^6}{(6612 \times 88\,733\,333) + (205\,000 \times 6\,666\,666.66)}$$

$$= 3.49 \text{ N/mm}^2$$

Similarly:

$$f_s = 205\,000 \times 100$$

$$\times \frac{9.39 \times 10^6}{(6612 \times 88\,733\,333) + (205\,000 \times 6\,666\,666.66)}$$

$$= 98.26 \text{ N/mm}^2$$

This can be checked using the modular ratio, which equals 31.0.

Applied bending stress in steel

$$3.46 \times 31.0 \times \frac{100}{110} = 91.50 \text{ N/mm}^2$$

To check the deflection use the modified timber inertia, $I_t + I_s \times$ Modular ratio

$$I_t = 88\,733\,333 \text{ mm}^4 \qquad I_s = 6\,666\,666.66 \text{ mm}^4$$

Therefore

Modified total $I_t = 88\,773\,333 + (6\,666\,666.66 \times 31)$

$$= 295\,399\,999 \text{ mm}^4$$

$$\text{Deflection} = \frac{5 \times (3.46 \times 4.66) \times 10^3 \times 4660^3}{384 \times 6612 \times 295.40 \times 10^6} = 10.86 \text{ mm}$$

$$< 0.003 \times 4660 = 13.98 \text{ mm}$$

### Bearing stresses

Maximum reaction = $9.30 + \dfrac{0.28 \times 4.66}{2.0} = 9.95$ kN

Applied bearing stress = $\dfrac{9.95 \times 10^3}{100 \times 2 \times 50} = 0.995$ N/mm$^2$

Allowable bearing stress = 2.20 N/mm$^2$

Timber and steel plate to be bolted together using 12 mm black bolts to BS 4190 at 400 mm centres on the neutral axis, as shown in Fig. 6.11.

## Design using steel channel section

As the steel beam will be part of a timber floor system it would be prudent to adopt the permissible stress design as outlined in BS 449 Part 2: 1969. The two channels will be bolted to timber joists as shown in Fig. 6.12 to facilitate the floor and ceiling board fixings.

The channels are not laterally restrained on their top flange, therefore the compressive stresses must be

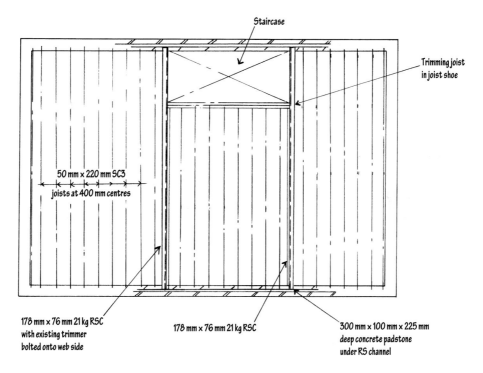

Fig. 6.12 Steel channel trimmers.

reduced accordingly:

$$\frac{D}{T} = 17.30 \qquad \frac{L}{r_{yy}} = \frac{4660}{22.50} = 207$$

From BS 449, Table 3: Permissible stress = 98 N/mm$^2$

Maximum bending moment = 6.02 kN m

Required section modulus $= \dfrac{9.39 \times 10^6}{98.0} = 95.816$ cm$^3$

A 178 mm × 76 mm × 21 kg/m RS channel gives a $Z$ value of 150.40 mm$^3$

$$\text{Deflection} = \frac{5 \times (3.46 \times 4.66) \times 10^3 \times 4660^3}{384 \times 205\,000 \times 1337 \times 10^4}$$

$$= 7.74 \text{ mm}$$

Use a 178 mm × 76 mm × 21 kg/m RS (rolled steel) channel section bolted to one of the existing trimmers, as shown in Figs 6.12 and 6.13.

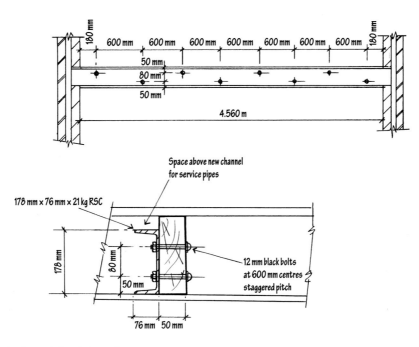

Fig. 6.13 RS channel bolted to existing timber trimmers.

### 6.2.2 Timber quality

Structural defects in timber floors can often result from the use of inferior-quality timber. Strength properties and durability are usually the prime considerations when specifying structural timber. In the past, most timber sizes were determined by experience and tradition using visual grading. More recently timber has been stress-graded using a non-destructive mechanical process. As a result, the amount of timber used in floors to support a given load has been reduced.

A specification for structural timber should state the following requirements:

● the strength class and grade stress;
● the cross-sectional sizes of the timbers, surface finish and timber lengths;
● the moisture content of the timber;
● details of any preservative treatment required.

### Strength class

The classification of timber is based on particular values of grade stress.

These strength classes are based primarily on bending strength, which means that any one species may occur in more than one strength class, depending upon the stress grade (see Table 6.5).

Most flooring timber used is softwood of strength class SC3 or SC4, and the species groups are Western Hemlocks, European Whitewoods, European Redwoods or Douglas Fir in accordance with BS 5268: Part 2.

### Grade stress

This is the stress that can be safely be permanently sustained by timber of a specific section size and of a particular strength class or species and grade.

Grade stress values are listed in BS 5268 : Part 2 for the BS 4978 grades and the North American grades of many structural softwoods.

**Table 6.7** Floor joists, SC3 grade timber: maximum clear span of joist (m). Partition loads have not been allowed for

| Size of joist (mm × mm) | Dead load, excluding the self-weight of the joist (kN/m²) | | | | | | | | |
| | Not more than 0.25 | | | More than 0.25 but not more than 0.50 | | | More than 0.50 but not more than 1.25 | | |
| | Spacing of joists (mm) | | | | | | | | |
| | 400 | 450 | 600 | 400 | 450 | 600 | 400 | 450 | 600 |
| 38 × 75 | 1.22 | 1.09 | 0.83 | 1.14 | 1.03 | 0.79 | 0.98 | 0.89 | 0.70 |
| 38 × 100 | 1.62 | 1.78 | 1.38 | 1.80 | 1.64 | 1.28 | 1.49 | 1.36 | 1.09 |
| 38 × 125 | 1.91 | 2.45 | 2.01 | 2.43 | 2.30 | 1.83 | 2.01 | 1.85 | 1.50 |
| 38 × 150 | 3.05 | 2.93 | 2.56 | 2.91 | 2.76 | 2.40 | 2.50 | 2.35 | 1.93 |
| 38 × 175 | 3.55 | 3.40 | 2.96 | 3.37 | 3.19 | 2.77 | 2.89 | 2.73 | 2.36 |
| 38 × 200 | 4.04 | 3.85 | 3.35 | 3.82 | 3.61 | 3.13 | 3.27 | 3.09 | 2.68 |
| 38 × 225 | 4.53 | 4.29 | 3.73 | 4.25 | 4.02 | 3.50 | 3.65 | 3.44 | 2.99 |
| 47 × 75 | 1.41 | 1.33 | 1.02 | 1.35 | 1.24 | 0.96 | 1.16 | 1.06 | 0.84 |
| 47 × 100 | 2.11 | 2.00 | 1.67 | 2.00 | 1.90 | 1.54 | 1.74 | 1.60 | 1.29 |
| 47 × 125 | 2.73 | 2.63 | 2.38 | 2.61 | 2.51 | 2.17 | 2.33 | 2.15 | 1.76 |
| 47 × 150 | 3.27 | 3.14 | 2.84 | 3.13 | 3.01 | 2.66 | 2.78 | 2.62 | 2.24 |
| 47 × 175 | 3.80 | 3.66 | 3.28 | 3.64 | 3.50 | 3.07 | 3.21 | 3.03 | 2.63 |
| 47 × 200 | 4.33 | 4.17 | 3.71 | 4.15 | 3.99 | 3.48 | 3.63 | 3.43 | 2.98 |
| 47 × 225 | 4.81 | 4.67 | 4.13 | 4.65 | 4.45 | 3.88 | 4.04 | 3.82 | 3.32 |
| 50 × 75 | 1.45 | 1.37 | 1.08 | 1.39 | 1.30 | 1.01 | 1.22 | 1.11 | 0.88 |
| 50 × 100 | 2.18 | 2.06 | 1.76 | 2.06 | 1.95 | 1.62 | 1.82 | 1.67 | 1.35 |
| 50 × 125 | 2.79 | 2.68 | 2.44 | 2.67 | 2.56 | 2.28 | 2.40 | 2.24 | 1.84 |
| 50 × 150 | 3.33 | 3.21 | 2.92 | 3.19 | 3.07 | 2.75 | 2.86 | 2.70 | 2.33 |
| 50 × 175 | 3.88 | 3.73 | 3.38 | 3.71 | 3.57 | 3.17 | 3.30 | 3.12 | 2.71 |
| 50 × 200 | 4.42 | 4.25 | 3.82 | 4.23 | 4.07 | 3.58 | 3.74 | 3.53 | 3.07 |
| 50 × 225 | 4.88 | 4.74 | 4.26 | 4.72 | 4.57 | 3.99 | 4.16 | 3.94 | 3.42 |
| 63 × 100 | 2.41 | 2.29 | 2.01 | 2.28 | 2.17 | 1.90 | 2.01 | 1.91 | 1.60 |
| 63 × 125 | 3.00 | 2.89 | 2.63 | 2.88 | 2.77 | 2.52 | 2.59 | 2.49 | 2.16 |
| 63 × 150 | 3.59 | 3.46 | 3.15 | 3.44 | 3.31 | 3.01 | 3.10 | 2.98 | 2.63 |
| 63 × 175 | 4.17 | 4.02 | 3.66 | 4.00 | 3.85 | 3.51 | 3.61 | 3.47 | 3.03 |
| 63 × 200 | 4.73 | 4.58 | 4.18 | 4.56 | 4.39 | 4.00 | 4.11 | 3.95 | 3.43 |
| 63 × 225 | 5.15 | 5.01 | 4.68 | 4.99 | 4.85 | 4.46 | 4.62 | 4.40 | 3.83 |
| 75 × 125 | 3.18 | 3.06 | 2.79 | 3.04 | 2.93 | 2.67 | 2.74 | 2.64 | 2.40 |
| 75 × 150 | 3.79 | 3.66 | 3.33 | 3.64 | 3.50 | 3.19 | 3.28 | 3.16 | 2.86 |
| 75 × 175 | 4.41 | 4.25 | 3.88 | 4.23 | 4.07 | 3.71 | 3.82 | 3.68 | 3.30 |
| 75 × 200 | 4.92 | 4.79 | 4.42 | 4.77 | 4.64 | 4.23 | 4.35 | 4.19 | 3.74 |
| 75 × 225 | 5.36 | 5.22 | 4.88 | 5.20 | 5.06 | 4.72 | 4.82 | 4.69 | 4.16 |

For domestic floor joists, Tables 6.7 and 6.8 are abstracts from the Approved Documents A1/2 for timbers of strength class SC3 and SC4.

## Sizes and surface finish

BS 4471 lists the range of lengths and cross-sectional areas for structural softwoods. Any specification should state clearly whether the timber is to be left sawn or whether it is to be planed or regularised to a finished size. Regularising is the process by which every piece of a batch of structural timber is sawn or machined to a uniform width. Both these processes result in a reduction in size from the basic size, and the maximum reductions are given in BS 4471.

One of the most common defects in timber floors occurs when the joists used have not been regularised, resulting in uneven and squeaky floors (Fig. 6.14). When fitting timber floor joists the carpenter should always check for excessive bow in the length of the joist, and reject those for use. If the bow is apparent but not excessive then he should ensure that the joists are all built with the slight bow on top. The worst situation occurs when the joists are not properly 'backed', and some are laid with bows in opposite directions, resulting in a poorly finished, uneven floor.

Timber sizes can vary appreciably if they have not been fully dried, or if the timbers have been allowed to become saturated. This drying-out will shrink the timber and reduce the strength of the section.

## Moisture content

This is defined as the amount of water in the timber expressed as a percentage of its oven dried mass. On building sites the moisture content can be readily measured using an electrical resistance moisture meter. It is important that these meters are regularly calibrated.

In 1995 new rules were introduced into the Standards for kiln-dried timber for specifying structural softwoods

*Table 6.8* Floor joists, SC4 grade timber: maximum clear span of joist (m). Partition loads have not been allowed for

| Size of joist (mm × mm) | Dead load, excluding the self-weight of the joist (kN/m$^2$) | | | | | | | | |
| | Not more than 0.25 | | | More than 0.25 but not more than 0.50 | | | More than 0.50 but not more than 1.25 | | |
| | Spacing of joists (mm) | | | | | | | | |
| | 400 | 450 | 600 | 400 | 450 | 600 | 400 | 450 | 600 |
|---|---|---|---|---|---|---|---|---|---|
| 38 × 75 | 1.34 | 1.26 | 1.09 | 1.29 | 1.22 | 1.05 | 1.17 | 1.11 | 0.93 |
| 38 × 100 | 2.02 | 1.91 | 1.66 | 1.92 | 1.82 | 1.58 | 1.70 | 1.62 | 1.42 |
| 38 × 125 | 2.65 | 2.55 | 2.28 | 2.53 | 2.43 | 2.15 | 2.25 | 2.14 | 1.89 |
| 38 × 150 | 3.17 | 3.05 | 2.77 | 3.03 | 2.91 | 2.65 | 2.72 | 2.62 | 2.37 |
| 38 × 175 | 3.69 | 3.55 | 3.22 | 3.53 | 3.39 | 3.08 | 3.17 | 3.05 | 2.76 |
| 38 × 200 | 4.20 | 4.04 | 3.68 | 4.02 | 3.87 | 3.52 | 3.62 | 3.48 | 3.15 |
| 38 × 225 | 4.70 | 4.54 | 4.13 | 4.52 | 4.35 | 3.95 | 4.07 | 3.91 | 3.54 |
| 47 × 75 | 1.49 | 1.41 | 1.22 | 1.43 | 1.35 | 1.17 | 1.29 | 1.23 | 1.07 |
| 47 × 100 | 2.23 | 2.12 | 1.84 | 2.11 | 2.00 | 1.75 | 1.87 | 1.77 | 1.56 |
| 47 × 125 | 2.84 | 2.73 | 2.48 | 2.72 | 2.61 | 2.37 | 2.44 | 2.34 | 2.07 |
| 47 × 150 | 3.40 | 3.27 | 2.97 | 3.25 | 3.13 | 2.84 | 2.93 | 2.81 | 2.55 |
| 47 × 175 | 3.95 | 3.80 | 3.46 | 3.78 | 3.64 | 3.31 | 3.41 | 3.28 | 2.97 |
| 47 × 200 | 4.50 | 4.33 | 3.95 | 4.31 | 4.15 | 3.78 | 3.89 | 3.74 | 3.39 |
| 47 × 225 | 4.94 | 4.81 | 4.43 | 4.79 | 4.66 | 4.24 | 4.36 | 4.20 | 3.81 |
| 50 × 75 | 1.54 | 1.45 | 1.26 | 1.47 | 1.39 | 1.21 | 1.33 | 1.26 | 1.10 |
| 50 × 100 | 2.30 | 2.18 | 1.90 | 2.17 | 2.06 | 1.80 | 1.92 | 1.82 | 1.61 |
| 50 × 125 | 2.90 | 2.79 | 2.53 | 2.77 | 2.67 | 2.42 | 2.50 | 2.40 | 2.13 |
| 50 × 150 | 3.46 | 3.34 | 3.03 | 3.32 | 3.19 | 2.90 | 2.99 | 2.87 | 2.61 |
| 50 × 175 | 4.03 | 3.88 | 3.53 | 3.86 | 3.71 | 3.38 | 3.48 | 3.34 | 3.04 |
| 50 × 200 | 4.59 | 4.42 | 4.03 | 4.40 | 4.23 | 3.85 | 3.97 | 3.81 | 3.89 |
| 50 × 225 | 5.02 | 4.88 | 4.52 | 4.86 | 4.73 | 4.33 | 4.45 | 4.28 | 3.89 |
| 63 × 100 | 2.51 | 2.41 | 2.12 | 2.40 | 2.28 | 2.01 | 2.11 | 2.01 | 1.78 |
| 63 × 125 | 3.12 | 3.01 | 2.74 | 2.99 | 2.88 | 2.62 | 2.69 | 2.59 | 2.35 |
| 63 × 150 | 3.73 | 3.59 | 3.27 | 3.58 | 3.44 | 3.13 | 3.22 | 3.10 | 2.82 |
| 63 × 175 | 4.33 | 4.18 | 3.81 | 4.16 | 4.00 | 3.65 | 3.75 | 3.61 | 3.28 |
| 63 × 200 | 4.86 | 4.73 | 4.34 | 4.71 | 4.56 | 4.16 | 4.28 | 4.12 | 3.74 |
| 63 × 225 | 5.29 | 5.15 | 4.81 | 5.13 | 4.99 | 4.66 | 4.76 | 4.62 | 4.20 |
| 75 × 125 | 3.30 | 3.18 | 2.90 | 3.16 | 3.05 | 2.77 | 2.85 | 2.75 | 2.50 |
| 75 × 150 | 3.94 | 3.80 | 3.46 | 3.78 | 3.64 | 3.32 | 3.41 | 3.28 | 2.99 |
| 75 × 175 | 4.57 | 4.41 | 4.03 | 4.39 | 4.23 | 3.86 | 3.97 | 3.82 | 3.48 |
| 75 × 200 | 5.06 | 4.93 | 4.59 | 4.91 | 4.78 | 4.40 | 4.52 | 4.36 | 3.97 |
| 75 × 225 | 5.51 | 5.36 | 5.02 | 5.34 | 5.20 | 4.86 | 4.96 | 4.82 | 4.45 |

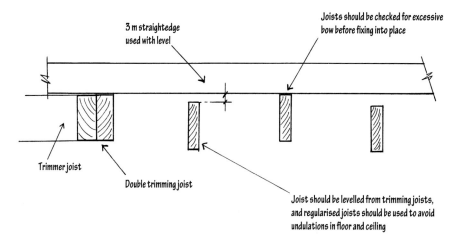

Fig. 6.14 Regularised timber joists.

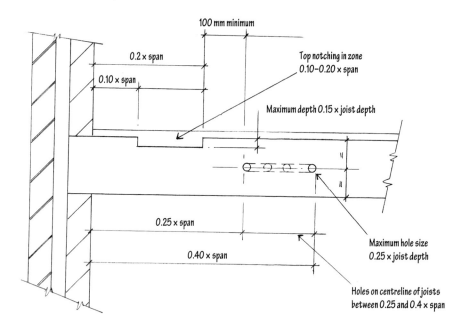

Fig. 6.15 Notching and drilling in timber joists.

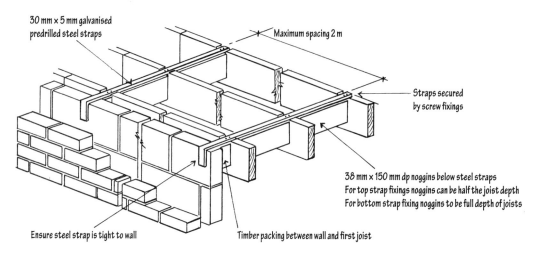

Fig. 6.16 Lateral restraint straps when floor joists are parallel to wall.

according to revised British Standards. These are intended to ensure that the moisture content of the timber is appropriate to the in-service conditions for which it has been specified. The requirement now is for all structural softwood of less than 100 mm thickness for internal applications such as floors and roofs to be graded at a moisture content of 20% or less. The timber must have a stress grade stamp indicating that it is DRY or KD (kiln dried).

In addition to improving the strength of timber, kiln drying improves dimensional stability, prevents transverse strutting from becoming loose, reduces distortion after fixing, and allows preservatives to penetrate into the wood pores more readily. It improves the durability of the timber, as fungi and moulds cannot live on dry timbers, as well as improving the surface for the use of paint or stain finishes, providing better bonding when using glues, and causing less risk of corrosion in metal components.

### Preservative treatments

In flat roofs and floors the possibility of fungal attack is very low if the waterproof layer remains intact and the roof or floor is well ventilated.

However, even when care is taken in the design, it is important to consider the risk of wetting in service for long periods.

BS 5268 Part 5: 1989 *Code of practice for the preservation of timber* provides guidance on assessing the risks of decay in particular building elements, and gives suitable preservative treatments for timbers.

Where a building is being converted it may be necessary for the existing timbers to be treated for woodworm infestation or House Longhorn Beetle. In

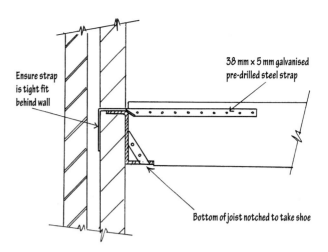

**Fig. 6.17** Metal restraint strap when using joist shoes.

such conversions it is always prudent to ensure that new timbers are treated with preservatives to guard against attack by wood-boring insects, or from fungal attack such as wet or dry rot infestation.

### 6.2.3 Bad workmanship

Bad workmanship practices can lead to major structural defects in timber floors, such as the following:

- Incorrect positioning of service notches and service holes. Figure 6.15 shows the permitted zones to ensure that the section modulus of the timbers is not reduced. Bottom notching of joists should never be done period.
- Lack of transverse strutting, especially on longer spans, results in excessive spring in a floor. Table 6.3

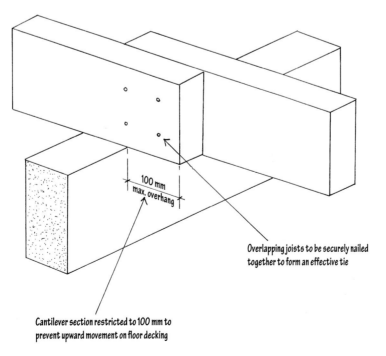

**Fig. 6.18** Excessive overlap on floor joists.

lists the NHBC requirements. The strutting can be solid or herringbone, and the gap between the last joist and the wall must be packed out as shown in Fig. 6.1.

- Failure to provide metal restraint straps from the walls into the top of floor joists where joists are parallel (Figs 6.16 and 6.17) will prevent the floor from achieving the necessary diaphragm action essential to stiffening the masonry structure.
- Excessive overlapping of floor joists on internal loadbearing walls can result in the ends of the joist being deflected upwards as the joist deflects. This is more severe on long spans, and can lead to severe distortion of the floor (Fig. 6.18).
- Inadequate connection of trimming joists around staircases. Failure to fully nail the proprietary joist shoes, the use of soft packings under overcut joist bottoms, and joists not wedged tight into the shoe can all lead to rotation of the joist, with resultant cracking to ceilings (Figs 6.19, 6.20 and 6.21).

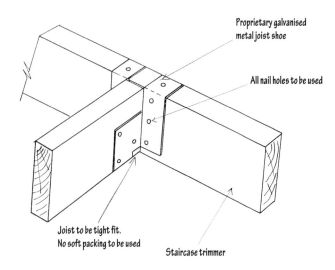

Fig. 6.19 Trimming joist connection.

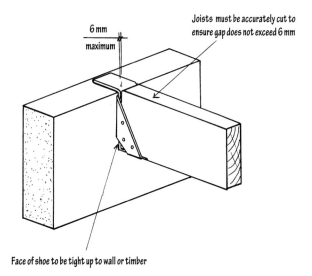

Fig. 6.20 Joist support shoe.

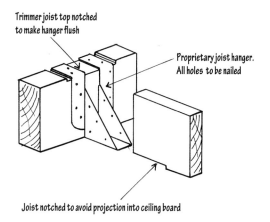

Fig. 6.21 Timber-to-timber joist hanger.

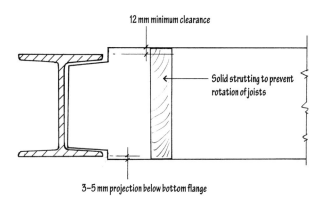

Fig. 6.22 Timber joists trimmed into steelwork.

- Inadequate trimming of joists into steel beam webs. To prevent twist occurring at the joist ends there should be a line of noggins. In addition, adequate clearances should be left above and below the beam flanges to prevent damage to boards and ceiling finishes (Fig. 6.22).

Where there is a combination of such defects arising from bad workmanship and poor materials the end product is a poorly constructed floor that is uneven and squeaky. It can also result in the house owner believing that there is a major problem with the floor.

## 6.3 Precast concrete floors

The housebuilding 'industry standard' is to build a concrete ground floor and a timber joist first floor. The concrete ground floor is generally a 100 mm thick ground-bearing slab laid on crushed stone infill.

One of the problems with this type of floor was the high number of failures as a result of poor compaction and subsequent settlement of the infill.

Other failures occurred as a result of sulphate attack when the infill materials were not chemically inert, and from floor heave following the rehydration of clay soils after the removal of trees and vegetation.

As a result of high insurance claims the NHBC introduced a mandatory requirement for ground floors

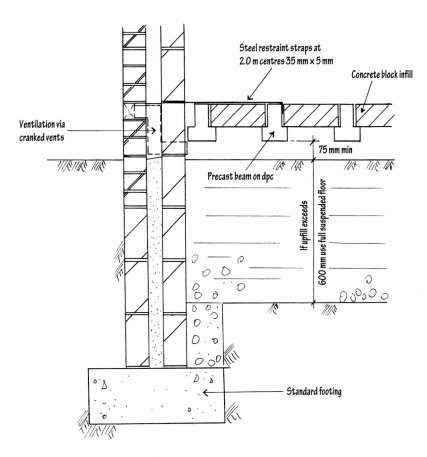

Steel restraint straps at
2.0 m centres 35 mm x 5 mm

Concrete block infill

Ventilation via
cranked vents

75 mm min

Precast beam on dpc

If upfill exceeds

600 mm use full suspended floor

Standard footing

*Fig. 6.23* Section through precast beam and block ground floor.

to be fully suspended when the upfill depth of stone was greater than 600 mm, as shown in Fig. 6.23.

For those builders who preferred a concrete floor to a timber joist floor the precast concrete industry produced a range of precast floor beam systems.

The basic cost difference between a ground-bearing slab and a precast floor is approximately £2/m². However, there are hidden savings with precast floors:

- speed of erection – no skilled operatives required;
- low-cost all-weather flooring system;
- immediate floor and working platform;
- good noise reduction between upper floors;
- No over-site concrete needed;
- clean dug site-fill used as infill, thereby saving costs of carting off-site;
- excellent fire resistance;
- easily provided void to cater for potential ground heave problems.

There are several types of precast floor that can be obtained from specialist concrete manufacturers:

- **Beam and block system**: This consists of a series of equally spaced precast, prestressed concrete beams with infill blocks laid between, as shown in Fig. 6.24. This is probably the most popular system, as the beams are easy to handle, and the builder can use house wall blocks as infill to save costs.
- **Hollow wide slab precast units**: These are used more by contractors on commercial buildings, offices and

factories. They are also popular in multi-storey housing developments (Fig. 6.25).

- **Solid precast planks with block infill**: The precast planks can be designed to take a structural screed and form a composite design, or the planks can support hollow blocks with in-situ tee-beams between (Fig. 6.26).
- **A solid plank composite floor**: This type of floor system, known as Omnia plank, consists of a precast plank with a structural lattice with hollow block inserts between the planks, as shown in Fig. 6.27.

A variation is to have the Omnia wide slab with hollow blocks or complete concrete infill used. These systems are more in use on sites where craneage is present.

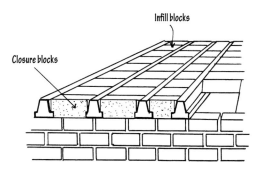

Infill blocks

Closure blocks

*Fig. 6.24* Precast concrete beam and block floor.

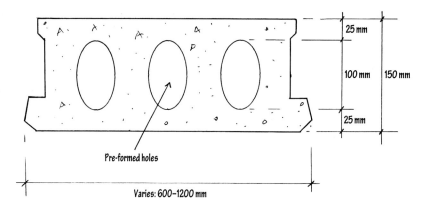

25 mm

100 mm   150 mm

25 mm

Pre-formed holes

Varies: 600–1200 mm

*Fig. 6.25* Wide slab, hollow, precast, prestressed concrete floor unit.

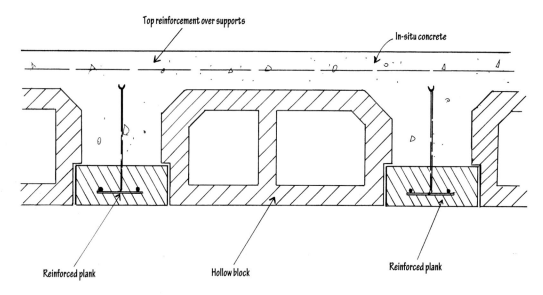

Top reinforcement over supports

In-situ concrete

Reinforced plank

Hollow block

Reinforced plank

*Fig. 6.26* Composite precast plank ribbed floor system with hollow infill blocks.

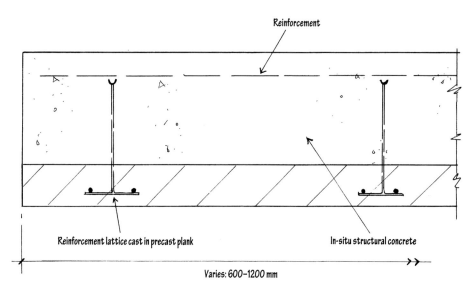

Reinforcement

Reinforcement lattice cast in precast plank

In-situ structural concrete

Varies: 600–1200 mm

*Fig. 6.27* Composite permanent precast plank floor with in-situ concrete infill.

### 6.3.1  Beam and block floor

Table 6.9 shows some typical spans for beam and block floors for guidance for domestic housing.

#### Infill blocks

These can be hollow or solid, and the standard size is 100 mm × 225 mm × 440 mm as used in wall constructions. The blocks are generally 3.50 N/mm$^2$ or 7.0 N/mm$^2$ compressive strength, manufactured to BS 6073: 1981.

For garage floors the block strength must be 7.0 N/mm$^2$ minimum, with a 60 mm structural screed on top with A98 mesh (Fig. 6.28).

#### Floor beams

Floor beams are cast in fixed steel moulds using vibrated concrete and prestressing techniques. They can be made and supplied in lengths up to 6 m, and the beams weigh approximately 32–34 kg/m run, depending on the beam configuration.

The beams are manufactured with a pre-camber as a result of the prestressing, and a good design is one in which the residual camber is very low after the floor is loaded. This camber generally works out at about 0.003 × span.

The concrete strength in the beams is generally about the 50 N/mm$^2$ range, and the concrete generally contains workability plasticisers to ensure that a dense concrete is achieved.

#### Installation

Once the masonry is up to level the floor beams are lifted into place. The beams should be lifted as near to each end as possible. The beams are placed onto a dpc, and must have a minimum bearing of 90 mm.

Closure blocks and infill blocks should be used, and any blocks built into the masonry must be bedded on

**Table 6.9** Load/span table for beam and block floors

| Beam centres (mm) | Self-weight (kg/m) | Loading conditions[a] (blocks of density 1300 kg/m³) | | | |
|---|---|---|---|---|---|
| | | Domestic W | Domestic + partition W + P | Domestic + partition W + W$_P$ | Garage W$_G$ |
| 502 | 90.9 | 4.98 | 4.35 | 2.95 | 4.44 |
| 389 | 76.3 | 5.58 | 4.96 | 3.25 | 4.76 |
| 277 | 61.7 | 6.00 | 5.86 | 3.54 | 5.18 |
| 606 | 132.8 | 6.00 | 5.58 | 4.30 | 5.54 |

[a] $W$ = self-weight + 50 mm screed at 1.20 kN/m$^2$ + domestic superload at 1.50 kN/m$^2$
$P$ = partition allowance (kN/m$^2$)
$W_P$ = 100 mm lightweight block partition, 2.30 m high, plastered both sides at 3 kN/m run across the span
$W_G$ = self-weight + 50 mm structural screed at 1.2 kN/m$^2$ + garage imposed load of 2.50 kN/m$^2$
All beams are prestressed and have an upward camber.

mortar. Any packing up must be done using hard packing such as tiles or slate.

The blocks should always be laid starting from the external wall, and laying in-board such that cut blocks are within the centre of the floor.

Once the blocks are infilled they must be grouted up with a coarse 3 : 1 sand and cement grout brushed in with a stiff brush. The blocks should be wetted up before this grouting is done. This grouting-up must be done before the floor is used for stacking blocks and bricks.

Where partition walls run parallel to the floor beams the precast manufacturer can design multiple beams to support the partition, as shown in Fig. 6.29.

#### Site preparation

The ground beneath the floor beams should be free from organic matter or vegetation. No oversite concrete is needed, but this type of floor must have through-ventilation of 600 mm$^2$ per metre length of wall. This is normally achieved by using telescopic crank ventilators. The minimum void space for normal sites is 75 mm,

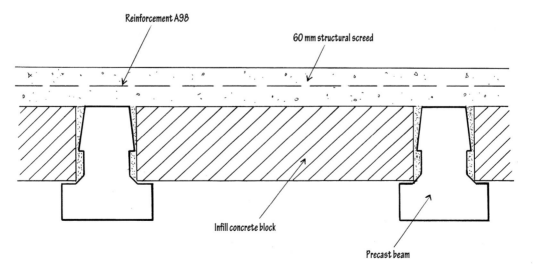

Reinforcement A98
60 mm structural screed
Infill concrete block
Precast beam

*Fig. 6.28* Screeded garage floor.

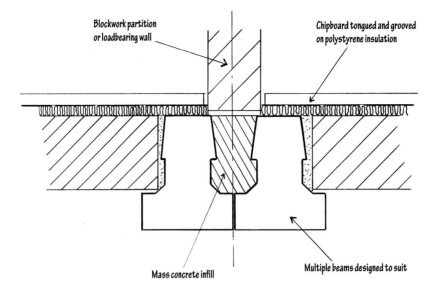

*Fig. 6.29* Detail of multiple beams to support partitions.

but where there are clay soils with a potential to heave then the void depth must be increased to 125 mm for low-plasticity clays, 150 mm for medium-plasticity clays and 225 mm for high-plasticity clays.

The ground below the floor should be free draining or drained, and provided this is the case then the ground can be below the external ground level. Figure 6.23 illustrates this point.

### Garage floors

For garage floors the imposed load to be used is 2.50 kN/m$^2$ and the alternative 9.0 kN point load. The beams must be covered with a structural screed 60 mm thick reinforced with A98 mesh reinforcement. The infill blocks must have a compressive strength of 7.0 N/mm$^2$.

### Upper floors

This floor system can also be used for upper floors provided crane facilities are available. The main requirements in this location are a fire resistance of 1 hour and adequate sound resistance.

For a 1 hour fire resistance the floor must have a bonded concrete screed.

Where infill blocks are built into the walls, the block strength must be 7.0 N/mm$^2$ minimum compressive strength or the strength of the wall blocks, whichever is the greater. These blocks must always be bedded on mortar (Fig. 6.30).

### Sound resistance

The revised Approved Document E of the Building Regulations 1991, which came into effect in 1992, now

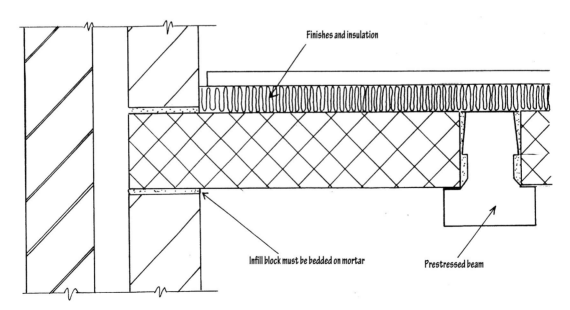

*Fig. 6.30* Upper floor infill blocks at main wall.

requires concrete separating floors to achieve a greater mass of 365 kg/m$^2$ for Type 1 floors or 300 kg/m$^2$ for Type 2 floors.

A Type 1 floor consists of a concrete base with a soft covering, where the resistance to airborne sound comes from the mass of the base, and the soft covering reduces the effect of impact sound.

A Type 2 floor consists of a concrete base with a floating layer, where the resistance to airborne sound comes from the mass of the base and finishes, and the resistance to impact sound depends on the resilient layer isolating the floating layer from the base.

For a Type 2 floor, the use of a beam and block floor with a concrete screed of 35 mm and a plasterboard ceiling 13 mm thick fixed to timber battens will meet the 300 kg/m$^2$ requirement using infill blocks with a density of 1800 kg/m$^3$.

For a Type 1 floor, the use of a 60 mm bonded screed with a plasterboard ceiling 13 mm thick fixed to timber battens will meet the 365 kg/m$^2$ requirement using infill blocks with a density of 1800 kg/m$^3$.

### 6.3.2 Hollow-core wide slab precast units

This type of unit is mainly used in multi-storey developments because of its speed of insulation, and because it is a relatively dry system.

The loadbearing walls supporting the units generally require the units to be placed onto a mortar levelling compound, as shown in Fig. 6.30.

This type of prestressed unit has several holes formed in the factory by inflated bags, which are withdrawn once the concrete has achieved its initial set.

The units lend themselves to being modified to allow for in-situ concrete and reinforcement to be installed for such situations as cantilevers, forming ties for continuity over walls where progressive collapse has to be catered for (Fig. 6.31).

The slabs can be overlaid with a concrete reinforced structural screed or have chipboard on polystyrene as a floating floor.

### 6.3.3 Prestressed plank composite floors

This type of floor construction combines the benefits of precast planks as permanent formwork and in-situ concrete infill to produce a composite structural floor.

This type of floor is most useful when designing floors that are required to provide ties and continuity where progressive collapse has to be catered for.

The prestressed reinforced plank is installed in the same way as wide-slab hollow units onto a mortar levelling compound. Some manufacturers use the Omnia system, which provides the main reinforcement in the planks in the form of a reinforced lattice with a top boom. This provides for a stiffer unit, and enables top reinforcement to be supported prior to placing the in-situ concrete.

Some manufacturers use hollow blocks to save weight, and the concrete ribs between the blocks are designed as a ribbed floor, as shown in Figs 6.25 and 6.26.

### 6.3.4 Specification for proprietary precast concrete floors

*1.0 General*

1.1   The 'Engineer' referred to in this Specification shall mean the Consulting Structural or Civil Engineer who has responsibility for the overall design of the building.

1.2   The Contractor referred to in this Specification shall mean the precast floor manufacturer. The general contractor for the building as a whole shall be referred to as the Main Contractor.

1.3   The terms 'Approved' and 'Approval' shall mean the written approval of the Consulting Engineer.

1.4   All British Standard Codes of Practice, Specifications or any other Building Standards referred to in this Specification shall be the latest version including all amendments published before the final tender date, unless otherwise stated.

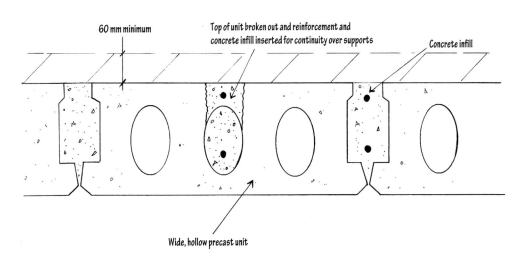

*Fig. 6.31* Continuity steel details for prestressed floor units.

1.5    The design, manufacture and installation works described in this Specification shall be carried out to the entire satisfaction of the Consulting Engineer. Clause 1.3 refers to the minimum standards of acceptance which may be supplemented or modified by this Specification.

1.6    No variance to this Specification or to the Clause 1.3 provisions will be allowed without the written agreement of the Consulting Engineer.

## 2.0 Design requirements to relevant British Standards

2.1    The Contractor-designed proprietary precast concrete floor shall be supplied by a specialist subcontractor approved by the job Architect or the Consulting Engineer.

2.2    The precast floor is to be designed to carry all the loads shown on the Engineer's drawings in addition to its self-weight.

2.3    All spans are to be taken as simply supported unless specifically noted otherwise on the Consulting Engineer's drawings. The precast units are to be designed in accordance with BS 8110 *Structural use of concrete*, which shall apply as the minimum standard of acceptance in respect of matters not covered in this Specification.

2.4    Any finishing screeds may not be taken as contributing to the structural strength of the finished precast floor.

2.5    Any specified structural screeds or other areas of in-situ infill are to be designed by the precast flooring Contractor unless shown otherwise on the Consulting Engineer's drawings.

2.6    All floors are to have a minimum dead weight of $220 \text{ kg/m}^2$.

2.7    All floors shall comply with the current Approved Documents in respect of both impact and airborne sound.

2.8    When used for upper floors the fire rating is to be a minimum of 1 hour. Finishes, screeds and the like shall not be considered as contributing to the fire resistance of the floor.

2.9    The size and location of service holes or ducts are to be as indicated on the Architect's or Consulting Engineer's drawings. The Contractor's attention is drawn to the need for trimming service holes positioned against unit bearing areas. All such trimming to be to the approval of the Architect and the Consulting Engineer.

## 3.0 Manufacture

3.1    No admixtures will be permitted unless the Consulting Engineer's permission is first obtained in writing.

3.2    High alumina cement shall not be used in the manufacture of the precast units.

3.3    All tolerances shall be in accordance with BS 8110 *Structural use of concrete*.

3.4    The Contractor's attention is drawn to the recommendations in BS 8110:Part 1: Clauses 6.2.5.2 chlorides in concrete; 6.2.5.3 sulphates in concrete; 6.2.5.4 alkali-silica reaction in concrete. No concrete is to be produced which does not comply with the recommendations contained within these clauses.

## 4.0 Finishes

4.1    All precast floors are to be suitably finished to receive any insulation or screed specified by the job Architect.

4.2    For upper floors the ceiling is to be fixed to the floor units using a proprietary clip fixing system to be agreed with the job Architect.

## 5.0 Delivery and storage

5.1    The precast floor manufacturer shall visit the site to determine and agree access for delivery and any cranage requirements. No extras will be allowed under any circumstances for the Contractor's lack of knowledge of the site or its surroundings.

5.2    Precast units will be off-loaded on site using appropriate lifting plant and units must be lifted as close to the ends of the beams as possible. Precast units must be stacked on level ground on timber bearers with timber packing between the floor beams.

## 6.0 Installation of precast units

6.1    The Contractor is to allow for all cranage, scaffolding and temporary propping required to safely erect and complete the floor including the provision of in-situ screeds, any in-situ make-up, and the grouting of the floor units.

## 7.0 Approvals

7.1    Design calculations and detailed assembly drawings are to be submitted to the job Architect for dimensional approval and to the Consulting Engineer and Approved Regulatory Bodies for structural approval prior to manufacture. Sufficient time must be allowed for in the manufacturer's programme for such approvals.

## 8.0 On-site modifications

8.1    Under no circumstances should precast floor units be modified on site without the written approval of the Consulting Engineer and the precast floor manufacturer.

## 6.4 In-situ suspended concrete floors

The use of in-situ suspended concrete floors in low-rise buildings for upper floors has declined in the last decade owing to the advantages offered by precast systems. There are, however, situations where an in-situ floor can provide the best technical solution: for example, when designing a building where the possibility of progressive collapse has to be catered for, i.e. buildings with five storeys or more.

In such cases a combination of in-situ structural toppings on a precast plank system used as a temporary shutter will produce at an economic cost a composite floor with the required robustness to prevent progressive collapse. Such a system easily permits the introduction of tying reinforcement into external walls and over internal walls to ensure structural integrity in the event of a gas explosion or similar incident.

In-situ reinforced concrete suspended floors have also become very popular on housing sites, where there are savings to be made if clean site-fill can be reused on site as make-up below floors. This method of construction will be covered in more detail in Chapter 7 Ground floors.

Major structural defects occurring in this form of construction can result from inadequate design, poor-quality materials and bad workmanship on site, often as a combined cause.

### 6.4.1 Design factors

All in-situ structural concrete should be designed in accordance with BS 8110 Part 1: 1997 *The structural use of concrete.*

The Design Engineer should ensure that the design criteria are based on the following.

● The finished structural element should have adequate fire resistance to meet the requirements of Part B of the Building Regulations Approved Documents. This will require consideration of suitable cover to reinforcement, and the type of surface finish to the underside and top of the floor.

There may be special considerations for buildings where the usage of the building may require more onerous precautions, as in the storage of inflammable materials for example.

● The imposed loads should be based on the building class as defined in BS 6399 Part 1: 1996. See Table 6.4.

● The correct allowance for weights of non-load-bearing partitions should be used (Table 6.1).

● Any walls supporting loads from above must be allowed for, as must the weight of finishes, such

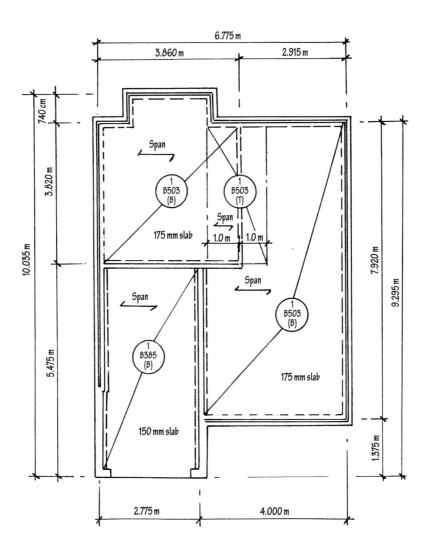

*Fig. 6.32* Plan showing reinforcement in in-situ reinforced concrete floor slab.

as screeds or timber boarding on insulation (Table 6.15).

● Where the floors of the building have to be designed to prevent progressive collapse, full compliance with Part A of the Approved Documents is required, and the detailed working drawings should clearly indicate the required details.

● Walls supporting the floors and any floor beams should be capable of carrying the loads without any overstressing. This may require appropriately sized padstones to be provided at beam reaction points, especially if the loadbearing blockwork is a light-weight thermal type with crushing strength of 2.8–3.50 N/mm².

● There should be clear and concise working drawings, which indicate the direction of the span of the floors, cover to reinforcements, trimming details around openings, and any relevant tying details into the main walls (Fig. 6.32).

The concrete design and concrete mix specification should comply with the relevant British Standards. The mix designs should take account of the strength and durability, and prescribed or designed mixes should comply with the tables in BS 8110 and BS 5328 (Tables 6.10. and 6.11).

For domestic housing projects the mixes listed in the NHBC Standards 2.1 Appendix B can be used by specifying the appropriate NHBC mix number (Table 6.12). Table 6.13 shows the exposure conditions for concrete mixes referred to in the NHBC standards.

### 6.4.2 Materials

The basic materials used in producing a reinforced concrete floor are steel reinforcement and ready-mixed concrete (water–aggregates–cement–admixtures).

#### Ready mix concrete

Many large projects may use concrete produced from a site batching plant, but most concrete used for low-rise developments is from external ready mix suppliers who are members of the Quality Scheme for Ready Mix Concrete (QSRMC), or similar schemes such as the BSI scheme. Such suppliers operate full quality control systems, which are intended to ensure that the concrete is delivered to the site as specified.

*Table 6.10* Guide to the selection of concrete mixes (from BS 5328)

| Element of structure | Designated mix | Standard mix | Recommended workability; nominal slump (mm) |
|---|---|---|---|
| *Foundations* | | | |
| Blinding and mass concrete fill | GEN 1 | ST 2 | 75 |
| Strip footings[a] | GEN 1 | ST 2 | 75 |
| Mass concrete foundations[a] | GEN 1 | ST 2 | 75 |
| Trench-fill foundations[a] | GEN 1 | ST 2 | 125 |
| Reinforced foundations[a] | RC. 35 | N/A | 75 |
| Foundations in Class 2 sulphate conditions | FND 2 | N/A | 75 |
| Foundations in Class 3 sulphate conditions | FND 3 | N/A | 75 |
| Foundations in Class 4A sulphate conditions | FND 4A | N/A | 75 |
| Foundations in Class 4B sulphate conditions | FND 4B | N/A | 75 |
| *Floors* | | | |
| House floors: unreinforced with screed or floating floor | GEN 1 | ST 2 | 75 |
| Garage floors: unreinforced | GEN 3 | ST 4 | 75 |
| Wearing surface: light foot and trolley traffic | RC 30 | ST 4 | 50 |
| Wearing surface: general industrial | RC 40 | N/A | 50 |
| Wearing surface: heavy industrial | RC 50 | N/A | 50 |
| *Other reinforced and prestressed concrete elements* | | | |
| Mild exposure condition | RC 30 | N/A | 75 |
| Moderate exposure condition | RC 35 | N/A | 75 |
| Severe exposure condition | RC 40 | N/A | 75 |
| Most severe exposure condition | RC 50 | N/A | 75 |

[a] In non-aggressive soils. Class 1 sulphate conditions

*Table 6.11* Minimum concrete cover for durability (from BS 8110: Part 1). Nominal cover to all reinforcement (including links) to meet durability requirements

| Condition of exposure (Clause 3.3.4) | Nominal cover (mm) | | | | |
|---|---|---|---|---|---|
| Mild | 25 | 20 | 20[a] | 20[a] | 20[a] |
| Moderate | – | 35 | 30 | 25 | 20 |
| Severe | – | – | 40 | 30 | 25 |
| Very severe | – | – | 50[b] | 40[b] | 30 |
| Extreme | – | – | – | 60[b] | 50 |
| Maximum free water/cement ratio | 0.65 | 0.60 | 0.55 | 0.50 | 0.45 |
| Minimum cement content (kg/m³) | 275 | 300 | 325 | 350 | 400 |
| Lowest grade of concrete | C30 | C35 | C40 | C45 | C50 |

[a] These covers may be reduced to 15 mm provided that the nominal maximum size of aggregate does not exceed 15 mm
[b] Where concrete is subject to freezing while wet, air entrainment should be used (Clause 3.3.4.2)
*Notes*
1. This table refers to normal-weight aggregate of 20 mm nominal max. size
2. For concrete used in foundations to low-rise construction see Clause 6.2.4.1

Table 6.12 General-purpose concrete mixes

| Location | Ground conditions | Designed mixes | | | Prescribed mixes | |
|---|---|---|---|---|---|---|
| | | Minimum concrete grade | Minimum cement content | Water/ cement ratio | NHBC mix no. | BS mix |
| Ground floor | Mild | RC 30 | 275 kg | 0.65 | 3 | C20 P |
| Ground floor | Moderate | RC 35 | 300 kg | 0.60 | 4 | C25 P |
| Ground floor | Severe exp. | RC 40 | 325 kg | 0.55 | 5 | C30 P |
| Upper floor | Mild | RC 30 | 275 kg | 0.65 | 3 | C20 P |

Table 6.13 Exposure conditions

| Environment | Exposure conditions |
|---|---|
| Mild | Concrete surfaces completely protected against weather or aggressive conditions except for brief periods of exposure to normal weather conditions during construction |
| Moderate | Concrete surfaces sheltered from severe rain or freezing wind when wet |
| | Concrete subject to condensation |
| | Concrete surfaces continuously under water |
| | Concrete in contact with non-aggressive soils (i.e. Class 1) |
| Severe | Concrete surfaces exposed to driving rain, alternate wetting and drying or occasional freezing or severe condensation |
| Very severe | Concrete surfaces exposed to sea water spray, deicing salts (directly or indirectly), corrosive fumes or severe freezing conditions while wet |
| Extreme | Concrete surfaces exposed to abrasive action, e.g. sea water carrying solids, flowing water with pH < 4.50, or machinery or vehicles |

## Reinforcement

In the UK it is usual to specify reinforcement for reinforced concrete elements by reference to the relevant British Standards. For reinforcement these are:

- BS 4449: 1997 *Specification for carbon steel bars for the reinforcement of concrete*
- BS 4466: 1989 *Specification for scheduling, dimensioning, bending and cutting of steel reinforcement for concrete*
- BS 4482: 1985 *Specification for cold reduced steel wire for the reinforcement of concrete*
- BS 4483: 1998 *Steel fabric for the reinforcement of concrete*

Most reinforcement in low rise developments is steel fabric reinforcement, which consists of drawn-down steel bars welded together to form square or rectangular grids.

Made with hard-drawn steel wire complying with BS 4482, the fabric is designated as a high-yield reinforcement. Table 6.14 shows the range of stock fabrics available for structural works.

The sheets of mesh are generally available in standard 4.80 m long by 2.40 m wide sizes.

Table 6.14 Stock steel fabrics for structure works

| Fabric ref. | Longitudinal wires | | | Cross-wires | | | Mass (kg/m²) |
|---|---|---|---|---|---|---|---|
| | Size | Spacing (mm) | Area (mm²/m) | Size | Spacing (mm) | Area (mm²/m) | |
| A98 | 5 | 200 | 87 | 5 | 200 | 98 | 1.54 |
| A142 | 6 | 200 | 142 | 6 | 200 | 142 | 2.22 |
| A193 | 7 | 200 | 193 | 7 | 200 | 193 | 3.02 |
| A252 | 8 | 200 | 252 | 8 | 200 | 252 | 3.95 |
| A393 | 10 | 200 | 393 | 10 | 200 | 393 | 6.16 |
| B196 | 5 | 100 | 196 | 7 | 200 | 193 | 3.05 |
| B283 | 6 | 100 | 283 | 7 | 200 | 193 | 3.73 |
| B385 | 7 | 100 | 385 | 7 | 200 | 193 | 4.53 |
| B503 | 8 | 100 | 503 | 8 | 200 | 252 | 5.93 |
| B785 | 10 | 100 | 785 | 8 | 200 | 252 | 8.14 |
| B1131 | 12 | 100 | 1131 | 8 | 200 | 252 | 10.90 |
| C283 | 6 | 100 | 283 | 5 | 400 | 49 | 2.61 |
| C385 | 7 | 100 | 385 | 5 | 400 | 49 | 3.41 |
| C503 | 8 | 100 | 503 | 5 | 400 | 49 | 4.34 |
| C636 | 9 | 100 | 636 | 6 | 400 | 70.80 | 5.55 |
| C785 | 10 | 100 | 785 | 6 | 400 | 70.80 | 6.72 |
| D49 | 2.5 | 100 | 49 | 2.5 | 100 | 49 | 0.77 |
| D98 | 5 | 200 | 98 | 5 | 200 | 98 | 1.54 |

*Table 6.15* Weights of floor finishes

| Finishes | Approximate weight (kN/m$^2$) |
|---|---|
| Asphalt flooring | 0.268 |
| Chipboard 22 mm thick, on timber battens with 25 mm thick mineral wool quilt insulation | 0.20 |
| Flexible PVC tiles (polyvinylchloride), 3.2 mm thick | 0.049 |
| Hardwood floor blocks | 0.079 |
| Lightweight floor screed per 25 mm thickness | 0.35 |
| PVC fibre reinforced tiles, 4.80 mm thick | 0.103 |
| Sand-cement screed per 25 mm thickness | 0.585 |

Some steel suppliers are now producing 'bespoke mesh' supplied in rolls, which is a system of various bar diameters at various spacings, the steel having been designed for specific areas of the floor. This cuts out waste on laps, and uses less reinforcement overall. The reinforcement is delivered to site in rolls, and is pre-fabricated to fit the areas being concreted. Distribution steel also comes in rolls and is placed as a separate transverse layer.

On some sites additional reinforcement in the form of bars may be required, and the two types in use are hot-rolled mild steel to BS 4449 (designated by suffix R on drawings) and cold-worked high-yield bars to BS 4461 (designated by suffix T on drawings).

### 6.4.3 Workmanship on site

Bad workmanship is often the cause of major structural defects in concrete floors despite the use of good-quality

concrete and reinforcement. The following standards must be adopted if serious structural defects are to be prevented:

- All reinforcement must have the concrete cover as specified on the engineer's drawings. This often requires the use of proprietary precast concrete support stools for bottom steel and pre-bent support chairs for top steel.
- The concrete ordered must be as specified by the engineer. Always check the delivery tickets before placing into the works. No water should be added to the supplied concrete mix to make it easier to place. If the concrete appears to be stiff a slump test should be taken, and if the slump test is too low, i.e. less than 50 mm, then the concrete should be sent back to the suppliers. All reinforced concrete should be well vibrated and properly cured using wet hessian or curing compounds sprayed on soon after floating-off is completed. This is most important in periods of hot weather, but in cold weather it is vitally important to provide additional protection to ensure that the new concrete is protected from the effects of frost. The temperature of any ready mix concrete delivered to site should never be less than 5 °C.
- Slump tests and cube tests should always be carried out on large sites. For suspended floors the concrete should have an average slump of about 50 mm. All concrete test cubes should be stored in a water tank at the correct temperature prior to their being crushed. A complete record should be maintained of all the concrete supplied to site, including the test results.
- Formwork and props to elevated concrete floors should remain in place until the concrete has

*Table 6.16* Reinforcement areas (mm$^2$) for groups of bars

| Diameter (mm) | Number of bars | | | | | | | | | | | |
|---|---|---|---|---|---|---|---|---|---|---|---|---|
| | 1 | 2 | 3 | 4 | 5 | 6 | 7 | 8 | 9 | 10 | 11 | 12 |
| 6 | 28 | 57 | 85 | 113 | 142 | 170 | 198 | 226 | 255 | 283 | 311 | 340 |
| 8 | 50 | 101 | 151 | 201 | 252 | 302 | 352 | 402 | 453 | 503 | 553 | 604 |
| 10 | 79 | 157 | 236 | 314 | 392 | 471 | 550 | 628 | 707 | 785 | 864 | 942 |
| 12 | 113 | 226 | 339 | 452 | 566 | 679 | 792 | 905 | 1020 | 1130 | 1240 | 1360 |
| 16 | 201 | 402 | 603 | 804 | 1010 | 1210 | 1410 | 1610 | 1810 | 2010 | 2210 | 2410 |
| 20 | 314 | 628 | 943 | 1260 | 1570 | 1890 | 2200 | 2510 | 2830 | 3140 | 3460 | 3770 |
| 25 | 491 | 982 | 1470 | 1960 | 2450 | 2950 | 3440 | 3930 | 4420 | 4910 | 5400 | 5890 |
| 32 | 804 | 1610 | 2410 | 3220 | 4020 | 4830 | 5630 | 6430 | 7240 | 8040 | 8850 | 9650 |
| 40 | 1260 | 2510 | 3770 | 5030 | 6280 | 7540 | 8800 | 10100 | 11300 | 12600 | 13800 | 15100 |
| 50 | 1960 | 3930 | 5890 | 7850 | 9820 | 11800 | 13700 | 15700 | 17700 | 19600 | 21600 | 23600 |

| Diameter (mm) | Spacings (mm) | | | | | | | |
|---|---|---|---|---|---|---|---|---|
| | 75 | 100 | 125 | 150 | 175 | 200 | 250 | 300 |
| 6 | 377 | 283 | 226 | 189 | 162 | 142 | 113 | 94 |
| 8 | 671 | 503 | 402 | 335 | 287 | 252 | 210 | 168 |
| 10 | 1050 | 785 | 628 | 523 | 449 | 393 | 314 | 262 |
| 12 | 1510 | 1130 | 905 | 745 | 646 | 566 | 452 | 377 |
| 16 | 2680 | 2010 | 1610 | 1340 | 1150 | 1010 | 804 | 670 |
| 20 | 4190 | 3140 | 2510 | 2090 | 1800 | 1570 | 1260 | 1050 |
| 25 | 6550 | 4910 | 3930 | 3270 | 2810 | 2450 | 1960 | 1640 |
| 32 | 10700 | 8040 | 6430 | 5360 | 4600 | 4020 | 3220 | 2680 |

achieved an adequate cube strength sufficient to support itself. Failure to observe this rule can result in excessive creep deflections months later, as removing props too soon causes early loading of the concrete and exacerbates the long-term creep process.

● Any construction joints should be formed in the middle third of a slab or beam.

● All installations of services etc. should be completed prior to concrete commencing. In addition the formwork should be checked for fallen debris, especially wire ties and nails.

● Admixtures should be used only where permitted by the engineer. The correct dosage should be maintained as outlined in the manufacturer's instructions.

● Do not load the concrete floor with building materials until it has achieved its 28-day cube strength. Early loading of a concrete floor can increase the effects of long-term creep as the bond between the reinforcement and concrete can be reduced.

● Check that all reinforcing steel is clean and free from loose rust or oils (especially mould oil) before concreting.

## Case history 6.3

### Introduction

A mill building was being converted into a two-storey development of flats for private sale. The existing mill floors were constructed of timber planking fixed to pitch pine joists. The floors were surveyed and opened up in selected areas to assess their condition.

Many of the joists had wet rot at their bearings into the main external walls, and as the design of the new housing layout required the provision of masonry walls the existing timber floor would require to be strengthened considerably.

There was also a problem with fire and noise at party wall positions, and it was therefore decided to remove the existing timber floor completely and replace it with an in-situ suspended concrete floor with finishes on top to provide improved sound resistance.

The new floor was designed to span two spans and support the internal timber stud partitions, which were non-loadbearing (Fig. 6.33). The existing external walls were solid natural sandstone, and it was decided to provide a new internal skin of insulating blockwork built of the existing footings, which would support the new floor.

### Design calculations

#### Design references

● BS 8110 *Structural use of concrete*: Part 1
● *Design data for rectangular beams and slabs to BS 8110: Part 1* by A.H. Allen
● BS 6399: 1996 Part 1 *Loading for buildings. Code of practice for dead and imposed loads*

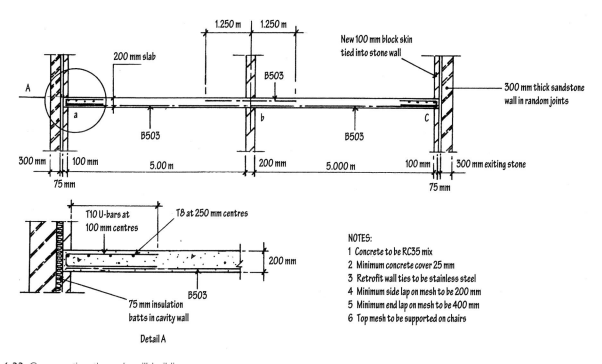

**Fig. 6.33** Cross-section through mill building.

### Loadings

|                                               | kN/m² |
| --------------------------------------------- | ----- |
| Imposed loading                               | $1.50 \times 1.60 = 2.40$ |

|                                               | kN/m² |
| --------------------------------------------- | ----- |
| Dead loading: stud partitions                 | 0.50  |
| chipboard, battens and insulation             | 0.20  |
| 200 mm concrete                               | 4.80  |
| 12.50 mm plaster                              | 0.25  |
|                               Total =         | $\overline{5.75} \times 1.40 = 8.05$ |

Design span $= 5.00 + 0.100 = 5.10$ m

The worst loading condition for centre support moment is with both spans fully loaded with their dead and imposed loads, using appropriate limit state factors of 1.40 for dead load and 1.60 for imposed loads.

Using moment distribution analysis: Distribution factors $= 0.50$

$$\text{FEM UDL} = \frac{10.45 \times 5.1^2}{L^2} = 22.65 \text{ kN m}$$

$$\text{FEM point load } a\text{–}b = \frac{Wab^2}{L^2} = \frac{5.71 \times 3.0 \times 2.10^2}{5.10^2}$$

$$= 2.90 \text{ kN m}$$

$$\text{FEM point load } b\text{–}a = \frac{Wa^2b}{L^2} = \frac{5.71 \times 2.10 \times 3.0^2}{5.10^2}$$

$$= 4.14 \text{ kN m}$$

| DF (a–b) | (b–a) 0.50 | 0.50 (b–c) | (c–b) |
| -------- | ---------- | ---------- | ----- |
| −22.65   | +22.65     | −22.65     | +22.65 |
| −2.90    | +4.14      | −4.14      | +2.90 |
| −25.55   | +26.79     | −26.79     | +25.55 |
| +25.55   |            |            | −25.55 |
| 0        | +12.75     | −12.75     | 0     |
| **0**    | **+39.54** | **−39.54** | **0** |

Span a–b bending moments at twentieth points from left to right (sagging is positive):

| 0   | 1    | 2    | 3    | 4     | 5     | 6     |
| --- | ---- | ---- | ---- | ----- | ----- | ----- |
| 0   | 5.08 | 9.47 | 13.2 | 16.20 | 18.60 | 20.30 |

| 7     | 8     | 9     | 10    | 11    | 12    | 13    |
| ----- | ----- | ----- | ----- | ----- | ----- | ----- |
| 21.30 | 21.60 | 21.20 | 20.20 | 18.50 | 15.70 | 11.20 |

| 14   | 15    | 16    | 17    | 18    | 19    | 20    |
| ---- | ----- | ----- | ----- | ----- | ----- | ----- |
| 5.98 | 0.081 | −6.49 | −13.7 | −21.7 | −30.3 | −39.6 |

Span b–c repeats but handed.

Worst case for mid-span moments is with one span fully loaded, 1.40 DL + 1.60 LL and one span loaded with $1.0 \times$ Dead load.

Span a–b bending moments at twentieth points from left to right (sagging is positive):

| 0   | 1    | 2     | 3     | 4     | 5     | 6     |
| --- | ---- | ----- | ----- | ----- | ----- | ----- |
| 0   | 5.49 | 10.30 | 14.40 | 17.90 | 20.70 | 22.80 |

| 7     | 8     | 9    | 10    | 11    | 12    | 13    |
| ----- | ----- | ---- | ----- | ----- | ----- | ----- |
| 24.20 | 24.90 | 25.0 | 24.40 | 23.10 | 20.80 | 16.70 |

| 14    | 15   | 16   | 17    | 18    | 19    | 20    |
| ----- | ---- | ---- | ----- | ----- | ----- | ----- |
| 11.90 | 6.40 | 0.25 | −6.58 | −14.1 | −22.3 | −31.1 |

Span b–c bending moments at twentieth points from left to right (sagging is positive):

| 0     | 1     | 2     | 3     | 4     | 5     | 6     |
| ----- | ----- | ----- | ----- | ----- | ----- | ----- |
| −31.1 | −25.4 | −20.1 | −15.1 | −10.5 | −6.26 | −2.41 |

| 7    | 8    | 9    | 10   | 11   | 12   | 13   |
| ---- | ---- | ---- | ---- | ---- | ---- | ---- |
| 1.07 | 4.17 | 6.11 | 7.42 | 8.36 | 8.93 | 9.12 |

| 14   | 15   | 16   | 17   | 18   | 19   | 20   |
| ---- | ---- | ---- | ---- | ---- | ---- | ---- |
| 8.94 | 8.39 | 7.46 | 6.16 | 4.48 | 2.43 | 0    |

With continuous suspended floors it is allowed to reduce the support moments by 20% and redistribute the moments onto the adjacent mid-span moments.

Therefore design for support moment of 80% of 39.54 kN m $= 31.68$ kN m

Mid-span adjusted moment $= 25 + 20\%$ of $39.54 = 25 + 7.92 = 32.92$ kN m

From Design Chart No. 1 (see Fig. 7.12 in Chapter 7):

$$\frac{M}{b \times d^2 \times f_{cu}} = \frac{31.68 \times 10^6}{1000 \times 170^2 \times 35} = 0.031$$

Therefore $K = 0.96$

Lever arm $z = 0.96 \times 170 = 163$ mm based on 25 mm cover and 10 mm bars

$$A_{st} = \frac{M}{0.87 f_y z} = \frac{31.68 \times 10^6}{0.87 \times 460 \times 163} = 485 \text{ mm}^2$$

Therefore use B503 fabric mesh in top of slab over central wall for 1.25 m each side.

From Design Chart No. 1:

$$\frac{M}{b \times d^2 \times f_{cu}} = \frac{32.92 \times 10^6}{1000 \times 170^2 \times 35} = 0.032 \quad K = 0.96$$

Lever arm $z = 0.96 \times 170 = 163$ mm

$$A_{st} = \frac{M}{0.87 \times f_y \times z} = \frac{32.92 \times 10^6}{0.87 \times 460 \times 163} = 504 \text{ mm}^2$$

Therefore use B503 in bottom of slab in both bays.

Check deflection:

BS 8110: Clause 3.4.6.7, Table 3.10

Percentage of steel is approximately 0.30% so modification factor $= 1.53$

Ratio of $\dfrac{\text{Basic span}}{\text{Effective depth}} = 26$ max

Allowable $\dfrac{\text{Basic span}}{\text{Effective depth}} = 26 \times 1.53 = 39.78$

Maximum allowable span $= 39.78 \times \dfrac{170}{1000}$

$= 6.760 \text{ m} > 5.10 \text{ m}$

*Use 200 mm thick slab, 35 N concrete reinforced with B503 mesh.*

## Points to remember

- **Statutory requirements:** All work must comply with all relevant Approved Documents and British Standards, Codes of Practice etc.
- **Structural design:** The design of all structural works and specifications should provide satisfactory performance. All structural design should be carried out by a chartered engineer or other suitably qualified persons in accordance with the current British Standards and Codes of Practice.

- **Materials:** All materials, building products and building systems must be suitable for their intended purpose.
- **Workmanship:** All construction work must be carried out in a proper, neat and workmanlike fashion and in accordance with manufacturer's instructions.

## References

*Building Research Establishment*

*Specifying timber*, BRE Digest 156 (1973).
*Remedial wood preservatives – use them safely*, BRE Digest 371 (1992).

*British Standards Institution*

BS 449–2:1969 *Specification for the use of structural steel in building.*
BS 648:1964 *Schedule of weights of building materials.*
BS 1881:1983–88 *Testing concrete.*
BS 4449:1997 *Specification for carbon steel bars for the reinforcement of concrete.*
BS 4461:1978 *Specification for cold worked steel bars for the reinforcement of concrete.*
BS 4466:1989 *Specification for scheduling, dimensioning, bending and cutting of steel reinforcement for concrete.*
BS 4471:1987 *Specification for sizes of sawn and processed softwood.*
BS 4482:1985 *Specification for cold reduced steel wire for the reinforcement of concrete.*
BS 4483:1998 *Steel fabric for the reinforcement of concrete.*
BS 4978:1996 *Specification for visual strength grading of softwood.*
BS 5268 *Structural use of timber*
  Part 2:1996 *Code of practice for permissible stress design, materials and workmanship*
  Part 5:1989 *Code of practice for the preservative treatment of structural timber.*

BS 5328 *Concrete*
  Part 1:1997 *Guide to specifying concrete.*
  Part 2:1997 *Methods for specifying concrete mixes.*
BS 5589:1989 *Code of practice for the preservation of timber.*
BS 5950 *Structural use of steelwork in building*
  Part 1:1990 *Code of practice for design in simple and continuous construction: hot rolled sections.*
BS 6073:1981 *Specification for precast concrete masonry units.*
BS 6178 *Joist hangers*
  Part 1:1990 *Specification for joist hangers for building into masonry walls of domestic dwellings.*
BS 6399 *Loading for buildings*
  Part 1:1996 *Code of practice for dead and imposed loads.*
BS 8103 *Structural design of low-rise buildings*
  Part 3:1996 *Code of practice for timber floors and roofs for housing*
  Part 4:1995 *Code of practice for suspended concrete floors for housing.*
BS 8110 *Structural use of concrete*
  Part 1:1997 *Code of practice for design and construction.*

The Building Regulations, Approved Documents, Parts A1, A2.

H.A. Allen, *Design data for rectangular beams and slabs to BS 8110: Part 1*, Palladian, London (1987).
J.A. Baird and E.C. Ozelton, *Timber designer's manual*, 2nd edn, Collins (1984).

### Useful addresses

Timber Research and Development Association (TRADA), Stocking Lane, Hughenden Valley, High Wycombe, Bucks HP14 4ND.

National House-Building Council (NHBC), Buildmark House, Chiltern Avenue, Amersham, Bucks HP6 5AP.

BRE Advisory Service, Building Research Establishment, Garston, Watford WD2 7JR.

# Chapter 7
# Ground floors

## 7.1 Introduction

Most low-rise commercial buildings are constructed with a concrete ground floor. Factory floors are usually reinforced concrete slabs laid using the long strip method on compacted stone sub-base material.

For low-rise housing and commercial buildings there are various options available to builders, and each type of floor has its own advantages. The types of ground floor construction used most in low-rise buildings are:

- timber joist floors on sleeper walls with ventilated air space (Fig. 7.1);
- full-span timber joist floors with ventilated air space;
- in-situ concrete ground-bearing slab (Fig. 7.2);
- in-situ fully suspended reinforced slab (Fig. 7.3);
- precast beam and block floor with ventilated void (Fig. 7.4).

## 7.2 Timber joist floors on sleeper walls

This is a long-standing traditional flooring method. It generally consists of 100 mm × 50 mm joists spanning between masonry honeycombed sleeper walls spaced at approximately 1800 mm centres. Floorboarding can be softwood timber, tongued and grooved boards, chipboard or plywood sheets. Insulation can be placed between the joists, suspended on proprietary netting fixed to the joists.

When using chipboard flooring it must be a moisture-resistant type C4 to BS 5669, and it must be laid with a 10 mm expansion gap at room perimeters between the flooring and the walls.

### 7.2.1 Design defects that can result

- Incorrect imposed loading used: domestic 1.50 kN/m², offices 2.50 kN/m².
- Incorrect size of joists specified for the span.
- Sleeper walls too far apart.
- Incorrect grade of timber specified.
- Inadequate design for infilled sites where the depth of infill below the oversite concrete that supports the sleeper walls exceeds 600 mm. In such situations sleeper walls built off the concrete oversite are not recommended, and joists must be placed full span between loadbearing walls.
- Inadequate design to cater for trees. Damage can result from clay shrinkage owing to the influence of existing trees or clay heave following removal of trees within or close to the building footprint.
- Partition loads not catered for in the design of the floor joists.

### 7.2.2 Materials defects that can result

- Structural timber not in accordance with British Standards: e.g. not kiln dried, excessive wane, too many knots.
- Imported infill material used below concrete oversite may not be chemically inert. Sulphate attack of the concrete oversite can be caused by the use of chemically unstable fills such as clinker, furnace bottom ash, colliery spoil and crushed bricks; these generally contain soluble sulphates, which can leach out and react with the concrete.
- The use of timber and flooring that has been damaged by water owing to poor storage on site.
- Regularised timber joists not specified. This can result in a very uneven floor.
- Failure to provide moisture-resistant flooring, i.e. Type C4 chipboard.
- Failure to use correct fastenings for chipboard flooring, i.e. ring shank nails or screws whose minimum length is 2.50 times the board thickness.

### 7.2.3 Workmanship defects that can result

- Inadequate compaction of sub-base below oversite concrete.
- Failure to provide a proper dampcourse below wallplates or joist ends.
- Floor joists not trimmed properly at openings in floor.

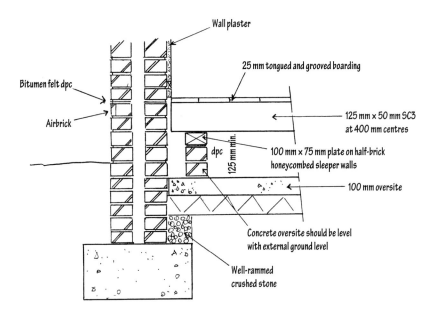

*Fig. 7.1* Timber joists on sleeper walls.

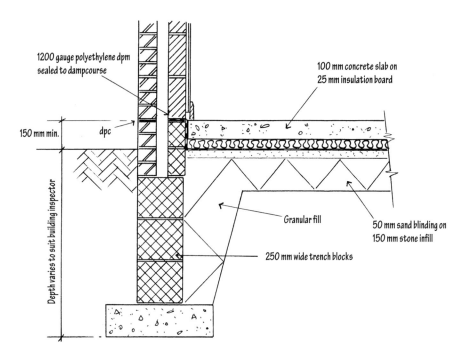

*Fig. 7.2* Concrete ground-bearing slab.

- Inadequate through-ventilation provided to the underfloor void.
- Floor joists not strutted properly, especially on long spans.
- Joists not doubled up when supporting partitions.
- Insufficient gap left around perimeter walls to chipboard and plywood flooring.
- Timber with excessive moisture content used, i.e. above 18%.
- Flooring fixed prior to building being substantially weathertight. Any ingress of driving rain can cause severe warping of the flooring and local damage.
- Failure to fit lateral restraint straps to joists running parallel with the external walls.

- Wet rot or dry rot arising from poor ventilation to underfloor void or badly constructed dampcourse below joist seatings.
- Settlement of the sub-base below oversite concrete, owing to the poor grading the sub-base stone having too many fines, or a high bulking factor due to high moisture content. Such materials result in significant subsidence as a result of the sub-base's drying out.

The Approved Documents require such floors to have a 125 mm minimum air space below the underside of the joists or 75 mm below the treated wall plate to prevent the risk of wood rot due to interstitial condensation or dampness. The joists can be laid onto a 100 mm × 50 mm

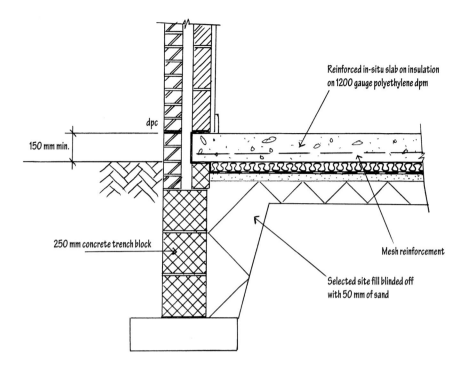

*Fig. 7.3* Reinforced in-situ concrete suspended floor slab.

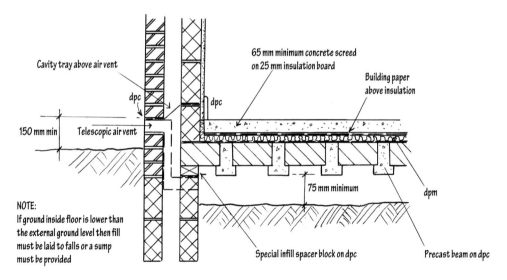

*Fig. 7.4* Precast beam and block screeded floor.

timber wall plate or can be built into the honeycombed sleeper walls, and the joists or wall plate must be laid on a dpc.

The void to such floors must be ventilated using airbricks that provide no less than 1500 mm² of open area per metre run of external wall. This can be achieved by using 225 mm × 150 mm clay airbricks at 1.50 m centres. These airbricks should not exceed 2.0 m spacings, and should be positioned within 450 mm of each end of any wall.

Every part of the air space below a timber floor must be thoroughly ventilated through openings on at least two opposite sides. This ventilation must extend through sleeper walls and other loadbearing walls built off foundations.

One of the most common defects encountered results from poorly ventilated air spaces. In some cases the air bricks have become blocked or covered over by external paving. This lack of air circulation can result in serious timber decay from fungal attack.

Another cause of timber decay in floor joists can arise when there is bridging via a path above the dampcourse due to earth piled against the wall, or to a raised pavement.

Fungal decay in timber is dependent on:

● condensation;
● rain penetration;
● contact with other damp materials, generally brick or block walls;
● contact with very humid air for relatively long periods.

## Case history 7.1

### Introduction

When decorating his ground floor lounge a houseowner noticed a small patch of dampness on the inside of an external cavity wall just above the skirting board level. He lodged a claim against his house warranty provider.

### Investigation

Close inspection of the surrounding walls revealed dampness in the ends of the floorboards at their junction with the skirting board on the external walls.

Examination of the external face of the cavity wall showed no dampness present. It was considered therefore that the presence of damp in the boards and internal wall pointed to the possibility that both the wall and floor joists were being penetrated by dampness.

It was established that the property was seven years old; the walls were of brick and block inner leaf, with 50 mm cavity. All the ground floor rooms had the traditional voided timber joist floor fixed to a timber wall plate on sleeper walls. Ventilation of the subfloor void was via air bricks through the external walls. Below the joists was a 100 mm thick concrete oversite.

Sections of the timber skirting and floorboarding were removed in the area of the observed damp patch. A close examination of the wall dampness showed that in one place it extended down to the mineral fibre dampcourse. By raking out the mortar joints it was possible to see that the dampcourse had been built without a lap, and there was a gap between the two pieces of dpc.

The ends of the joists supporting the floorboards and the wall plate and brick sleeper wall underneath were very wet, and the wall plate was showing signs of fungal decay. It was also noticed that the wall plate was bedded

in mortar directly onto the sleeper wall, and was wet throughout its length. The sleeper wall was also very damp, and was built off a poor-quality concrete oversite (Fig. 7.5).

### Conclusion and recommendations

It was established that the cause of the dampness in the internal walls and floorboarding was due to rising damp from the ground below.

The dampness in the wall was due to the failure of the dampcourse in the inner leaf of the external wall by virtue of its being butted up and not being continuous. The dampness in the floorboarding and timber wall plate was due to a failure to provide a dampcourse below the timber wall plate on the sleeper wall.

Further investigation showed that this was occurring only on this one sleeper wall, but it was also found that not all of the external airbricks extended through the inner leaf into the air space. This was the cause of the musty smell below the floor.

### Remediation works

It was decided to repair the badly built dampcourse by adding an airbrick and providing a proper overlap to the dampcourse sections. The next step was to remove the skirting boards to allow the wall to dry out. The timber wall plate that was showing signs of rot was replaced with a new treated wall plate onto a fully bedded dampcourse.

While the section of flooring was up, the poor areas of oversite concrete were broken out and replaced with a 1:3:6 site-mixed concrete.

Once the existing floor joists had dried out, new floorboards were fixed and the skirtings were replaced.

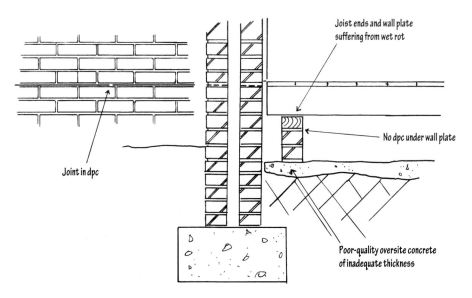

*Fig. 7.5* Defects in floor and wall construction as built.

## 7.3 In-situ concrete ground-bearing slabs

This is probably the most common type of ground floor construction used by housebuilders. The concrete floor is generally placed onto a compacted crushed stone sub-base with a 1200 gauge polyethylene dpm linked into the wall dampcourse.

The slab can be provided with a power-floated finish, concrete screed, asphalt finish or timber boarding on treated battens onto a quilt insulation or similar insulation. For domestic properties it is usually 100 mm thick.

If a concrete screed is used as a finish it must have a minimum thickness of 50 mm or, in the case of an unbonded screed, 65 mm.

### 7.3.1 Design defects that can result

● Inadequate design for infilled sites where the depth of infill below the concrete slab exceeds 600 mm. In such situations the floor should be a fully suspended reinforced concrete slab or a precast floor spanning between loadbearing walls.
● Inadequate design to cater for trees. Damage can result from clay shrinkage due to the influence of existing trees or clay heave following removal of trees within or close to the building footprint.
● Inadequate slab thickenings under non-loadbearing blockwork partition.

### 7.3.2 Materials defects that can result

● The use of poor-quality concrete can result in a defective floor, especially in a garage, where there is a higher degree of exposure and wear.
   The NHBC provides minimum standards for concrete floors (Table 7.1), and requires a GEN 1 mix for floors with permanent finishes, i.e. screed or floating floor, and a GEN 2 mix where no permanent finish is to be used, i.e. carpet or cushion floor. For garage ground-bearing floors without reinforcement, a GEN 3 mix is recommended.
● Existing or imported sub-base materials below the slab may contain soluble sulphates, which could attack the concrete if no polyethylene dpm is present.

### 7.3.3 Workmanship defects that can result

● Inadequate compaction of sub-base fills below the floor slab.
● Failure to construct the floor dpm properly and link it into the dampcourse in the main walls.
● Adding water or workability additives to a concrete mix to make it easier to place. This can result in concrete with a low strength and poor resistance to frost.
● Placing concrete over frozen sub-base materials. This can result in floor heave when the sub-base thaws out, and subsequent cracking of the floor.
● Failure to provide adequate protection to a newly cast floor from the elements. Floor slabs poured in

Table 7.1 Minimum concrete specifications for general purposes

| Structural element | Designated mix | | Characteristic strength (N/mm²) | Minimum cement content (kg/m³) |
|---|---|---|---|---|
| Unreinforced ground-bearing slabs | GEN 1 | Permanent finish | 10 | 175 |
| | GEN 2 | No permanent finish | 15 | 200 |
| Garage slab: ground bearing | GEN 3 | | 20 | 220 |
| Dwelling floor fully suspended and reinforced concrete | RC 35 | | 35 | 300 with W/C ratio of 0.60 |
| Garage floor fully suspended reinforced slab | RC 35 | | 35 | 300 with W/C ratio of 0.60 |

The above mixes are the minimum cement contents for non-aggressive soil conditions, i.e. where soluble sulphates are Class 1 or less
Recommended workability: a nominal slump of 75 mm is recommended for slabs without reinforcement. For RC 35 mixes a 50 mm nominal slump is recommended.

hot, windy or cold weather should be covered with damp hessian or damp sand, or sprayed with a proprietary curing compound, to prevent the surface from drying out too rapidly or freezing.

In very cold weather concrete floors should be constructed in accordance with the *NHBC Standards* Chapter 1.4 'Cold weather working'. These standards recommend that the minimum temperature of ready-mixed concrete when delivered should be 5 °C, and the newly placed concrete must be properly protected from the elements by polyethylene sheeting or a curing compound.

The three most common defects to ground floors are caused by:

● inadequate compaction of sub-base infill material below the floor slab or oversite concrete;
● excessive depth of upfill to floors, or floors being laid on existing deep fills that have a potential for ongoing self-weight or collapse settlement;
● use of fill materials that are not chemically inert, and which have a potential for reactive chemical changes.

### 7.3.4 Sub-base fills

The specification and compaction of sub-base fill materials is most important. Prior to 1980 50% of NHBC claims repair costs were as a result of faulty fills below ground floors. Where upfilling is required this should consist of clean crushed concrete, crushed stone, hoggin or other approved hard materials, having the following properties:

● The fill material should be regular in shape, and should pass a 150 mm diameter ring. It should be reasonably graded such that there are sufficient proportions of finer material to ensure that it can be well consolidated into a compact homogeneous layer.

- It should be free from organic matter, non-combustible, and free from excessive dust. It should also have a low soluble sulphate content, and be free from other chemicals that could have a deleterious effect on concrete.
- The fill should be chemically inert and physically stable (wet or dry), and should not contain any clay or shales.
- Demolition fills should be avoided, as they contain deleterious materials such as gypsum plaster, timber, or bricks from old flues.
- Fills containing colliery waste, demolition materials, furnace bottom ash, steel or iron slags, expansive shales and soft chalks are generally not suitable as structural fills below floor slabs, and are best not used. Apart from the risk of sulphate damage to the floor slab these types of fill pose a danger to the foundations and underbuilding works if the soluble sulphate contents are higher than the concrete or mortar can tolerate.

### 7.3.5 Compaction of fills

Where the depth of upfill to a floor slab is less than 600 mm it is acceptable to use a normal ground-bearing floor construction. The compaction of the fill below the slab or oversite should be carried out in discrete layers not exceeding 200 mm depth, and the most suitable compaction plant is a small vibrating smooth roller such as a Bomag 600, each layer placed having four to six passes of the roller. Vibrating plate compactors are not as efficient as vibratory rollers except in confined spaces or when compacting narrow trenches.

It is no use placing fill without consolidating it efficiently. Defects will result when the full depth of fill is deposited at one time in the actual footprint of the floor, spread and rolled from the intended finished level. The tipping of fill direct from a lorry into a floor area should not be allowed as this is one of the major causes of slabs settling. Badly filled and compacted ground can only lead to trouble, and careful site supervision is needed to ensure good compaction standards. Table 7.2 gives details of the type of plant and layer thicknesses for the compaction of granular fills.

### 7.3.6 Depth of fills

The depth of upfill below a ground-bearing floor slab should not exceed 600 mm from the formation level of natural strata to the underside of the slab. This is an NHBC requirement.

In certain situations the fills present may be existing and may have been in place for many years. While it could be argued that such fills could support the weight of a thin concrete slab, there is a potential for subsidence or other problems to occur, depending on the nature of the fill and the materials it contains.

For many years it was fashionable to use vibrocompaction for new houses built over the footprints of previous backfilled cellars. Generally precast floors were used, but on occasions ground-bearing slabs were constructed on the basis that the fills had been improved up to a specified bearing capacity. Subsequently a lot of dwellings built this way developed problems of floor subsidence and invasive dry rot.

Examination of the sub-base fill below the floors of many of the claims investigated revealed decaying timbers in the cellar backfill, with mycelium fungus

Table 7.2 Compaction requirements for granular fills

| Type of compaction plant | Weight category | Number of passes for layers not greater than | | |
|---|---|---|---|---|
| | | 110 mm | 150 mm | 225 mm |
| Smooth wheeled roller | Mass/metre width of roller | | | |
| | 2700–5400 kg | 16 | Unsuitable | Unsuitable |
| | Over 5400 kg | 8 | 16 | Unsuitable |
| Pneumatic-tyred roller | Mass/wheel | | | |
| | 4000–6000 kg | 12 | Unsuitable | Unsuitable |
| | 6000–8000 kg | 12 | Unsuitable | Unsuitable |
| | 8000–12000 kg | 10 | 16 | Unsuitable |
| | Over 12 000 kg | 8 | 12 | Unsuitable |
| Vibrating roller | Mass/metre width of roller | | | |
| | 700–1300 kg | 16 | Unsuitable | Unsuitable |
| | 1300–1800 kg | 6 | 16 | Unsuitable |
| | 1800–2300 kg | 4 | 6 | 10 |
| | 2300–2900 kg | 3 | 5 | 9 |
| | 2900–3600 kg | 3 | 5 | 8 |
| | 3600–4300 kg | 2 | 4 | 7 |
| | 4300–5000 kg | 2 | 4 | 6 |
| | Over 5000 kg | 2 | 3 | 5 |

growth present (generally accompanied by a strong mushroomy smell), and owing to the highly voided nature of the fills, collapse settlement had taken place. Such claims proved to be very difficult to resolve because the main structural foundations were also supported on the same fills where they passed over or close to the cellar footprint. In such situations, removing the existing fills is not possible, and an appropriate barrier membrane must be inserted to prevent spore migration, with a new in-situ suspended floor laid on top. It will still be necessary to check the fills for soluble sulphates.

Where the fills are deep, foundations are passing through onto firm natural strata and the fills contain no timber or other organic matter, it is often sufficient to use a fully suspended ground floor. Where this is the solution it is important that the fills are tested for sulphates and chemical stability to prevent damage to the foundations and underbuilding.

### 7.3.7 Chemical stability of fills

Before the introduction of polyethylene damproof membranes, sulphate attack often occurred as a result of unsuitable fills used below oversite concrete supporting sleeper walls or concrete ground floor slabs. However, soluble sulphates still attacked foundations, concrete blockwork and mortar where the floor dpms were present, and the overall effects must always be looked at when a new development is to be constructed or a claim for damage is being investigated.

The presence of sulphates in the natural ground or fills does not always mean that sulphate attack of concrete

## Case history 7.2

A major developer was proposing to construct dwellings on a large redevelopment site, which required the demolition of many four-storey flats. The flats had been erected in the early 1960s, and were system built using precast panels and precast floors linked together. They were being demolished as a result of water penetration and chloride attack on the concrete panels.

Drawings made available indicated that the ground floors had been constructed as ground-bearing slabs, 200 mm thick. The main external precast walls and party walls were supported off a ring beam spanning between deep pad foundations onto a hard shaley clay. Following removal of the superstructure, it was proposed by the developers that because of the good condition of the ground floor slabs they wished to retain them. As the footprint of the new dwellings was to match the old, the new party walls would not be coincident with the existing party wall ground beams but would be constructed off the existing slab. The developers' logic was that it was such a good floor, and it had been down for 20 years without any problems, that it would be more economical to use it. These proposals were examined by NHBC and rejected on the grounds that further investigations were required of the slabs and the fill below them.

In addition they were also concerned about placing foundations blind onto unknown ground conditions. The developers were not happy about having to carry out invasive trial pit investigations through such good-quality concrete. After much discussion the developers agreed to do a trial foundation on the line of a new party wall by cutting the existing slab each side with a Stihl saw, removing the concrete, and excavating through the fills. This excavation was inspected by the NHBC engineer, and the findings confirmed the NHBC concerns. The excavations showed:

- The concrete slab was of good quality, but it varied in thickness from 250 mm down to 50 mm in many places over the length of the trench.

- The fill below the floor slab consisted of approximately 1.50 m of whole-brick hardcore, highly voided, topped off with a black ashy material. During excavation of the trench, the fills at the top remained fairly compact owing to the effect of constraint provided by the slab over. As the depth increased beyond 800 mm the fill collapsed dramatically, leaving large voids under the slab at each side.
- No damproof membrane was evident.
- Investigation of the front and rear walls revealed the edge beams to be integral with this slab, with no reinforcement evident.

It was therefore concluded that the main concrete panels had actually spanned from pad to pad, acting as a deep beam.

The excavations showed clearly the inherent risks to the building if party walls were built off the existing floor, and the developer went along with new proposals that placed footings through the fill onto the natural shaley clays, and with the provision of a new in-situ reinforced slab laid on a polyethylene dpm. The old concrete slab was removed because of the risk of sulphates attacking it and heaving up the new slab.

This case history clearly illustrates the potential dangers in building blind over previous development. While the existing slab had remained in good condition for many years thanks to the point-to-point contact of the hard fills, subsidence was always a possibility in the long term. Had the original proposals been accepted it is highly probable that major foundation damage would have resulted, which would have far exceeded the savings to be made by the developer, not to mention the subsequent distress of any future houseowners.

will occur. It is the water- or acid-soluble sulphates that need to be determined, as these are dissolved out to form an aqueous solution that attacks the hardened cement in concrete and mortars.

Sulphates can occur in natural ground, mainly in the typical London clays, Oxford, Kimmeridge clays and Keuper marls, and the main salts are calcium sulphates (gypsum or selenite), magnesium sulphates (Epsom salts) and sodium sulphates (Glauber's salt). The factors that influence the rate of attack are:

● the type of sulphate present;
● the acidity of the ground;
● the level of the groundwater and its mobility;
● the type of concrete used and its cement content.

It follows therefore that in dry natural strata, without standing groundwater, it is highly unlikely that any significant chemical attack will take place on good-quality concrete. In natural undisturbed, unfissured clays the movements of groundwater will be very slow, resulting from rain percolation.

However, where there is a water table present it is essential to determine whether it is static or mobile. Particular care should be taken in any site investigation to establish the groundwater regime. Since groundwater levels can fluctuate there is the risk of aqueous sulphates, and the acidity of the groundwater will lead to the lime proportion of the cement in concrete being attacked.

Samples of soil and groundwater should therefore be taken from relevant depths and tested for acidity (pH) and soluble sulphate content. Any precautionary measures should be based on an adequate number of tests. Too few tests can result in inadequate measures or over-conservative measures being adopted.

Where industrial fills such as ash, furnace bottom ash, crushed brick hardcore or colliery spoil are evident the sulphates in these fills can be diluted out by the

*Table 7.3* Concrete requirements in soils

| Class | Concentration of sulphate and magnesium | | | | | Cement type[a] | Minimum cement content $(kg/m^3)^{b,c}$ | Maximum free water |
| | In soil or fill | | | In groundwater (g/L) | | | | |
| | By acid extraction | By 2 : 1 water/soil extract (g/L) | | | | | | |
| | % $SO_4$ | $SO_4$ | Mg | $SO_4$ | Mg | | | |
| 1 | <0.24 | <1.20 | | <0.40 | | A–L | d | 0.65 |
| 2 | If >0.24 classify on the basis of a 2 : 1 extract | 1.2–2.3 | | 0.4–1.40 | | A–G / H / I–L | 330 / 280 / 300 | 0.50 / 0.55 / 0.55 |
| 3 | | 2.3–3.7 | | 1.4–3.0 | | H / I–L | 320 / 340 | 0.50 / 0.50 |
| 4 | | 3.7–6.7 | <1.2 | 3.0–6.0 | <1.0 | H / I–L | 360 / 380 | 0.45 / 0.45 |
| | | 3.7–6.7 | >1.2 | 3.0–6.0 | >1.0 | H | 360 | 0.45 |
| 5 | | >6.7 | <1.2 | >6.0 | <1.0 | As for Class 4 plus surface protection | | |
| | | >6.7 | >1.2 | <6.0 | >1.0 | | | |

[a] Types of cement:
A  Portland cement to BS12
B  Portland blastfurnace slag cements to BS146
C  High-slag blastfurnace slag to BS4246
D  Combinations of Portland cements to BS12 and BS6699
E  Portland pfa cements to BS6588
F  Combinations of Portland cements to BS12 and pfa to BS3892 Part 1
G  Pozzolanic pfa cement to BS6610
H  Sulphate-resisting cement to BS4027
I  High-slag blastfurnace cement to BS4246
J  Combinations of Portland cements to BS12 and BS6699
K  Portland pfa cements to BS6588
L  Combinations of Portland cements to BS12 and pfa to BS3892 Part 1
[b] Cement content includes pulverised fuel ash (pfa) and slag
[c] Cement contents relate to 20 mm nominal maximum size aggregate. In order to maintain the cement content of the mortar fraction at similar values, the minimum cement contents given should be increased by 40 $kg/m^3$ for 10 mm nominal maximum size aggregate, and may be decreased by 30 $kg/m^3$ for 40 mm nominal maximum size aggregate, as described in Table 8 of BS 5328 : Part 1.
[d] The minimum value required in BS 5328 : Part 1 : 1990 and BS 8110 : 1985 is 275 $kg/m^3$ for unreinforced structural concrete in contact with a non-aggressive soil. A minimum cement content of 300 $kg/m^3$ (BS 8110) and maximum free water/cement ratio of 0.60 is required for reinforced concrete. A minimum cement content of 220 $kg/m^3$ and maximum free water/cement ratio of 0.80 is permissible for Grade 20 concrete when using unreinforced strip footings and trench fill for low-rise buildings in Class 1.

groundwater leaching out the salts over the long term, and the resulting aqueous solution will attack concrete foundations and blockwork walls.

Table 7.3 from BRE Digest 363 gives recommendations for dealing with soluble sulphates for various acidic conditions in the ground or groundwater.

Figures 7.6 and 7.7 illustrate the typical failure patterns in concrete floors and external walls when Portland cement concrete is attacked by sulphates, a prime source of which are products containing gypsum. The effect of the sulphates is to cause an increase in volume of the attacked concrete. The chemical reaction causes the slab to expand and arch upwards from the hardcore below. This expansion puts a considerable horizontal thrust on the external walls of the building, and often the brickwork moves on the dampcourse. One other sign of damage caused by sulphate attack is the presence of vertical cracking at the corners of walls, the cracking actually passing through the bricks in a straight line about six to eight courses above dampcourse level. This is usually accompanied by vertical cracks at regular centres below the dampcourse.

Remedial works in such cases are not possible, and the only solution is to remove the floor and the fill. If the external walls have been pushed out too far, then demolition is the only answer.

### 7.3.8 Expansion of fills

There are some fills that are chemically unstable, and which often have a potential to heave under certain conditions. They include materials such as blastfurnace slag from old slag heaps, and colliery spoils which contain iron pyrites and other chemicals that can oxidise out sulphides and calcites in the fills to form gypsum. During these chemical reactions the fills can undergo signicant volumetric changes. Even clean natural Whitbian shales quarried in the Cleveland area have been known to be very expansive when recompacted.

*Blastfurnace slags*

Slag from the manufacture of iron is a hard strong material and, if inert, is suitable for use as a fill material.

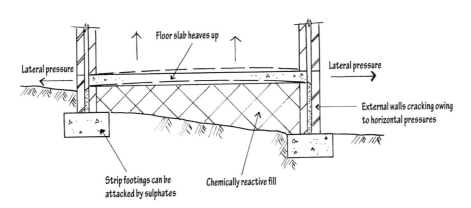

*Fig. 7.6* Sulphate attack on concrete floor.

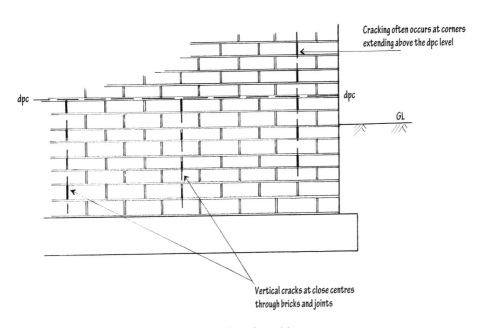

*Fig. 7.7* Vertical cracking of external walls as a result of sulphate attack on floor slab.

The slags can contain sulphates, but these are mainly calcium sulphates, which have a limited solubility in water. When used as fill the slags are free-draining, and the materials should therefore be tested using the water extract method, as proposed in BRE Digest 174, to determine the soluble sulphate content. Any test in which the amount of $SO_3$ is less than 1.0 g/L may be taken as falling in Class 1 of Digest 174.

Slags from the steelmaking process should not be used as fills, as they can contain constituents such as free lime, free magnesia or refractory bricks from the furnace, and these are known to be unstable.

Old slag banks should be thoroughly tested before being used as fill below buildings. Air-cooled blastfurnace slag is essentially a manufactured crystalline rock. Part of the function of a blastfurnace slag in the ironmaking process is to preferentially attract the sulphur present in the iron melt.

In slag that has been 'air cooled' to produce a crystalline product, the sulphur is at first present as sulphides. It is now recognised that long exposure in old slag tips allows these sulphides to progressively oxidise to sulphates.

Under specific circumstances the chemical reaction taking place can produce a mineral similar to tricalcium sulphoaluminate hydrate (ettringite).

The formation of this ettringite is associated with a large increase in volume. It is somewhat analogous to sulphate attack on concrete, and has similar consequences in that volumetric instability results. Clearly this is a very important consideration for a material that could be placed below a building floor.

In such slags it is therefore vitally important that proper testing is carried out to confirm whether the reactive activity has ceased. This is best done by testing for:

- water-soluble sulphates, as described in BS 1047;
- total sulphates;
- gypsum content;
- calcite;
- ettringite;
- plus an accelerated expansion test.

In addition, X-ray diffraction tests for mineral phases, thermogravimetric analysis and differential thermal analysis will provide considerable information on the amount of ettringite already formed, and on the amount of residual gypsum still available for the oxidic reaction process. These tests will also detect other chemicals that can influence the pH value of the material, and which can influence the efficacy of the accelerated expansion test.

The accelerated expansion test was developed by George Thomas of Thomas Research Services Ltd in Scunthorpe, Lincolnshire. George Thomas is a leading authority on iron and steel slags. The test itself is a direct test for slag expansion in which the accelerated results are set alongside real-time expansion tests extending over many years.

It is not a cheap or easy test, and from collection of the samples to completion of the test results can take up to 3–4 weeks. The main purpose of the test is to measure the likely maximum residual expansion available from the reactive changes occurring in the slag laid, i.e. from sulphite to sulphate (gypsum) to ettringite. Only the change from gypsum to ettringite is significant.

Sometimes the testing has to be carried out when the slag materials are being placed, and samples should be taken at the ratio of one sample for every 500 m$^3$ placed. It is advantageous to test only crushed-down slags, and test results at 7 day and 14 day intervals can be useful when an ongoing construction operation is taking place. Sufficient time must therefore be allowed between the layer thicknesses of placed crushed slag in order that the materials laid can be validated.

While such slags can be shown to be stable by expert testing and analysis it must be recognised that there is always some risk in using materials from industrial processes, and it is not always possible to be 100% certain of the slag's behaviour. The mere disturbance of a slag bank and crushing of the slag and recompaction can modify the local chemical balance, and in these cases post-testing of the slags is a must to establish the significance of this effect.

In conclusion, ettringite is the formation of tricalcium sulphoaluminate hydrate. Sulphate already taken up in ettringite no longer represents an expansion hazard. Calcite is a product of 'weathering' caused by long-term exposure of the slag heap, and where the lime is very slowly leached out as calcium hydroxide, which is carbonated to calcium carbonate. There is no calcite in freshly made slag, so the calcite content is a rough index of the amount of weathering that has taken place. It should be noted that there is no significant sulphate, gypsum or ettringite present in freshly made blastfurnace slags. All these constituents are the result of long-term exposure in the slag heap.

### Colliery spoil (coal measure mudstones)

Known commercially as 'minestone' (unburnt) or red shale (burnt), colliery spoils have a notorious record for causing damage when used in contact with concrete. Most of the recorded cases investigated resulted in sulphate attack caused by the oxidation of the sulphides and pyritic constituents during the burning process. The soluble sulphates then reacted with the concrete. However, there are now many recorded instances where there have been failures resulting from the expansion of such fills in the UK, at Washington in County Durham, Crigglestone, near Wakefield, and Lofthouse, near Wakefield.

### Whitbian shales

As-dug Whitbian shales of the Upper Lias were quarried extensively in the Great Ayton locality in Cleveland, and there were over 10 000 houses affected in the Teeside locality in the 1970s, where these shales were used as underfloor fill. The presence of such fills below a floor did not always result in heave, and in many cases whole estates were blighted even when no damage had resulted.

The presence of some moisture is, however, necessary for the oxidation processes to take place and to enable the sulphates produced to migrate. Both the chemical and bacterial oxidation processes depend on access to oxygen, and the effect of quarrying and crushing the shales accelerated the reactions.

*Cleveland ironstone shales*

Similar to the as-dug Whitbian shales, these waste products from the ironstone mining industry in Cleveland

were very expansive. However, the materials were stored in large colliery spoil heaps for many years, and considerable weathering and reaction took place. The effects of these shales were not as severe as those of the clean, as-dug shales, presumably as a result of the weathering process in the tips.

The types of expansive fill described above should not be used below house floors unless written permission has been given by the NHBC.

## Case history 7.3

### Introduction

A large estate near Wakefield was constructed over a previous colliery spoil tip. Foundations for the houses were 2.0 m deep concrete trenchfill down onto a firm to stiff brown clay. The ground floor construction consisted of a 150 mm thick, fully suspended reinforced concrete slab laid on a 1200 gauge polyethylene dpm. The interior of the dwellings developed cracking to floors and partitions, and the insurance claim was investigated by the author. The property was 8 years old.

### Investigation

From the crack patterns to the internal non-loadbearing walls and upward movement of the ground floor it was clear that the ground floor slabs were heaving up. The worst cracking was in those areas where there was less dead weight reaction, and where internal non-loadbearing walls had been built off the slab. There was no damage evident in the external walls internally or externally. The upward heave was confirmed by a grid-levelling check of the ground floors using a water level.

It was decided to open up the ground floor, and a trial pit 2.20 m deep was hand dug down to the natural clays. The reinforced concrete slab was in good condition, and the polyethylene dpm was also intact. The trench-fill concrete foundations were also seen to be in a sound

condition. The builder's records had confirmed that the concrete mix used for the foundations had a minimum cement content of 220 kg/m³ Portland cement.

Samples of the colliery spoil were taken for chemical analysis, and these results are shown in Table 7.4.

### Conclusions

As there was a polythene dpm present between the concrete slab and the colliery spoil it was considered that sulphate attack was not a cause of the movement. Examination of the concrete cores taken at the sampling points confirmed the concrete to be sound. The test results indicated negligible sulphides, high percentages of total sulphates (Class 3 and 4 of BRE Digest 363), and significant proportions of calcium carbonate. Table 7.5 (after R.K. Taylor) shows some of the volumetric increases for the various minerals in Coal Measure mudrocks due to internal chemical processes.

In simple terms the principal chemical reactions taking place were due to:

1  iron sulphides oxidising in damp conditions to form ferrous and ferric sulphate plus sulphuric acids;
2  the sulphuric acids then reacting with the calcium carbonate (calcite) to form calcium sulphate (gypsum);
3  the ferrous sulphates reacting with the clay minerals present to form jarosite.

*Table 7.4* Chemical test results

| Location of sample | Sample no. | Total sulphides (%) | Total sulphates (%) | Calcium carbonate (%) |
|---|---|---|---|---|
| Lounge | 1 | <0.01 | 1.35 | 6.30 |
| | 2 | <0.01 | 0.60 | 1.20 |
| | 3 | <0.01 | 1.88 | 0.60 |
| | 4 | <0.01 | 1.25 | 0.60 |
| Hallway | 5 | <0.01 | 0.55 | 11.80 |
| | 6 | <0.01 | 0.85 | 7.80 |
| Kitchen | 7 | <0.01 | 0.93 | 7.50 |
| | Average values | <0.01 | 1.05 | 5.11 |

*Table 7.5* Crystalline solid expansion due to mineral oxidation

| Original mineral | Mineral formed | Volume increase (%) |
|---|---|---|
| Pyrite (FeS₂) | Jarosite | 115 |
| | Melanterite | 536 |
| | Anhydrous ferrous sulphate | 350 |
| Calcite (CaCO₃) | Gypsum | 103 |
| | Bassanite | 189 |
| Julite (clay mineral) | Jarosite | 8 |

*Fig. 7.8* Removal of colliery spoil from below ground floors.

These reactions resulted in crystalline expansion of the colliery spoil fills.

It was apparent that sulphides in the colliery shale fills had oxidised expansively to form sulphates and sulphuric acids, and that these had then reacted with the calcium carbonates present to form gypsum or bassanite. This type of reaction can produce more than double the volume of the original constituents. With a thickness of 2 m of colliery shale and with a potential for 5.11% of reactive materials the increase in volume would result in a theoretical total heave of 102 mm.

### Remediation works

In view of the potential for further long-term expansion it was decided that the only practical solution was to remove the ground floor slab, remove all the colliery spoil from below the building footprint, replace with clean inert fills and reconstruct a new suspended ground floor slab. It was also considered prudent to remove the colliery spoil from the outside of the main external walls to minimise any lateral pressures on the house walls during the temporary excavation of the inside fills, and a 1200 gauge polyethylene sheet was placed against the external concrete face prior to backfilling with well-compacted clean dolomite to increase its resistance to sulphate attack. The inside excavation was backfilled with crusher-run stone, and a new suspended reinforced concrete floor was placed on a 1200 gauge polyethylene dpm. This floor was supported on new brick walls built off the trench-fill foundations.

*Fig. 7.9* External trench around external walls.

## 7.4 Fully suspended in-situ reinforced concrete floors

With the increased cost of landfill many builders now use site fill as a temporary shutter, and construct reinforced concrete slabs reinforced with mesh where the infills are deeper than 600 mm. On certain sites where existing fills are present the ground floors can be in-situ suspended, with the main foundations on piles or deep trench-fill concrete.

Designs can be based on a simply supported condition or fully continuous over loadbearing walls. The fully continuous mode will require the provision of top reinforcement placed for a minimum distance of one-quarter of the span each side of the wall. Provided the design is correctly carried out, a 35 N concrete is provided and the mesh is given the correct cover, then no problems should result.

These floors can be provided with a power-floated finish, or be provided with concrete screeds or timber on insulation floating finishes.

Tables 7.8 and 7.9 show typical loading conditions and design requirements for various spans.

### Case history 7.4

#### Introduction

Several years after the construction of a large housing estate, problems of sinking floors resulted in many house owners making claims to the house warranty insurance provider. On investigation it was found that the cause of the ground floor subsidence was inadequate compaction of stone upfill and drying-out of the stone upfill. In some situations excessive subsidence of approximately 35 mm was measured, and investigation after the floors were removed highlighted that the stone fills had been placed on top of existing site clays, which had been disturbed and softened by site traffic.

It was decided to remove all the floors and replace them with fully suspended in-situ slabs using the existing stone fill as a temporary shutter. Around the external walls an additional 100 mm block wall was constructed of the footings to support the new slabs.

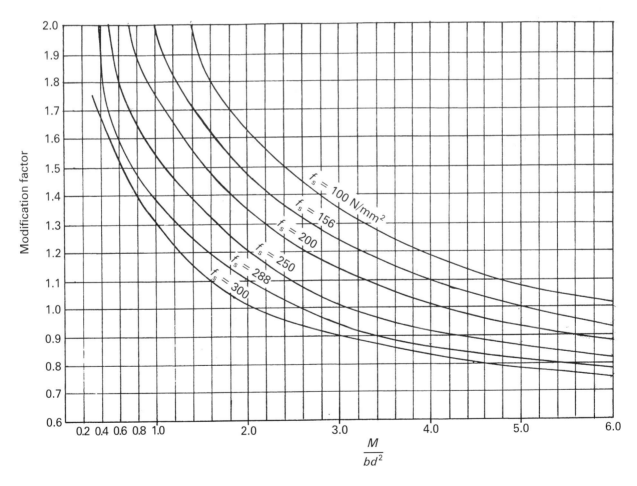

*Fig. 7.10* Modification factor chart (BS 8110).

To rationalise the construction it was decided to use one fabric reinforcement, B503, in a continuous mode, and to cater for the various spans by increasing the slab thicknesses accordingly to keep the slab span-to-depth ratios within the recommended limits to ensure acceptable deflections.

## Slab design

Excessive deflections of ground floor slabs can cause damage to finishes, and as a slab is a slender structural element the restrictions on the span-to-depth ratios become more important.

Minimum effective depth

$$= \frac{\text{Span}}{\text{Basic ratio} \times \text{Modification factor}}$$

The modification factor is based on the actual area of reinforcement used at the tension zone, and Fig. 7.10 shows the modification factors for the tension steel plotted in the form of a graph for ease of use for the various service stress conditions. It can be seen from the graph that a lower service stress, $f_s$, gives a higher modification factor, and hence a thinner slab can be used. The service stress is reduced by providing an area of tension steel greater than that required to resist the ultimate design moment.

$$F_s = \frac{2f_y}{3} \times \frac{A_{s\ reqd}}{A_{s\ prov}}$$

where no redistribution of the moments has taken place.

The span-to-depth ratios can be checked using the service stress appropriate to the characteristic stress shown in Table 7.6. The basic span-to-depth ratios are shown in Table 7.7.

The design philosophy will be based on a two-span continuous slab, spanning in one direction. This covers the majority of domestic floors, and the general rule is to apply the following coefficients, which will give the maximum moment cases for various live load situations:

- hogging moment over central support taken as $0.125\ wl^2$
- positive moment mid-span taken as $0.096\ wl^2$

Floor slabs have a 15 mm asphalt finish, partitions will be timber stud, plastered both sides, and slab thicknesses from 150, 175, 200 and 225 mm will be used.

| Loadings: | (kN/m²) |
|---|---|
| 15 mm asphalt finishes | 0.30 |
| stud partitions | 0.50 |
| slab self-weight 0.15 × 24 | 3.60 |
| Total = 4.40 × 1.40 = 6.16 | |

Imposed loading    $1.50 \times 1.60 = 2.40$
Total $= 8.56$ kN/m²

Design span in lounge $= 3400 + 100 = 3500$ mm

Effective depth $= h - 40 - 4$ and breadth, $b = 1000$ mm

Maximum bending moment $= 8.56 \times 0.125 \times 3.50^2$
$= 13.10$ kN m

Effective depth, $d = 106$ mm

$$\frac{M}{bd^2} = \frac{13.10 \times 10^6}{1000 \times 106^2} = 1.16$$

**Table 7.7** Basic span-to-depth ratios (BS 8110)

| Span mode | Rectangular sections |
|---|---|
| Cantilever | 7 |
| Simply supported | 20 |
| Continuous | 26 |

**Table 7.6** Modification factor for tension reinforcement

| Service stress (N/mm²) | | M/bd² | | | | | | | | |
|---|---|---|---|---|---|---|---|---|---|---|
| | | 0.50 | 0.75 | 1.0 | 1.50 | 2.0 | 3.0 | 4.0 | 5.0 | 6.0 |
| | 100 | 2.0 | 2.0 | 2.0 | 1.86 | 1.63 | 1.36 | 1.19 | 1.08 | 1.01 |
| | 150 | 2.0 | 2.0 | 1.98 | 1.69 | 1.49 | 1.25 | 1.11 | 1.01 | 0.94 |
| | 156 | 2.0 | 2.0 | 2.0 | 1.66 | 1.47 | 1.24 | 1.10 | 1.0 | 0.94 |
| $f_y = 250$ | 167 | 2.0 | 2.0 | 1.91 | 1.63 | 1.44 | 1.21 | 1.08 | 0.99 | 0.92 |
| | 200 | 2.0 | 1.95 | 1.76 | 1.51 | 1.35 | 1.14 | 1.02 | 0.94 | 0.88 |
| | 250 | 1.90 | 1.70 | 1.55 | 1.34 | 1.20 | 1.04 | 0.94 | 0.87 | 0.82 |
| | 288 | 1.68 | 1.50 | 1.38 | 1.21 | 1.09 | 0.95 | 0.87 | 0.82 | 0.78 |
| | 300 | 1.60 | 1.44 | 1.33 | 1.16 | 1.06 | 0.93 | 0.85 | 0.82 | 0.76 |
| $f_y = 460$ | 307 | 1.56 | 1.41 | 1.30 | 1.14 | 1.04 | 0.91 | 0.84 | 0.80 | 0.76 |

Modification factor $= 0.55 + \dfrac{477 - f_s}{120(0.90 + M/bd^2)} \leqslant 2.0$

where $M$ is the design ultimate moment at mid span or at the support.

From Fig. 7.11 with

$$f_{cu} = 35 \text{ then } \frac{100A_s}{bd} = 0.30$$

Figures 7.11 and 7.12 are design charts to be used for the determination of lever arm factors and areas of reinforcement based on the characteristic strength of the concrete.

Therefore area of steel,

$$A_s = \frac{0.30 \times 1000 \times 106}{100} = 318 \text{ mm}^2$$

Use B503 in top of slab over support walls

Service stress

$$f_s = \frac{2f_y}{3} \times \frac{A_{s\,reqd}}{A_{s\,prov}} = 0.66 \times 460 \times \frac{318}{503} = 192.0 \text{ N/mm}^2$$

From Fig. 7.10, by interpolation, modification factor = 1.73

Span-to-depth ratio $= 26 \times 1.73 = 44.98$, which is greater

than $\dfrac{3500}{106} = 33.0$

**Bottom steel**

Maximum bending moment $= 0.096 \times 8.56 \times 3.50^2$
$$= 10.06 \text{ kN m}$$

$$\frac{M}{bd^2} = \frac{10.06 \times 10^6}{1000 \times 106^2} = 0.895$$

From Fig. 7.11 $\dfrac{100A_s}{bd} = 0.26$

$$A_{s\,reqd} = \frac{0.26}{100} \times 1000 \times 106 = 275.60 \text{ mm}^2$$

Use B503 in bottom of slab

Service stress, $f_s = \dfrac{2}{3} \times 460 \times \dfrac{275.60}{503} = 168.02 \text{ N/mm}^2$

From Fig. 7.10, modification factor = 2.0,

therefore span-to-depth ratio $= 26 \times 2.0 = 52 > 33$

Provide B503 in bottom and over supports for a distance equal to one quarter of the span each side. See Tables 7.8 and 7.9 for various span conditions.

Figure 7.13 shows the span to slab thickness parameters using the B503 mesh, and Fig. 7.14 shows a typical slab layout and reinforcement details.

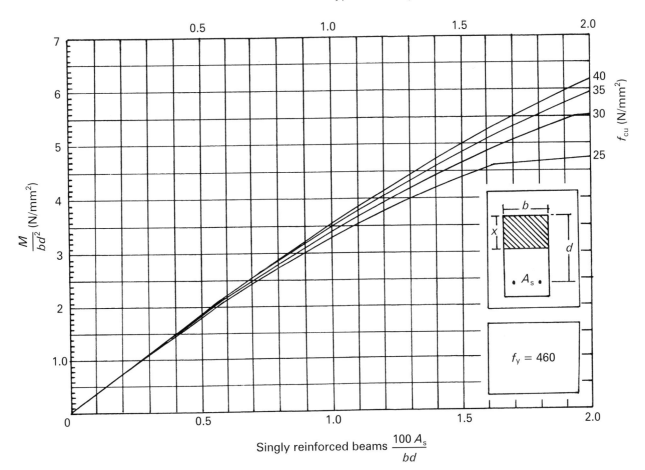

Singly reinforced beams $\dfrac{100 A_s}{bd}$

*Fig. 7.11* Design Chart no. 9.1 for reinforced concrete floor slabs.

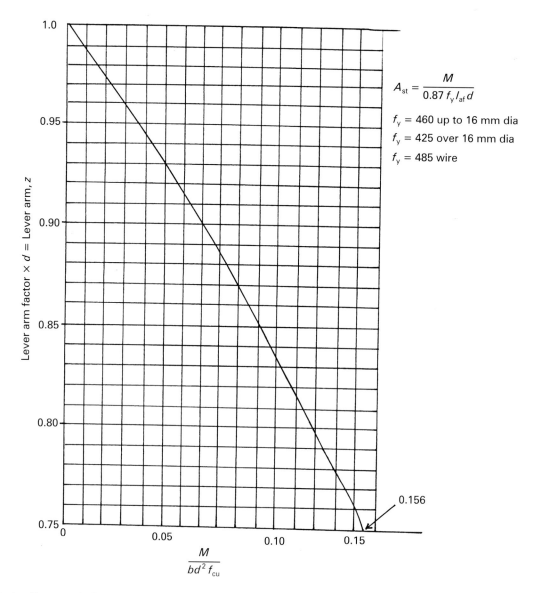

**Fig. 7.12** Design Chart no. 1. Graphical determination of lever arm factor: lever arm curve for limit state design. When $M/bd^2f_{cu}$ exceeds 0.156, compression steel is required.

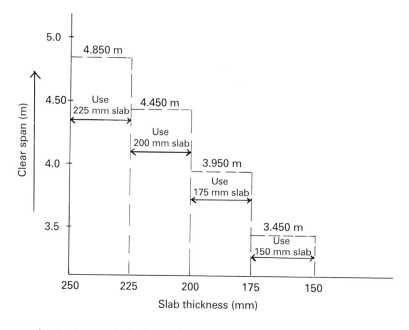

**Fig. 7.13** Maximum clear spans for simply supported slabs carrying stud partitions and reinforced with B503 mesh reinforcement.

**Table 7.8** Floor slab loadings

| Floor thickness (mm) | Dead loading, $g_k$ | | | | Imposed loading, $q_k$ | Design loading, $1.4g_k + 1.6q_k$ |
|---|---|---|---|---|---|---|
| | Slab | Finishes | Partitions | Total | | |
| 150 | 3.60 | 0.30 | 0.50 | 4.40 | 1.50 | 8.56 |
| 175 | 4.20 | 0.30 | 0.50 | 5.00 | 1.50 | 9.40 |
| 200 | 4.80 | 0.30 | 0.50 | 5.60 | 1.50 | 10.24 |
| 225 | 5.40 | 0.30 | 0.50 | 6.20 | 1.50 | 11.08 |

**Table 7.9** Bending moments and span-to-depth ratios

| Slab depth (mm) | Effective depth (mm) | Design span (m) | Bending | | | | | Deflection | | | |
|---|---|---|---|---|---|---|---|---|---|---|---|
| | | | udl $(kN/m^2)$ | Md | $\dfrac{Md}{bd^2}$ | $\dfrac{100A_s}{bd}$ | $A_s$ $(mm^2)$ | $A_s$ $(mm^2)$ | $f_s$ $(N/mm^2)$ | Mod. factor | Actual span: depth ratio |
| 150 | 106 | 3.40 | 8.56 | 12.3 | 1.10 | 0.29 | 307 | 503 | 187 | 1.75 | 45 |
| 150 | 106 | 3.50 | 8.56 | 13.1 | 1.16 | 0.31 | 328 | 503 | 200 | 1.67 | 43 |
| 150 | 106 | 3.55 | 8.56 | 13.4 | 1.19 | 0.32 | 339 | 503 | 206 | 1.63 | 42 |
| 150 | 106 | 3.60 | 8.56 | 13.8 | 1.23 | 0.33 | 349 | 503 | 212 | 1.58 | 41 |
| 175 | 131 | 4.05 | 9.40 | 19.2 | 1.12 | 0.29 | 379 | 503 | 231 | 1.56 | 40 |
| 175 | 131 | 4.075 | 9.40 | 19.5 | 1.13 | 0.30 | 393 | 503 | 239 | 1.52 | 39 |
| 175 | 131 | 4.10 | 9.40 | 19.7 | 1.15 | 0.31 | 406 | 503 | 247 | 1.48 | 38 |
| 200 | 156 | 4.55 | 10.24 | 26.5 | 1.08 | 0.28 | 436 | 503 | 265 | 1.44 | 37 |
| 200 | 156 | 4.60 | 10.24 | 27.0 | 1.11 | 0.29 | 452 | 503 | 275 | 1.38 | 36 |
| 225 | 181 | 4.95 | 11.08 | 33.9 | 1.03 | 0.27 | 489 | 503 | 297 | 1.32 | 34 |
| 225 | 181 | 5.00 | 11.08 | 34.6 | 1.05 | 0.28 | 506 | 503 | 308 | 1.27 | 33 |

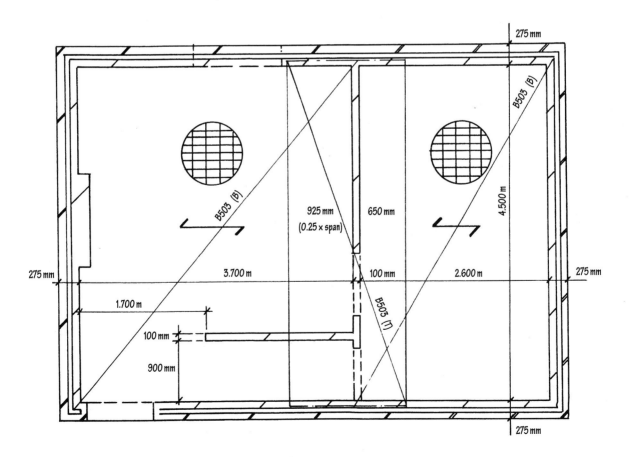

**Fig. 7.14** Typical slab layout and reinforcement details.

## Case history 7.5

### Introduction

The owner of a detached dwelling in Outibridge, Sheffield, made an insurance claim as a result of cracking to the external brickwork below the ground floor bay window. The engineer acting for the builder had examined the cracking and concluded that it was due to the external wall being subjected to a lateral pressure from the retained fill below the ground floor slab. The cavity wall had been filled with concrete up to ground level, and as the depth of infill was only 1.10 m it was questionable whether the wall was adequate.

### Investigation

With the floor coverings rolled back it was evident that the floor was cracked, and water level readings indicated a slight crowning at the centre of the floor. Close examination also revealed that:

● the fireplace surround was actually higher than when installed, as indicated by the plaster lipping approximately 3 mm;

● the floor slab at the rear patio window was slightly higher than the portion of slab cast over the external wall, and a significant crack was evident in what had been a monolithic floor.

It was the author's view that the movements indicated were a result of upward heave, and bearing in mind that the estate was close to iron and steel manufacturing sites the possibility of sulphate attack or slag expansion came to mind.

The floor was therefore broken out locally, and a sample of the underfloor sub-base materials was removed for testing. Visual examination showed the materials to be a crushed slaggy material. The two samples were sent off for chemical analysis, and the results are shown in Table 7.10.

These chemical tests confirmed that the sub-base materials were slag from an ironworks slag bank, and the material was chemically reactive. The high ferric

oxide and calcium had reacted with the sulphates present to form gypsum. In addition the free magnesia and free lime will have reacted with the other agents present.

Subsequent checks with the local authority highways department confirmed that a crushed slag fill had been used under the adopted highway, and the source of the material was traced back to a local supplier who processed the slags for spent mineral products and sold the material left as sub-base. The building inspector for the site had also rejected this fill when it had been placed under a garage floor. It was clearly an isolated case, but one that should not have occurred.

### Conclusions and recommendations

As the soluble sulphates were too high for the concrete used in the foundations it was decided to remove all the deleterious fill from below the floor and replace the concrete floor on clean crushed stone. There was a recommendation that the wall to the bay window be checked once the fills had been removed. Local repairs and repointing were carried out to the external brickwork.

### Schedule of works

1   Remove, store and refit internal doors.
2   Seal off staircase with plywood and polythene to reduce dust.
3   Strip out kitchen, store units in garage for refitting on completion.
4   Allow for removing and refitting plumbing and electrical services to utilities.
5   Drain down central heating system.
6   Take down fireplace and hearth. New hearth to be fitted later.
7   Allow for protection to ground floor doors, radiators and patio doors.
8   Break out existing floor slab and cart off site.
9   Excavate out all the deleterious fill below the floor down to the natural strata. This to include slag fill down the foundation easement.
10  Where depth of fill to go back exceeds 600 mm then provide lean-mix concrete up to 600 mm below the floor slab.
11  Provide and compact in thin layers not exceeding 150 mm DoT Type 1 granular sub-base materials. This fill to be blinded off with sand.
12  Provide and lay a 1200 gauge polyethylene dpm with adequate laps and turned up 200 mm at the edges.
13  Provide and lay 100 mm concrete floor slab with rough tamped finish using a GEN 3 mix.
14  Provide a minimum 50 mm sand : cement screed and make provision for replacing service pipes in

*Table 7.10* Chemical test results

| Range of chemical tests carried out | Sample no. 1 Hard slaggy material | Sample no. 2 Light sandy material |
|---|---|---|
| Total sulphate content (SO$_3$) (%) | 2.0 | 1.60 |
| Water-soluble sulphate content, 2 : 1 extract SO$_3$ (g/L) | 0.10 | 0.10 |
| Calcium content as CaO (%) | 26.60 | 2.10 |
| Magnesium content as MgO (%) | 6.1 | 0.50 |
| Iron content as Fe$_2$O$_3$ (%) | 17.20 | 17.20 |
| Magnesium content as MnO$_2$ (%) | 4.90 | 3.20 |
| Chromium content as Cr$_2$O$_3$ (%) | 2.90 | 0.90 |
| Aluminium content as Al$_2$O$_3$ (%) | 2.60 | 2.30 |

ducts in screed. Ensure that all copper pipes are protected with tape.

15  Carry out repairs to external wall below bay window.

16  Replace any damaged skirting boards.

17  Refit kitchen units and carry out full decoration to ground floor walls and ceilings.

18  Connect up and test all services and central heating system.

## Case history 7.6

### Introduction

The owner of a large bungalow on an estate on the outskirts of Hull was experiencing cracking of internal partitions, and movement at the junction with the ceiling. The dwelling was approximately 14 years old at the time of the investigation.

### Investigation

Examination of the main external walls showed no signs of any subsidence cracking, and it was concluded that damage as a result of foundation movement could be ruled out. It was noticeable that the external walls were showing vertical cracking below the dampcourse, and at the gable junction this vertical cracking had extended in a straight line above the dampcourse for approximately three brick courses.

The floor was opened up in several places. This revealed 100 mm × 50 mm timber joists on a dpc onto a honeycombed sleeper wall. The sleeper walls were built off the concrete oversite. The concrete oversite was seen to be badly cracked in quite a few areas. This oversite concrete was then broken out, and was seen to be 100 mm thick concrete laid directly onto brick hardcore fill. This hardcore fill consisted of whole bricks, and was very voided. Samples of the bricks and surrounding matrix were removed for chemical analysis.

The samples were tested for soluble sulphates using the 2 : 1 water : soil extract method; the soluble sulphates present were 1.20 g/L. This value would have required a concrete with a Class 2 mix and 330 kg/m$^3$ of cement. The oversite concrete was a 1 : 3 : 6 mix with approximately 220 kg/m$^3$ of cement. Examination of the con-

crete showed a marked deterioration at the interface with the hardcore fill. Examination of the mortar and concrete foundations also showed some signs of attack but not as severe.

### Conclusions and recommendations

Following examination of the test results on the fills it was decided that removal of these fills was the only long-term solution. It was clear that initially the sleeper walls had subsided as the poorly compacted fills settled under their own weight. This would have been the reason for the large gaps at the ceiling junctions with the internal walls. Once the sulphate attack passed its initial phase the oversite was trying to expand, and then eventually started to lift. The initial gap would have masked this movement, but once the initial gap had closed up, the further ongoing movement caused cracking in the upper floor partitions.

A scheme was prepared to remove the concrete oversite and hardcore fill and replace with clean, inert stone. The ground floor partitions were temporarily hung from the first floor joists using timber battens. This allowed the ground floor area to be completely removed at one go. All the fill down the sides of the foundation easement was also removed off site.

The 100 mm thick oversite concrete was replaced on a 1200 gauge polyethylene dpm, and the original floor joists were replaced on new sleeper walls. New tongued and grooved boarding was fitted.

The minor cracking around the external walls was resolved by removing the cracked bricks and replacing them with matching bricks, followed by repointing the bed joints.

## 7.5 Precast beam and block floors

This type of floor system has become popular with house builders and contractors for ground floor construction for the following reasons:

- It is quick to erect.
- There is no need for a concrete oversite; this allows surplus selected site fill to be used below the floor.
- No shuttering is involved.
- Once erected, the floor forms a quick and useful working platform.
- Services are easily accommodated.

- Foundation loads are reduced, enabling foundation widths to be reduced.

The types of structural defect that can occur have already been covered by the section on upper floors, but the main requirement when used as a ground floor is the provision of a minimum 75 mm ventilated air space below.

The precast beams are manufactured by prestressing thin-gauge, high-tensile steel wires in the concrete, and these wires only have 15 mm concrete cover. In ground floor environments, dew pointing from condensation or dampness can occur on the bottom of the beams and cause corrosion in the thin wires. The BBA certificate therefore requires the ventilated void to be provided to prevent this from happening. The precast beams should always be placed on a layer of dampcourse at their supports.

Before using the floor as a working platform, all the infill blocks should be grouted up using a 1 : 6 cement : sharp sand mix well brushed into all joints.

Where floor blocks are built into the main walls there are two important points to remember:

● The blocks are of the same strength as the wall blockwork.

● The floor block is bedded onto the wall with mortar. Dry bedding of such blocks is not acceptable in any circumstance.

There are other types of precast unit that can be used for ground floors, such as precast hollow wide slab, precast Omnia decking with in-situ toppings, and precast beam and block using dense polystyrene blocks instead of concrete.

All these systems have the advantage of speedy construction, plus the use of a dryer system.

## Points to remember

● Is the depth of infill to a floor in excess of 600 mm? If so, use a fully suspended floor construction.
● Is a void required below the floor to allow for heave, and is the air space ventilated properly?
● Has the fill below the floor been tested to confirm that it is chemically inert?
● Have precautions against sulphate attack been catered for in the floor concrete, the foundations and the underbuild mortar?
● Are the blocks in the floor the same strength as that required for the wall blocks?
● Have the floor blocks built into the wall been laid on a mortar bed?
● Have the precast beams and timber wall plates been laid on a dampcourse?

## References

### Building Research Establishment

Defect Action Sheets 121 (1988), 65 (1985) and 36 (1983).
*Cracking in buildings*, BRE Digest 75 (1966).
*Sulphate and acid resistance of concrete in the ground*, BRE Digest 363 (January 1996).

### British Standards Institution

BS 12 : 1996 *Specification for Portland cement.*
BS 146 : 1996 *Specification for Portland blastfurnace cements.*
BS 648 : 1964 *Schedule of weights of building materials.*
BS 1047 : 1983 *Specification for air-cooled blast furnace slag aggregate for use in construction.*
BS 1377 : 1990 *Methods of test for civil engineering purposes.*
BS 3892 *Pulverized-fuel ash*
  Part 1 : 1997 *Specification for pulverized-fuel ash for use with Portland cement.*
BS 4027 : 1996 *Specification for sulphate-resisting Portland cement.*
BS 4246 : 1996 *Specification for high slag blastfurnace cement.*
BS 4482 : 1985 *Specification for cold reduced steel wire for the reinforcement of concrete.*
BS 5328 *Concrete*
  Part 1 : 1997 *Guide to specifying concrete*
  Part 2 : 1997 *Methods for specifying concrete mixes.*
BS 5930 : 1999 *Code of practice for site investigations.*
BS 6399 *Loading for buildings*
  Part 1 : 1996 *Code of practice for dead and imposed loads.*
BS 6588 : 1996 *Specification for Portland pulverized-fuel ash cements.*
BS 6610 : 1996 *Specification for pozzolanic pulverized-fuel ash cement.*
BS 6699 : 1992 *Specification for ground granulated blastfurnace slag for use with Portland cement.*
BS 8103 *Structural design of low-rise buildings*
  Part 1 : 1994 *Code of practice for stability, site investigation, foundations and ground floor slabs for housing.*
BS 8110 *Structural use of concrete*
  Part 1 : 1997 *Code of practice for design and construction*
  Part 3 : 1985 *Design charts for singly reinforced beams, doubly reinforced beams and rectangular columns.*

### Institution of Structural Engineers

*Subsidence of low-rise buildings* (1994)

### National House-Building Council

*NHBC Standards* (1999):
  Chapter 4.2 Building near trees
  Chapter 5.1 Substructure and ground bearing floors
  Chapter 5.2 Suspended ground floors.

# Chapter 8
# Main walls

## 8.1 Introduction

This chapter will deal with the type of structural defect resulting from the main loadbearing walls' failure to perform satisfactorily. Foundation failures or subsidence movements are generally reflected in the superstructure walls, and these types of defect will be covered in the chapter dealing with foundation failures.

Modern masonry construction for low-rise buildings generally consists of external cavity walls built in facing brick, and an inner blockwork leaf with a 75–100 mm cavity. Party walls or internal loadbearing walls are generally built in blockwork. This form of construction has been shown to be economical up to four storeys high. Above this height the additional robustness required to carry the higher loads makes the construction uneconomic and impracticable, as such buildings allow little scope in the way of safety factors for bad workmanship, walls built out of plumb, or the use of substandard materials. There can be little doubt that the quality of brickwork now accepted as normal falls far below the recommendations of the relevant British Standard Codes of Practice. Such structural masonry must be built and supervised to a higher specification to ensure accurate dimensions and quality control of strength.

The other main factor that has a significant effect on the design is the requirement that buildings above five storeys in height must be designed to withstand any incidence or accidental explosion that could cause progressive collapse. This makes a masonry building more expensive than a building that has fewer than five storeys, owing to the need to build more robustness and flexibility into the design, notwithstanding the additional costs of the higher specification and supervision needed.

Structural defects can occur in loadbearing walls as a result of:

- design faults;
- problems with materials;
- inadequate workmanship.

On many occasions the defects caused can be due to a combination of the above factors.

## 8.2 Design of walls

Masonry like any other structural material, requires an understanding of its strengths and weaknesses in order that an economic design can be produced in accordance with accepted statutory requirements.

Masonry's main weakness is its low tensile resistance, and it is therefore not well suited to catering for eccentric forces or lateral loads, which require the insertion of sufficient bracing and shear walls to transmit such loads into the foundations.

### 8.2.1 Dimensional stability

The dimensions of masonry elements in a building are continually changing. Bricks expand because of moisture absorption, and mortars contract as they harden. Unfortunately there are still many designers who are not aware of or choose to ignore the effects of masonry movements, and it is not uncommon to see long terraced buildings with no movement joints. Modern construction uses cement-based mortars rather than the lime : sand mortars used in the early twentieth century. This results in less mass in the wall, but such thin structural elements are not as tolerant of movement, as they do not have the long-term plasticity given by lime mortars.

Ways must therefore be found of accommodating thermal, shrinkage and moisture expansive movements in masonry elements.

The four main causes of dimensional change are:

- deformation as a result of vertical compressive loading;
- movements caused by temperature changes;
- movements arising from changes in the moisture content of the brickwork;
- movements caused by chemical reactions between the bricks and mortar.

### 8.2.2 Deformation or cracking due to loading

When a masonry pier or wall is subjected to a vertical compressive load it shortens by a minute amount. This ratio between any change of length and the original length is known as *strain*, and its amount is directly related to the stress on the element: the greater the stress the greater the strain, if the masonry behaves elastically.

### 8.2.3 Movement caused by temperature changes

All building materials alter their dimensions where there are temperature fluctuations. Brickwork has a coefficient of expansion of approximately $6 \times 10^{-6}$ per °C. This means that for every one degree rise in temperature a 12 m length of wall will expand by $6 \times 10^{-6} \times 12 \times 10^3 = 0.072$ mm. A temperature rise of 40 °C would therefore give rise to an unrestrained expansion of $40 \times 0.072 = 2.88$ mm. Such movements in strong bricks built in a strong cement mortar with varying degrees of restraint will most likely result in some cracking of the wall.

Movement in the vertical direction is different from horizontal movements because of the changes in restraint. It is most important to consider the aspect of a wall, as the surface temperature of a south-facing brick wall can be as high as 50 °C. It is also important to remember that the external leaf of a cavity wall will be less restrained in comparison to the inner leaf, which may be infilling a framed structure. Figure 8.1 shows the rates of thermal expansion per degree C for various building materials. For brick-faced buildings that exceed 12 m in length it is advisable to provide vertical movement joints in the outer leaf to minimise the risk of major cracking, and Table 8.1 lists joint requirements for various materials.

### 8.2.4 Movement caused by moisture expansion

Loss or gain of moisture dimensionally affects masonry, especially clay bricks. Properly fired clay bricks undergo only very small reversible moisture movements, and these are generally negligible. However, clay brickwork can suffer as a result of 'permanent' long-term moisture expansion, and such movements are time related. Research has shown that:

● bricks that are kiln fresh undergo considerably greater expansion than those that have been allowed to mature slowly for several weeks before being used;
● the only practical way to cater for the long-term phenomenon is to provide adequate movement joints in the structure.

Table 8.1 Recommended movement joint spacing

| Material | Joint thickness (mm) | Recommended spacing (m) |
|---|---|---|
| Clay brickwork | 15 | 12–15 maximum |
| Calcium silicate bricks (sand limes) | 10 | 7.50–9 maximum |
| Concrete blocks | 10 | 6.0 |
| Concrete bricks | 10 | 6.0 |
| Any masonry in a parapet wall situation | 10 | Half of the above spacing and 1.50 in from corners |

The above joint thicknesses are the amounts of movement likely to result, and the joint material used should be compressible. Joint fillers such as cellular polyurethane, foam rubber and flexible cellular polyethylene are acceptable. To ensure that the above movements can take place the joint width may need to be thicker to accommodate the compressibility characteristic of the filler material used.

The spacing of the first movement joint from a return wall should not be more than 50% of the above spacing.

When differing materials are used, tied together, special consideration needs to be given to cater for the differential effects of the different materials.

Movement joints must run the full height of the structure, and extend below the dampcourse down to the foundation.

Where masonry forms panels and cladding to framed structures then the movement joints should be provided in accordance with BS 5628 : Part 3.

All external movement joints should be filled with a compressible filler and sealed off with a mastic sealant.

At movement joints wall ties are required at each side of the joint at vertical spacing of 300 mm or each block course.

**To reduce cracking**
● Avoid mixing bricks and blocks or blocks of varying densities together.
● Always provide a joint where dissimilar materials abut.
● Consider the benefits of introducing masonry bed joint reinforcement under and above window and doorway openings.

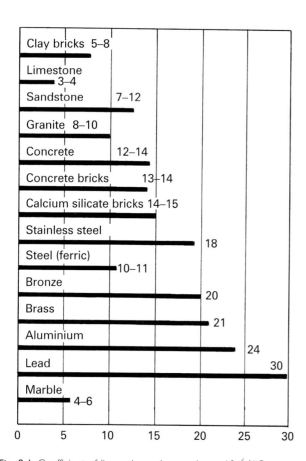

*Fig. 8.1* Coefficient of linear thermal expansion $\times 10^{-6}/$°C.

## Case history 8.1

### Introduction

While this case history relates to a multi-storey building, it has been included on the basis that the defects referred to can also occur on buildings up to four or five storeys if horizontal and vertical movement joints are not properly detailed at floor levels. Cracking had developed on the main external brickwork cladding on three eight-storey blocks of flats in Gateshead. The defect took the form of vertical cracking at the corners of the gable walls, extending up through several floors at different levels. The cracking had passed through the bricks and mortar. In addition, the brick slip facings at floor levels had tented owing to the extreme load transfer from above, and were falling off, posing a danger to the public.

### Investigation

The building had been constructed using an industrialised building technique known as concrete no-fines. When used for multi-storey buildings the concrete columns in the structural walls were cast monolithic with the no-fines concrete walls by using sliding metal formwork within the timber shuttering. This required the building to be constructed in single-storey-height pours. Ties were cast in the formwork to fix to the brick cladding. On completion of the walls, the beams and in-situ floors were cast. The outer brick skin was supported on a nib,

which was monolithic with the floor beam, and as the floors were not taken through onto the elevation it was required to use brick slips at the beam positions, as shown in Fig. 8.2.

A plumb and level survey of the damaged elevations was carried out. This survey formed the basis for the wall tie assessment, and provided a valuable guide to the best areas to use the borescope. Following this survey, a detailed inspection of each gable end was carried out from a maintenance hoist, using a borescope at various points to check on the condition and spacing of the wall ties.

Drawings of the construction were also available and proved to be very useful.

### Conclusions and recommendations

The plumb survey highlighted that there was likely to be a problem with wall ties. While the outer brick cladding was found to be reasonably plumb, the inner no-fines walls in several instances were found to be out of plumb to such a degree that the galvanised ties cast in the wall during construction did not reach the outer leaf, and in many situations those ties that did reach did not have the required 50 mm penetration.

Removal of bricks in selected sections at the corners revealed that the concrete nibs supporting each storey height of walling had been knocked off, either during the stripping-out of the formwork or by the bricklayers to make walling-through easier. This load transference resulted in high compressive stresses being developed at the lower storeys, which had resulted in vertical stress fractures.

The spalling of the brick slips was caused by the moisture expansion of the brickwork combined with 'compression shrinkage' of the concrete frame under load. This transfer of load from the frame through to the wall below caused the brick slips to be compressed and to buckle. No horizontal movement joints had been provided to the main structure at any floor level.

As the walls had a potential for further movement, and there was a risk of injury to the public from falling masonry slips, it was clear that large-scale remedial measures were required to solve the problems.

The main requirement was to provide a horizontal movement joint at each floor level such that only the wall loads from each storey were carried by the floor beams. From examination of the drawings it was possible to provide stainless steel angles with a long stalk bolted to the face of the existing floor beam and above the concrete nib, as shown in Fig. 8.3.

Before the needling of brickwork could be carried out, it was essential that the cavity walls were strengthened by the insertion of retrofit wall ties. These were inserted at column positions using standard stainless steel retrofit steel tie rods. For the no-fines wall a special tie was devised by Hilti Ltd using an epoxy injection system. Ties

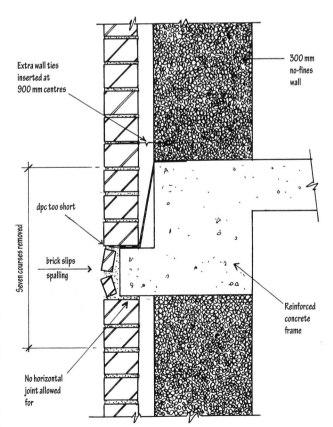

Extra wall ties inserted at 900 mm centres

300 mm no-fines wall

dpc too short

Seven courses removed

brick slips spalling

Reinforced concrete frame

No horizontal joint allowed for

*Fig. 8.2* Existing support detail.

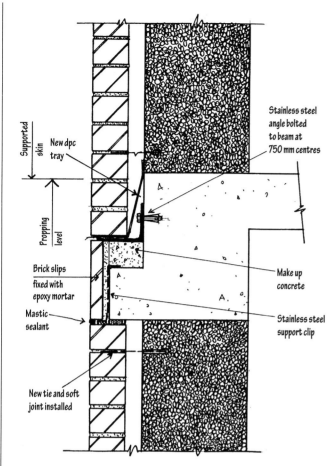

*Fig. 8.3* New support detail.

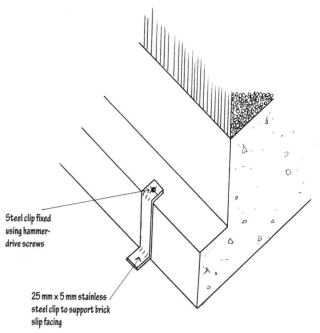

*Fig. 8.4* Brick slip support detail.

installed were tested at random for pull-out values during the installation programme.

One major problem occurred in that the brick supplier did not manufacture the same thickness of brick in the required matching colour. The original brick was 73 mm thick, whereas the newer metric bricks were 65 mm thick. To overcome this problem it was decided to use brick-on-edge slip bricks onto an epoxy cement backing with a mechanical fixing arrangement for added strength, as shown in Fig. 8.4. The special toggle clip was hung from the top of the beam nib, and the turned-up portion at the bottom supported the brick slip in addition to the bond from the epoxy mortar backing.

The remedial scheme required the removal of seven courses of imperial bricks at each floor level. This required the outer leaf of brickwork to be supported on needles, and the scaffolding was specifically designed to support these needles. Steel ledgers were installed at

915 mm centres, bolted to the concrete floor beam, with hardwood timber fox-wedged packings driven in from the side between the ledgers and brickwork. The small reaction at the scaffold end of the ledger was approximately 0.45 kN, and this was allowed for in the scaffold design.

It was therefore important to ensure that the scaffold was erected to a tight vertical tolerance using good-quality steel tubing with sufficient bracing. The works were started from the roof level, and at any one time operatives were working at three floor levels, as the various stages of work were carried out in a specified sequence of providing needles, removing brickwork, installing angles and rebuilding the new brickwork with its soft joint and horizontal tie fixing below the beam.

## Framed buildings

In a framed building, or where the floors rest on the internal skin of the structure, it is therefore important to provide for a compressible joint horizontally below each floor level so that each floor beam carries only the storey above, and the problems of brick moisture expansion and frame shortening are catered for in such a way that the differential effects are minimised.

### 8.2.5 *Movement caused by chemical reactions*

The most-documented form of damage involving chemical reactions in brickwork is that caused by soluble sulphates in the bricks. These sulphates are leached out when the wall becomes excessively wet, and the acidic solution formed reacts with a constituent of the set

cement known as tricalcium aluminate, $C_3A$, which produces a calcium compound that is expansive.

In severe cases this sulphate attack can completely disrupt the brickwork, causing cracking and movement, both vertically and horizontally.

In one case investigated by the author in Huddersfield, the expansion at each mortar bed joint had resulted in

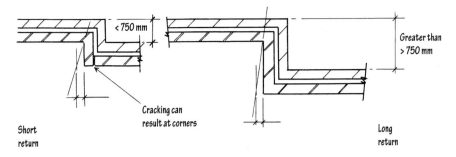

*Fig. 8.5* Rotational cracking to return walls.

excessive vertical expansion, and this had resulted in the roof trusses rotating. This chemical attack had been caused by a combination of high exposure to driving rain and a high soluble sulphate content in the bricks. The movement was non-reversible, and the only solution was to re-skin the house using a different brick. Had the original bricks been laid using sulphate-resistant mortar, as recommended by the brick manufacturer, the defects would not have arisen.

The defect is identified by cracking of the mortar bed joint in the centre. While it is not a common defect, and is slow to develop, the risk is greatly reduced if the mortar is made using sulphate-resisting Portland cement to BS 4027, where the $C_3A$ content is limited to a maximum of 3.50%.

The other source of chemical attack is from soluble sulphates in the ground or groundwater. Fills containing ash, brick rubble, colliery shales etc. generally contain soluble sulphates, and where such conditions are known the use of sulphate-resistant cement in mortar and foundation concrete is recommended.

Where the levels of soluble sulphates are Class 3 or higher, then the use of strong 1:3 sulphate-resistant mortar mixes is the only solution: the denser the mortar mix, the greater the resistance to the sulphate solution.

### 8.2.6 Designing for movement

Where differential movements are likely to occur it is essential that the design caters for the movements from the outset.

Cracking of masonry due to horizontal movement is more common than that due to vertical movement, owing to the additional restraint in the vertical plane.

When investigating the thermal and moisture movement type of defect the following points should be looked for:

● Oversailing of the end of the wall on the dampcourse. Cracking rarely passes below the dampcourse, owing to the restraint from the foundation and ground friction effects. However, it is always advisable to extend any movement joint in the superstructure below the dpc into the foundation walling and down to the foundation.

● Long terraces with many doors and windows are susceptible to thermal expansion, especially on south-facing walls. Long walls with short returns are also likely to develop some cracking at the short return, and Fig. 8.5 shows the rotation problem with short returns. Movement joints should be based on Table 8.1

## Case history 8.2

The owner of a flat in Newcastle had lodged a claim under his house insurance warranty for cracking to an internal wall. The wall, which was cracked close to the external junction, was not a loadbearing wall, and was built using 140 mm blockwork bonded into the external wall blockwork. The concrete blockwork forming the inner leaf of the external wall had been built without any provision for contraction movement. The manufacturer's recommendations for the type of block used were that contraction joints should be provided at a maximum of 6.0 m.

The internal walls were dry-lined; a section of the dry-lining was removed, and the cracking was observed in the blockwork. The most significant point recorded was that the main external wall had a 215 mm step back at the position of the crosswall, and the length of return was 9.30 m. The length of the non-loadbearing crosswall was

6.35 m. It was clear that contraction of the outer leaf blockwork had caused rotation at the strong, short return and hence the cracking. While this type of cracking was not considered to be a major structural defect under the terms of the house warranty, the damage was repaired as a gesture of goodwill because the damage had resulted from the builder's not having followed the guidance given in the warranty provider's standards. These recommended that block manufacturer's requirements be followed in regard to movement joints at 6.0 m centres.

Repair works required the removal of the dry-lining to both sides of the internal wall and a short section of the wall was rebuilt using 7 N blocks in a 1:1:6 mortar. Where the blockwork was toothed into the existing the bed joints were reinforced with masonry reinforcement. The dry-lining was renewed, and the walls were decorated.

## Case history 8.3

### Introduction

This case history highlights the defects that can result in new brickwork when walls are built using bricks that have not been allowed to mature for a sufficient period after manufacture, i.e. 'kiln fresh' bricks.

On a residential site in Yorkshire several dwellings were found to be suffering from unsightly cracking in the external brickwork. The cracking had developed several days after the bricklaying had been finished.

### Investigation

A detailed examination of the houses was carried out, and the ground conditions were examined. It was concluded that the cracking was not the result of any foundation movement. This was based on the fact that no internal cracking was evident, and the cracking to the external walls did not extend below the dampcourse. The foundations were also shown to be suitable for the ground conditions.

### Conclusions

Samples of the mortar were taken and analysed, and were found to contain acceptable ratios of sand to cement. In view of these findings it was decided to involve the brick manufacturer, and to check the building records and delivery tickets for the bricks. The manufacturer's records confirmed that the bricks supplied to the site had not been allowed to mature for a sufficiently long period after coming out of the kiln. Because the bricks were kiln fresh they had been subject to unrestrained expansion due to absorption of moisture soon after having been laid. Checking the firing dates with the actual building programme showed that the bricks had been walled within one week from firing. It was clear that the manufacturer had sent out bricks that were kiln fresh, and the site agent had not allowed sufficient time to elapse before using them (it is good practice to let bricks stand on site for at least 7 days before use). In this instance the manufacturer and the builder had joint responsibility for the problems that had arisen. The manufacturer replaced the bricks at no cost to the builder, and the builder rebuilt those sections of wall that had been damaged.

---

- Long lengths of walling between floors, with the brickwork above and below perforated with doors and windows, are very vulnerable to cracking due to the differential effects along the elevation. The vertical panels of brickwork between the windows are subjected to tensile and compressive forces when the long section of wall above expands.

When clay bricks are removed from the kiln after firing they are generally allowed to cool off and mature for a period before being delivered. During this maturing period they take up moisture from the atmosphere, and initially expand at a high rate. This expansion then continues at a slower rate, and then diminishes to negligible amounts at about 100 days after firing. Figure 8.6 shows the expansion/time relationship of brick moisture absorption. Maximum moisture absorption varies from 0.2% to 0.50% in the UK.

Figure 8.7 shows the type of cracking that can be caused by brick thermal expansion and moisture expansion if long terraces are built without movement joints. In this situation the short lengths of wall between doors and windows are subjected to tensile and compressive forces due to the expansion or contraction of the differing lengths of walling. In addition the long lengths of walling at the dampcourse level are more restrained than the long lengths higher up.

Bricks tested at the BRE expanded at a rate of 0.80 mm/m in the first 8 years; 50% of this expansion took place in the first week.

It is therefore important to consider the effects of different expansion rates when retaining walls or walls below ground are constructed using two types of brick. In long walls with cross headers with an engineering brick behind the facing brickwork, the facing brickwork will develop vertical tension cracks as the backing skin takes up moisture at a higher rate. Brickwork can be reinforced to minimise the tension cracking, and movement joints should be used in long walls. Where possible the use of cross headers should be avoided, and flexible wall ties should be provided instead.

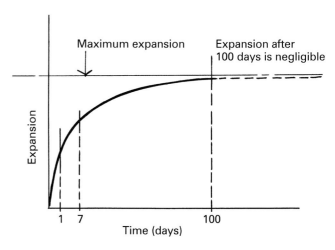

*Fig. 8.6* Unrestrained moisture absorption of clay bricks after firing.

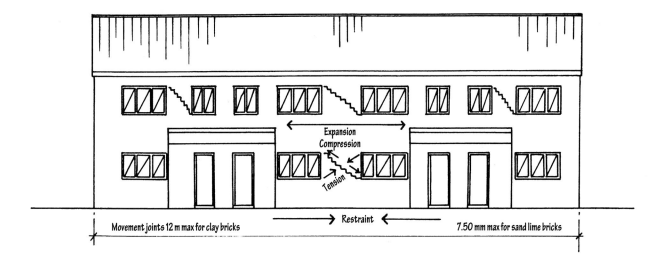

*Fig. 8.7* Movement in a long terrace block.

### Calcium silicate bricks (sand limes)

Sand lime bricks are made from a mixture of sand and hydrated lime, moulded under pressure and hardened off in autoclaved ovens using steam, which helps to chemically bond the constituents. While this type of brick has good resistance to soluble sulphates from the ground, the designer needs to take care when specifying them that the drying shrinkage is catered for. Unlike clay bricks, which expand, sand lime bricks shrink slightly when they dry out, and failure to provide movement joints could result in unsightly cracking.

The drying shrinkage of bricks made to BS 187 : Part 2 should not exceed 0.025% of the original wet length, except for bricks of Classes 2B and 3B, for which the limit is 0.035%.

For Class 1 bricks there is no limit, but they should only be used for internal walls in dry conditions, above dampcourse levels.

Table 8.2. lists the various classes of calcium silicate brick and their uses.

Because of their lime content, calcium silicate bricks should never be used where there are fills containing slags derived from iron or steel processes. Such slags can have

*Table 8.2* Classes of calcium silicate brick and their uses

| Class of brick | Recommended use | Minimum average crushing strength (N/mm²) | Maximum average drying shrinkage (%) |
|---|---|---|---|
| I | Suitable only for internal non-facing use above ground level dpc where an average strength of 7 N/mm² is required. Group 4 or 5 mortars only should be used with Class I | 7 | No requirement |
| 2A | Suitable for use where an average strength of 14 N/mm² is required for facing work, whether external or internal, provided none of the conditions that necessitate the use of bricks in Class 3A, 3B, 4 or 5 apply | 14 | 0.025 |
| 2B | Suitable for use as for Class 2A except that this class should not be used when structural reasons or mortar durability require mortars stronger than group 4 | 14 | 0.035 |
| 3A | Suitable for situations where brickwork is liable to be continuously saturated with water, or brickwork is exposed repeatedly to temperatures below freezing when saturated, and for any unprotected brickwork below ground | 20.50 | 0.025 |
| 3B | Suitable for use where an average strength of 20.50 N/mm² is required, and for the same purposes as Class 3A, except that this class should not be used where the drying shrinkage of the brickwork may cause cracking or when stronger mortars than group 4 are required for strength and durability. This class is suitable for brickwork piers, unrestrained or partially restrained infill panels of relatively short length, and any work below dpc at ground level | 20.50 | 0.035 |
| 4, 5 | Suitable for use where high crushing strengths are needed | 27.50–34.50 | 0.025 |

*Table 8.3* Mortars for use with sand lime bricks

| Class of sand lime brick | Type of construction | No risk of frost during construction | Freezing may occur during construction |
|---|---|---|---|
| 1 | Internal walls<br>Inner leaf of cavity walls<br>Backing to external solid walls | 5 or 4<br>5 or 4<br>4 | 4 or 3<br>3 |
| 2A, 2B | External walls, including the outer leaf of cavity walls, facing to solid construction<br>Work below dpc level but more than 150 mm above ground | 4 | 3 |
| 3A, 3B | Work below ground or within 150 mm above ground<br>Parapet walls (not rendered)<br>External free-standing walls<br>Parapet walls (rendered) | 3[a]<br>3<br><br>4 | 3[a]<br>3<br><br>3 |
| 4 | Sills and copings<br>Earth retaining walls | 2<br>2[a] | 2<br>2[a] |
| 5 | Appropriate for calculated loadbearing brickwork when a minimum crushing strength of 34.50 N/mm$^2$ is required | To be specified by the designer to give the required strength and durability with adequate workability | |

[a] Where sulphates are present in the groundwater use sulphate-resistant cement

high concentrations of magnesium sulphates or ammonium sulphates, which attack the calcium silicate bricks. Mortar mixes are also very important; Table 8.3 shows the recommended mortar mixes to use for sand lime bricks.

It is important to remember, when using sand lime construction, that the mortar should always be weaker than the bricks. Joints in long walls should not exceed 7.50 m centres.

### 8.2.7 Defects caused by compression failure

BS 5628 Part 1: 1992 deals with unreinforced masonry construction, and lays down the methods of design and construction. The most important factor in the design and construction of a masonry wall is the thickness of walls. Walls need to be correctly sized to ensure that the direct compressive stresses are not greater than the allowable stresses in the Code. The slenderness ratio of the wall is therefore very important. In essence, design resistance must exceed design load by a recognised factor.

## Case history 8.4

### Introduction

This case history illustrates two important points:

● the need to ensure that structural elements supporting large areas of roof, such as large purlins or multiple girder trusses, impose concentrated loads at their bearings;
● where low-strength thermal blockwork with compressive strengths of 2.80 N and 3.50 N is being used for main walls, the need to check the stress below any concentrated loadbearing positions to see whether a padstone or spreader is required.

In this case the house owner had complained about vertical cracking to the internal skin of the external wall.

### Investigation

This involved removing plaster from the cracked section. It was found that the cracking extended right through the blockwork; it was possible to insert a feeler gauge

through the wall. Examination of the blockwork revealed that it was a lightweight thermal autoclaved block with a basic compressive strength of 2.80 N/mm$^2$.

Figure 8.8 shows the roof truss layout; it was clear from the site measurements that the cracking occurred below the triple girder trusses used in the hip-end construction. The triple girder truss supported short mono trusses and the sloping hip rafters. The triple girder truss was fixed to a 100 mm × 50 mm deep timber wall plate.

### Calculations

From the roof truss calculations the computer printout showed a worst-case reaction of 20 kN. Applying a load factor of 1.50 gave an ultimate design load of 30 kN.

Width of triple girder truss = 3 × 35 = 105 mm

Width of wall subjected to the bearing = (2 × 50) + 105

$$= 205 \text{ mm}$$

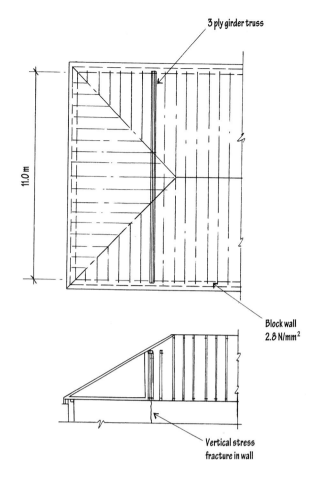

*Fig. 8.8* Roof plan: details of girder truss.

The mortar was judged to be a 1:1:6 cement:lime:sand, i.e. Type 3.

Allowable compressive strength of blocks = 2.80 N/mm$^2$

Direct bearing stress on top of wall $= \dfrac{W}{A} = \dfrac{30 \times 10^3}{100 \times 205}$

$$= 1.46 \text{ N/mm}^2$$

where $W =$ ultimate load, and $A =$ area of section.

BS 5628 Part 1 allows higher stresses to be used in the vicinity of concentrated loads depending on the type of bearing onto the wall.

In this case the bearing was a type 2 bearing, which permits an increase of 1.50.

Local allowable design stress $= \dfrac{1.50 \times f_k}{\gamma_m}$

where $f_k =$ characteristic strength of the blockwork; $\gamma_m =$ partial safety factor for materials and workmanship (Table 4, BS 5628).

From Table 2(d), BS 5628: $f_k = 2.80$ N/mm$^2$

From Table 4, BS 5628: $\gamma_m = 3.50$

Local allowable design load/unit area, $g_A$

$$= \frac{1.50 \times 2.80}{3.50} = 1.20 \text{ N/mm}^2$$

This was clearly less than the actual compressive stress, and it was considered necessary to install a dense concrete padstone.

Length of padstone $= \dfrac{W}{t \times g_A} = \dfrac{30 \times 10^3}{100 \times 1.20} = 250$ mm

A 300 mm long × 215 mm high × 100 mm thick padstone was installed during the repairs to the blockwork wall. This required the girder truss to be propped at the node point nearest to the wall, with an extra prop close to the wall. The props were taken down onto the ground floor slab. Following the repairs the props were removed after 7 days.

## Case history 8.5

### Introduction

A four-storey residential development was built in Harrogate with car parking situated on the ground floor. Shortly after the building was completed it was noticed that cracking had developed on the inside blockwork to the garage piers.

### Investigation

Figure 8.9 shows the rear elevation of the building. The precast floors all spanned front to back, and there were two central corridor walls supporting the floors. The as-built blockwork strengths from ground floor to first floor were 10 N/mm$^2$, and from second floor to roof were 7.0 N/mm$^2$. The trussed rafter roof spanned front to rear, and consisted of concrete tiles on battens and felt, the trusses spanning onto the front and rear walls.

At ground floor level the rear garage openings were 3.00 m wide, and steel beams carried the masonry above, supported on concrete padstones on the 1500 mm wide piers. Examination of the ground floor blockwork to the front wall of the building revealed no cracking, and calculations showed that the 10 N strength blockwork was adequate for the gable, party and rear walls. It was therefore necessary to check the design of the rear piers between garage units.

### Calculations

As most of the ground floor piers were showing similar cracking it was necessary to check the load resistance of

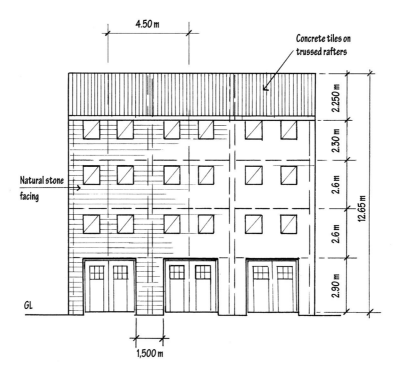

4.50 m

Concrete tiles on
trussed rafters

Natural stone
facing

2.250 m

2.30 m

2.6 m

2.6 m

12.65 m

2.90 m

GL

1,500 m

**Fig. 8.9** Rear elevation.

the rear piers. As many of the flats were unoccupied it was considered realistic to consider only dead loads when doing the assessment of the floor loadings. It was also considered unnecessary to apply any stresses due to wind loading.

### Geometric properties of rear 1500 mm piers

Thickness of blockwork leaf, $t_1 = 100$ mm
Thickness of stone-faced outer leaf, $t_2 = 125$ mm
$t_{ef}$ = effective thickness of pier
$h_{ef}$ = effective height of pier
$k$ = effective thickness coefficient (Table 5, BS 5628)
$e_x$ = eccentricity of loading at top of pier
$\lambda$ = slenderness ratio = 0.75 × storey height
$\beta$ = capacity reduction factor (Table 7, BS 5628)
$\gamma_m$ = partial safety factor for materials
$\gamma_f$ = partial safety factor for load
$f_k$ = characteristic compressive strength of masonry
$N$ = ultimate design vertical load
$N_r$ = ultimate design load resistance of pier

### Loadings

| Roof: Dead load | Serviceability loads | Ultimate loads |
|---|---|---|
| | (kN/m²) | (kN/m²) |
| concrete tiles | 0.60 | |
| battens and felt | 0.05 | |
| trusses | 0.15 | |
| insulation | 0.05 | |
| plasterboard and skim | 0.15 | |
| Total = | 1.0 × 1.40 = | 1.40 |

Imposed loading

| snow loading | 0.60 | | |
|---|---|---|---|
| storage | 0.25 | | |
| Total = | 0.85 × 1.60 | = | 1.36 |
| | Total | = | 2.76 |

Precast floors: Dead load

| 150 mm precast prestressed units | 2.20 | | |
|---|---|---|---|
| 50 mm sand : cement screed | 1.20 | | |
| plastered ceiling | 0.20 | | |
| stud partitions | 0.50 | | |
| Total = | 4.10 × 1.40 | = | 5.74 |

Imposed loading    =   1.50 × 1.60   =   2.40

### External walls

| | (kN/m²) | | (kN/m²) |
|---|---|---|---|
| 125 mm natural stone facings | 3.50 × 1.40 | = | 4.90 |
| 100 mm concrete blockwork | 1.50 × 1.40 | = | 2.10 |
| plaster coat + skim | 0.25 × 1.40 | = | 0.35 |
| Total | = 5.25 × 1.40 | = | 7.35 |

Roof loading:   $\dfrac{12.50}{2.0} \times 2.76 \times 4.50$   $= 77.625$ kN

Floor loading:   $\dfrac{6.25}{2.0} \times 5.74 \times 3 \times 4.50$   $= 243$ kN

Walls: inner leaf
Area of blockwork supported $= (5.20 + 2.30) \times 4.50$
$$= 33.75 \ m^2$$
less windows $6 \times 1.0 \times 0.80 \quad = \underline{4.80}$
net area $\qquad\qquad\qquad = 28.95 \ m^2$

Weight of inner leaf (including plaster) $= 28.85 \times 2.45$
$$= 70.68 \ kN$$

design load, $N$, on piers at first-floor level
$$= 77.625 + 243 + 70.68 = 391 \ kN$$

As the floor slabs were providing enhanced restraint to the inner leaf

$$h_{ef} = 0.75 \times 2600 = 1950 \ mm$$

Because the pier was bonded into the party wall the pier had an enhanced stiffness,
$t_p/t = 3.0$, and from Table 5, BS 5628, stiffening coefficient $= 2.0$

therefore $t_{ef} = \dfrac{2}{3}(125 + 2 \times 100) = 216 \ mm$

Slenderness ratio, $\lambda = \dfrac{1950}{216} = 9.0 \qquad e_x = \dfrac{t}{2} - \dfrac{t}{3} = \dfrac{t}{6}$

From Table 7, BS 5628, for $e_x = 0.167$ $\beta = 0.77$

$N_r =$ stress $\times$ area loaded

$$= \frac{\beta \times f_k \times t_1 \times b}{\gamma_m} = \frac{0.77 \times f_k \times 100 \times 1500}{3.50}$$

$N_r = (33 \times 10^3) f_k$, which is $> N$

therefore $f_k = \dfrac{391 \times 10^3}{33 \times 10^3} = 11.84 \ N/mm^2$

From Table 2(d), BS 5628, using mortar type 3 (1 : 1 : 6), for $f_k = 11.24$ the pier blockwork strength would need to be 20 $N/mm^2$ if using blockwork.

It was clear that the pier blockwork was overstressed and needed to be strengthened. Calculations were also carried out to check the remaining ground floor walls.

*Front walls*

The full height to ground floor equals $(3 \times 2.60) + 2.30 = 10.10 \ m$

| Total load/metre run | (kN/m) |
|---|---|
| Inner block wall $= 10.10 \times 2.45$ | $= 24.73$ |
| Floors $3 \times 8.14 \times \dfrac{6.75}{2}$ | $= 76.51$ |
| Roof $2.76 \times \dfrac{12.50}{2}$ | $= \underline{17.25}$ |
| Total | $= 118.49 \ kN/m$ |

$$h_{ef} = 2600 \times 0.75 = 1950 \ mm$$

$$t_{ef} = \frac{2}{3}(100 + 125) = 150 \ mm$$

Slenderness ratio $= \dfrac{1950}{150} = 13 \qquad e_x = 0.16 \qquad \beta = 0.76$

Using 10 $N/mm^2$ blockwork, $f_k = 8.20$

Ultimate design load resistance of wall, $N_r$

$$= \frac{\beta \times t_1 \times b \times f_k}{\gamma_m}$$

$$= \frac{0.76 \times 100 \times 1000 \times 8.20}{3.50 \times 10^3} = 178 \ kN/m$$

The 10 N strength blocks were therefore adequate for the ground floor walls.

**Conclusion**

It was decided that replacing the blockwork piers with engineering bricks would be too expensive, and a scheme for strengthening the piers was adopted using steel channel sections each side of the pier and welded to the steel beams above. The steel channels were taken down onto the pier foundation using a nominal base plate, and after installation the channels were bolted to the wall at 1.0 m centres. This steel channel was also considered as a suitable protection to the blockwork pier in the event of a vehicle impact. The steel beam was already transferring load onto the blockwork pier via the padstone. It was therefore necessary to support the steel beam on Acrow props and remove the concrete padstone. The steel channel was then positioned and

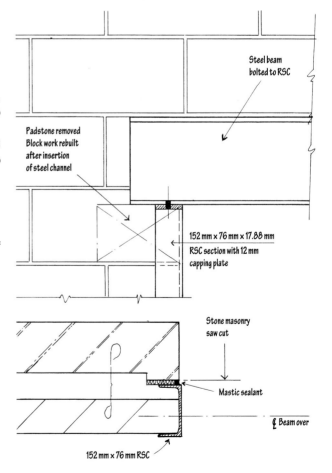

Fig. 8.10 Channel fixing details.

bolted to the steel beam bottom flange, as shown in Fig. 8.10. Once the base plate grout had attained its strength the Acrow props were removed, and the blockwork was then built back up to the underside of the steel beam.

### Final design load down pier

Including imposed loads on floors

$$= 391 + 3 \times 2.40 \times \frac{6.25}{2} \times 4.50 \qquad = 492.52 \text{ kN}$$

Additional storey of blockwork $= 2.45 \times 2.60 \times 4.50$

$$= \underline{28.66}$$
$$\text{Total} = 520.91$$

$$\text{Load carried by blockwork} = \frac{1500}{4500} \times 520.91$$

$$= 173.63 \text{ kN}$$

As the ultimate design load resistance/metre $= 178$ kN this is adequate.

Load carried by each channel $= 173.63$ kN

For practical purposes try $152 \times 76 \times 17.88$ kg/m. RS channel

Design to BS 5950: Effective length, $L_e$
$$= 0.85 \times 3.55 = 3.00 \text{ m (Table 24)}$$

Radius of gyration, $r_{yy} = 22.30$ mm

Slenderness ratio, $\lambda$

$$= \frac{L_e}{r_{yy}} = \frac{3000}{22.30} = 134.50$$

From Table 27(c), BS 5950: $p_c = 81$ N/mm$^2$

$$\text{therefore } P_c = p_c \times A_g = \frac{81 \times 22.80 \times 10^2}{10^3}$$

$$= 184 \text{ kN} > 173.63 \text{ kN} - \text{OK}.$$

where $p_c$ = compressive strength (N/mm$^2$), $P_c$ = compressive resistance of column (kN), and $A_g$ = gross sectional area of steel section.

### 8.2.8 Laterally loaded masonry

The foregoing examples dealt with defects related to the direct compressive strengths of masonry arising from vertical loads from floors, walls and roofs. However, buildings also have to be capable of adequately resisting lateral forces from wind, and from accidental damage (generally buildings of five storeys or more).

Wind pressure is assumed to act statically on a traditional brick structure. The magnitude of these wind forces is dependent on:

- location of the site;
- local topography;
- the mean hourly wind speed;
- the height of the building;
- length of the building.

The wind loadings on buildings are calculated from BS 6399 Part 2: 1997.

While masonry is strong in compression it is very weak in tension, and masonry structures must therefore be designed to resist these lateral forces adequately. Figure 8.11 shows a low-rise building subjected to wind loading, and the way the lateral load is transferred down into the foundations via the floors and shear walls.

When designing for wind forces two cases must be considered:

- the strength of the individual masonry elements;
- the overall stability of the building.

There are not many recorded instances of low-rise masonry buildings that have failed because they were unable to resist lateral wind pressures. Quite often it is excessively high winds that cause storm damage to the roofs and cladding of buildings.

Before going into the strength of individual masonry elements it is worth studying the overall stability aspect in more detail. In steel-framed factory buildings and commercial buildings, resistance to lateral wind forces is provided by diagonal steel cross-bracing, or by the brick panels parallel to the direction of the wind. Quite often steel-framed buildings require to have the steel bracing fitted prior to any brickwork being built. In masonry buildings the key to good design is to ensure that there is sufficient robustness available in the structure to transfer the loads down into the foundations. This will be achieved by ensuring that the floors and roof to the building are sufficiently robust to act as horizontal stiff diaphragms. This often requires floors that span parallel to shear walls to be provided with built-in 'strong point' connections, as shown in Fig. 8.12.

For flats, hostels, hotels and residential buildings there are generally sufficient walls available, because of the subdivision into units, to provide sufficient lateral stiffness in both directions. For buildings of three storeys or more the provision and positioning of stairwells, lift shafts and corridor walls will play an important part in providing the lateral rigidity.

The most suitable arrangement of walls is one that results in a cellular wall system, as shown in Fig. 8.13.

For structures where the front and rear elevations are primarily just clad in non-loadbearing materials, it is essential to introduce corridor walls and lift shafts to compensate for the lack of stiffness. For such cross-wall construction it is prudent to have some added robustness to cater for building inaccuracies or accidental damage. Robustness in a building cannot be achieved purely by provision of a suitable wall layout. Floor systems must have adequate bearings on the supporting walls, and

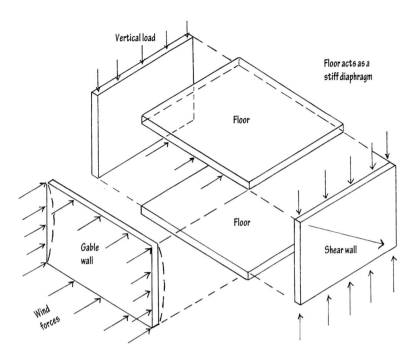

*Fig. 8.11* Wind transfer.

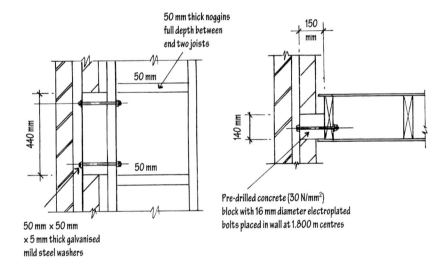

*Fig. 8.12* Timber floor strong point connections.

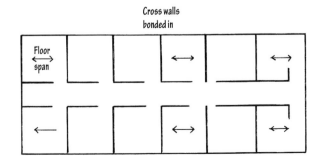

*Fig. 8.13* Robust cellular wall layout.

must have lateral ties into walls parallel to the direction of span of the floor.

Cast in-situ reinforced concrete floors with a two-way spanning capacity will provide a very stiff diaphragm action compared with timber joist floors or precast plank floors, which are a series of individual structural elements. Such floors must be tied together in such a way that any lateral forces cannot unzip the floor from the surrounding walls.

When considering lateral stability, the provision of movement joints through the structure to accommodate thermal, foundation or seismic movements must be taken into consideration, i.e. by the provision of double columns or walls at the joint. Roofs must be provided with sufficient diagonal and longitudinal bracing, tied into the supporting walls, to ensure that the whole roof acts as a stiff plate.

In some situations it may not be possible to provide the stiffness required by just using unreinforced masonry. Quite often the introduction of steel reinforcement into a masonry panel can extend its limits, as in large sports halls, buildings affected by seismic forces, and exhibition buildings designed with fin wall constructions, for example. The general method adopted is to provide the reinforcement into pockets or cavities, which are grouted up as construction progresses. Such designs are heavily dependent on good, sound workmanship if the as-built structure is to model the theoretical design. For buildings that need to be designed to prevent progressive collapse from an explosion incident, reinforcement used in masonry and floors generally proves to be very cost-effective.

### 8.2.9 Designing for wind

A short low-rise building requires more stiffness per metre than a longer building. The normal design approach is to consider the walls and piers as a series of vertical cantilevers, the wind loads being translated into the top of the individual cantilevers in proportion to their flexural stiffness. This design approach, while fairly conservative, is used widely because of its simplicity. In practice the true stiffness of a building is highly indeterminate, because there are many factors that are not taken into account, such as the stiffening effects of non-loadbearing partitions, the effects of fixity, the actual length of return walls to be considered, pre-compression loads, mortar strength and workmanship.

## Case history 8.6

### Introduction

This case history relates to a modern two-storey dwelling, built in traditional brick and blockwork construction. Some cracking had been observed by the owners on the front elevations, and following an investigation by the insurance warranty provider the damage was considered to be non-structural. While the cement : sand ratios from several mortar samples varied considerably, none of the mortar sampled had cement : sand ratios less than 1 : 6.

The owner decided to seek independent advice from a consulting engineer, and the engineer's report indicated that the cracking could have resulted from wind forces on the gable. As the fenestration to the front elevation was more than two-thirds of the total elevation, it was necessary to check the stability of the dwelling and the individual piers to the front elevation.

Figure 8.14 shows the ground floor plan and Fig. 8.15 the gable elevation. Figure 8.16 shows the front elevation, and Fig. 8.17 the first floor room layout and direction of floor spans. The ground floor was a concrete ground-bearing slab, cast independent of the main walls.

By inspection the rear and gable elevations complied with the criteria for stability, and only the front elevation needed to be considered.

When the aggregate width of the openings on an elevation exceeds two-thirds of the length of the wall, the individual piers must be checked by calculation.

Length of elevation = 8.00 m,
summation of openings
$= 2.275 + 1.725 + 0.915 + 0.915 = 5.830$
Ratio of openings in elevation length

$$\frac{5.830}{8.00} = 0.73 > 0.66$$

### Design loadings:

#### Roof

|  | (kN/m²) |
|---|---|
| Dead loads: | |
| concrete roof tiles | 0.55 |
| rafters | 0.25 |
| insulation | 0.020 |
| services | 0.050 |
| ceiling board and plaster | 0.15 |
| | Total = 1.02 |
| Imposed loading: | |
| snow | 1.0 |
| storage | 0.25 |
| | Total = 1.25 |

#### First floor

|  | (kN/m²) |
|---|---|
| Dead loads: | |
| flooring | 0.09 |
| joist, noggins etc. | 0.095 |
| ceiling | 0.150 |
| finishes | 0.035 |
| | Total = 0.370 |
| Imposed loading: | |
| Domestic | 1.50 |

#### Walls and partitions

| | |
|---|---|
| 75 mm stud + plasterboard | 0.75 |
| 100 mm concrete blockwork | 1.370 |
| 100 mm brickwork | 2.15 |
| plaster per face | 0.22 |
| plasterboard + finish | 0.15 |

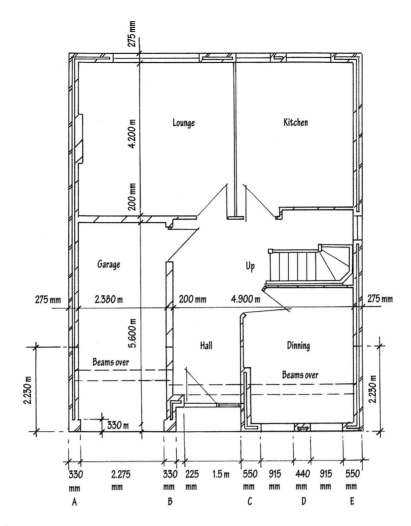

Fig. 8.14 Ground floor plan.

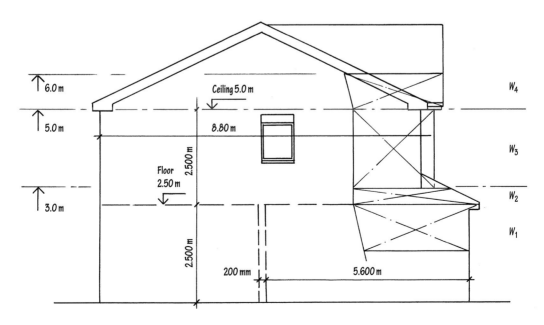

Fig. 8.15 Gable elevation.

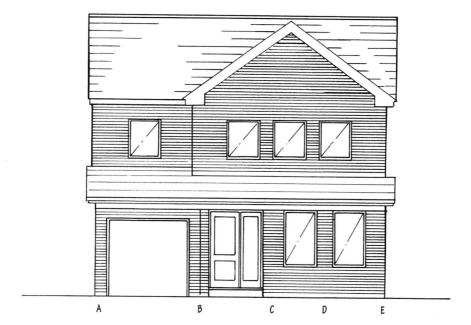

*Fig. 8.16* Front elevation.

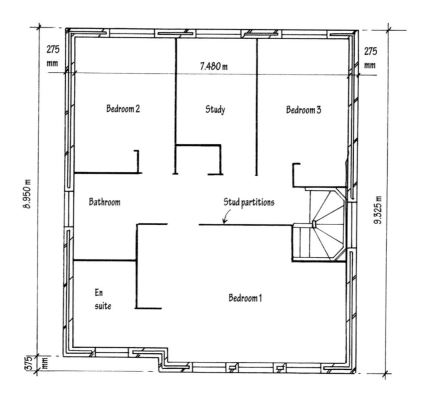

*Fig. 8.17* First floor plan.

## External masonry details as-built

Brickwork: Crushing strength of bricks taken as 10 N/mm$^2$

Characteristic compressive stress, $f_k$ taken as 4.10 N/mm$^2$ in Class 3 mortar (Table 2, BS 5628, Pt 1)

Flexural strength of brickwork, $f_{kx} = 0.40$ N/mm$^2$ (Table 3, BS 5628, Pt 1)

Blockwork: Crushing strength of blocks 10 N/mm$^2$ and $\gamma_m = 3.50$ (Table 4, BS 5628)

Characteristic compressive stress, $f_k$, taken as 3.50 N/mm$^2$ in Class 3 mortar

Flexural strength of blockwork, $f_{kx} = 0.25$ N/mm$^2$ (Table 3, BS 5628)

## Wind pressures

Basic wind velocity, $v = 46$ m/s. $S_1 = S_2 = 1.0$

| Height of element (m) | Ground roughness $S_2$ Class B | $V_s = v \times S_1 \times S_2 \times S_3$ | $q = kV_s^2$ (kN/m²) $k = 0.613$ |
|---|---|---|---|
| 0–3 | 0.67 | $46 \times 0.67 \times 1.0 = 31$ | 0.590 |
| 3–5 | 0.74 | $46 \times 0.74 \times 1.0 = 34$ | 0.710 |
| 5–10 | 0.93 | $46 \times 0.93 \times 1.0 = 42.78$ | 1.122 |

where $S_1$, $S_2$, $S_3$ are wind load factors used from Code of Practice CP3, Chapter V, Part 2; $V_3 =$ wind speed; and $q =$ wind force (N/mm²).

This wind load, acting on the gable end, is transmitted through the floors and roof to act as a horizontal force at window heads, and this force is in proportion to the relative stiffness of the piers. Five piers – A, B, C, D and E – will take the wind force, and these piers have a return length of 2.230 m.

(kN)

$$W_1 = \frac{1}{2} \times \left( \frac{5.7}{2.0} + \frac{3.90}{2.0} \right) \times \frac{2.50}{2.0} \times 0.590 \quad = 1.770$$

$$W_2 = \frac{1}{2} \times \left( \frac{5.7}{2.0} + \frac{3.9}{2.0} \right) \times (3 - 2.50) \times 0.590 \quad = 0.708$$

$$W_3 = \frac{1}{2} \times \left( \frac{4.3}{2.0} + \frac{2.9}{2.0} \right) \times (5.0 - 3.0) \times 0.71 \quad = 2.550$$

$$W_4 = \frac{1}{2} \times \left( \frac{5.1}{2.0} + \frac{2.9}{2.0} \right) \times (6.125 - 5.0) \times 1.12 = \underline{2.52}$$

Total $W_f = 7.54$

| Section no. | Area of section (m²) | $y$ (mm) | $Ay^2$ | $I_{self} = \dfrac{bd^3}{12}$ | Total $I \times 10^9$ |
|---|---|---|---|---|---|
| 1 | $1.9 \times 0.10$ $= 0.19$ | 93 | $1.64 \times 10^9$ | $\dfrac{1.9 \times 0.10^3}{12}$ $= 0.158 \times 10^9$ | 1.798 |
| 2 | $1.9 \times 0.10$ $= 0.19$ | 82 | $1.27 \times 10^9$ | $\dfrac{1.9 \times 0.10^3}{12}$ $= 0.158 \times 10^9$ | 1.428 |
| 3 | $0.33 \times 0.33$ $= 0.109$ | 22 | $0.05 \times 10^9$ | $\dfrac{0.33 \times 0.33^3}{12}$ $= 0.9 \times 10^9$ | 0.95 |
| | Total $= 0.489$ | | | | Total = 4.176 |

$$\text{Approximate stiffness} = \frac{I + Ay^2}{L_e} \qquad I = \frac{bd^3}{12} + Ay^2$$

where $A =$ area of the section, and $y =$ distance from the centroid of the section to the centre of the segment being considered.

### Consider pier A

Area of pier $= 488\,900$ mm²     Centroid $= 143$ mm

(see table bottom left).

### Consider pier B

Area of pier $= 552\,400$ mm²     Centroid $= 196$ mm

(see table bottom right).

| Section no. | Area of section (m²) | $y$ (mm) | $Ay^2$ | $I_{self} = \dfrac{bd^3}{12}$ | Total $I \times 10^9$ |
|---|---|---|---|---|---|
| 1 | $1.28 \times 0.20$ $= 0.256$ | 44 | $0.495 \times 10^9$ | $\dfrac{1.28 \times 0.2^3}{12}$ $= 0.853 \times 10^9$ | 1.348 |
| 2 | $0.6 \times 0.10$ $= 0.062$ | 94 | $0.547 \times 10^9$ | $\dfrac{0.62 \times 0.1^3}{12}$ $= 0.051 \times 10^9$ | 0.598 |
| 3 | $0.1 \times 0.355$ $= 0.035$ | 133.5 | $0.632 \times 10^9$ | $\dfrac{0.1 \times 0.355^3}{12}$ $= 0.372 \times 10^9$ | 1.004 |
| 4 | $0.1 \times 0.280$ $= 0.028$ | 174 | $0.847 \times 10^9$ | $\dfrac{0.1 \times 0.28^3}{12}$ $= 0.182 \times 10^9$ | 1.029 |
| 5 | $0.275 \times 0.1$ $= 0.0275$ | 309 | $2.625 \times 10^9$ | $\dfrac{0.275 \times 0.1^3}{12}$ $= 0.022 \times 10^9$ | 2.640 |
| 6 | $0.345 \times 0.1$ $= 0.0345$ | 84 | $0.243 \times 10^9$ | $\dfrac{0.345 \times 0.1^3}{12}$ $= 0.028 \times 10^9$ | 0.271 |
| 7 | $0.330 \times 0.330$ $= 0.109$ | 31 | $0.104 \times 10^9$ | $\dfrac{0.33 \times 0.33^3}{12}$ $= 0.988 \times 10^9$ | 1.092 |
| | Total = 0.552 | | | | Total = 7.982 |

## Pier C

Area of pier $= 465\,500$ mm$^2$    Centroid $= 161$ mm
Total $I = 7.70 \times 10^9$

## Pier D

Area of pier $= 103\,000$ mm$^2$    Centroid $= 220$ mm
Total $I = 1.854 \times 10^9$

## Pier E

Area of pier $= 522\,000$ mm$^2$    Centroid $= 170$ mm
Total $I = 8.76 \times 10^9$

### Summation of stiffness of piers

The effective height of piers A and B is taken as $1.0h$. Piers C, D and E are short columns built integral with brickwork above and below windows, and the effective height is considered to be $0.5h$. This will reflect the true stiffness, and these piers will attract the highest proportions of the lateral force.

$$\frac{I}{L_e} = \frac{4.17}{2.50} + \frac{7.98}{2.50} + \frac{7.70}{0.5 \times 1.4} + \frac{1.86}{0.5 \times 1.4} + \frac{8.76}{0.5 \times 1.4}$$

$$= 1.668 + 3.192 + 11.0 + 2.657 + 12.514 = 31.03$$

Total horizontal force, $W_f = 7.54$ kN, is shared out to each pier in proportion to its stiffness:

$$\text{Force on pier A} = \frac{1.668}{31.03} \times 7.54 = 0.405 \text{ kN}$$

$$\text{pier B} = \frac{3.192}{31.03} \times 7.54 = 0.776 \text{ kN}$$

$$\text{pier C} = \frac{11.00}{31.03} \times 7.54 = 2.672 \text{ kN}$$

$$\text{pier D} = \frac{2.657}{31.03} \times 7.54 = 0.646 \text{ kN}$$

$$\text{pier E} = \frac{12.514}{31.03} \times 7.54 = \underline{3.041} \text{ kN}$$

$$= 7.540 \text{ kN}$$

### Check pier A

$W_a = 0.405$ kN

Vertical loading on inner leaf:                                    (kN)

Dead load:

$$\text{Main roof } \frac{8.80}{2} \times \frac{2.90}{2} \times 1.10 = 7.01$$

$$\text{Roof over garage } 1.36 \times \frac{2.90}{2} \times 1.10 = 2.16$$

$$\text{First floor brickwork } \frac{2.90}{2} \times 2.50 \times 3.65 = 13.23$$

First floor brickwork $1.0 \times 2.75 \times 1.50$      $= 4.12$
Pier at ground floor $2.230 \times 2.50 \times 1.50$      $= 8.36$

$$\text{Pier at first floor } 1.02 \times \frac{2.38}{2} \times 0.37 = \underline{0.45}$$
$$\text{Total} = 35.33$$

Inner leaf, imposed loads:                                    (kN)

$$\text{Roof/loft } \frac{8.80}{2} \times \frac{2.90}{2} \times 1.25 = 7.97$$

$$\text{Roof at garage entrance } 1.36 \times \frac{2.90}{2} \times 1.0 = 1.97$$

$$\text{First floor load } 1.0 \times 1.50 \times \frac{2.38}{2} = \underline{1.78}$$
$$\text{Total} = 11.72$$

Basic design stresses: $g_k$, $q_k$ and $w_k$

$$g_k = \frac{35.33 \times 10^3}{2.230 \times 0.1 \times 10^6} = 0.158 \text{ N/mm}^2$$

$$q_k = \frac{11.72 \times 10^3}{2.230 \times 0.1 \times 10^6} = 0.05 \text{ N/mm}^2$$

$$w_k = \frac{0.405 \times 2.50 \times 0.187 \times 10^9}{4.17 \times 10^9} = 0.045 \text{ N/mm}^2$$

#### Case 1

$$1.4g_k + 1.6q_k = 1.4 \times 0.158 + 1.6 \times 0.05$$
$$= 0.30 \text{ N/mm}^2$$

#### Case 2

$$1.4g_k + 1.4 w_k = 1.4 \times 0.158 + 1.4 \times 0.045$$
$$= 0.284 \text{ N/mm}^2$$

#### Case 3

$$0.9g_k - 1.4w_k = 0.9 \times 0.158 - 1.4 \times 0.045$$
$$= 0.08 \text{ N/mm}^2$$

#### Case 4

$$1.2g_k + 1.2q_k + 1.2w_k$$
$$= 1.2 \times 0.158 + 1.2 \times 0.05 + 1.2 \times 0.045$$
$$= 0.304 \text{ N/mm}^2$$

Vertical loading on outer leaf:                                    kN

Dead loads:
    First floor brickwork $2.275 \times 2.275 \times 2.16$   $= 11.18$
    Pier at ground floor level
                $2.275 \times 2.50 \times 2.16$   $= \underline{12.28}$
                                Total $= 23.46$

Basic design stresses:

$$w_k = \frac{0.405 \times 2.50 \times 0.187 \times 10^9}{4.17 \times 10^9} = 0.045 \text{ N/mm}^2$$

$$g_k = \frac{23.46 \times 10^3}{2230 \times 100} = 0.105 \text{ N/mm}^2$$

No imposed load down outer leaf, $q_k = 0$

Case 1                                    $N/mm^2$

$1.4g_k + 1.6q_k = 1.4 \times 0.115 + 0$        $= 0.161$

Case 2

$1.4g_k + 1.4\,w_k = 1.4 \times 0.115 + 1.6 \times 0.045 = 0.233$

Case 3

$0.9g_k - 1.4w_k = 0.9 \times 0.115 - 1.4 \times 0.045 = 0.0405$

Case 4

$1.2g_k + 1.2q_k + 1.2w_k$
$= 1.2 \times 0.115 + 0 + 1.2 \times 0.041$       $= 0.187$

This pier is therefore adequate for compression, and no tension was present.

### Check pier B

$W_b = 0.776$ kN

Vertical loading on inner leaf:                    (kN)

Dead load:

  Roof and loft

$$\left( \frac{8.80}{2} \times \frac{2.90}{2} + \frac{9.10}{2} \times \frac{1.725}{2} \right) \times 1.02 \quad = 10.50$$

Roof over porch

$$\left( \frac{3.203}{2} \times 1.325 + \frac{2.00}{2} \times 0.95 \right) \times 1.02 \ = 3.12$$

Brickwork to first floor

$$\frac{2.90 + 1.725}{2} \times 2.50 \times 3.65 \qquad = 21.10$$

First floor $\dfrac{2.38 + 1.725}{2} \times 1.0 \times 0.37$    $= 0.75$

Pier at ground floor level
$0.552 \times 2.50 \times 21.50$                $= 29.67$
                                     Total $= 65.14$

Imposed loads:

  Roof and loft $13.62 \times \dfrac{1.25}{1.02}$        $= 16.69$

  First floor $0.75 \times \dfrac{1.50}{0.37}$        $= 3.04$
                                     Total $= 19.73$

Basic design stresses:                        $(N/mm^2)$

$$w_k = \frac{0.776 \times 2.50 \times 0.196 \times 10^9}{7.98 \times 10^9} \qquad = 0.047$$

$$g_k = \frac{65.14 \times 10^3}{0.552 \times 10^6} \qquad = 0.118$$

$$q_k = \frac{19.73 \times 10^3}{0.552 \times 10^6} \qquad = 0.035$$

Case 1

$1.4g_k + 1.6q_k$
$= 1.4 \times 0.119 + 1.6 \times 0.035$     $= 0.222\ N/mm^2$

Case 2

$1.4g_k + 1.4w_k$
$= 1.4 \times 0.119 + 1.4 \times 0.047$     $= 0.231$

Case 3

$0.9g_k - 1.4w_k$
$= 0.9 \times 0.119 - 1.4 \times 0.047$     $= 0.042$

Case 4

$1.2g_k + 1.2q_k + 1.2w_k$
$= 1.2 \times 0.119 + 1.2 \times 0.035$
$+ 1.2 \times 0.047$             $= 0.240$

The pier is adequate in compression, but case 3 shows that the stresses are approaching a tension condition. Allowable tension for blockwork of $0.25/3.50 = -0.071$. Pier B is therefore adequate.

### Check pier C

$W_c = 2.672$ kN

Vertical loading on inner leaf:                    (kN)

Dead load:

  Main roof $\dfrac{9.10}{2} \times \dfrac{3.375}{4} \times 1.02$       $= 3.91$

  Canopy roof $\dfrac{3.375}{4} \times 0.90 \times 1.02$       $= 0.774$

  Brickwork to first floor $\dfrac{3.375}{4} \times 3.74 \times 1.50 = 4.73$

  First floor $\left( \dfrac{3.0}{2} + \dfrac{1.80}{2} \right) \times 0.37$       $= 0.888$

  Pier at sill level $\dfrac{0.465}{2} \times 1.40 \times 12$       $= 3.906$
                                     Total $= 14.208$

Inner leaf, imposed loads:                      (kN)

  Roof $\dfrac{3.91 \times 1.25}{1.02}$             $= 4.79$

  Canopy roof $\dfrac{0.774 \times 1.25}{1.02}$       $= 0.948$

  Floor $\dfrac{0.888 \times 1.50}{0.37}$           $= 3.60$
                                     Total $= 9.338$

Basic design stresses:                        $(N/mm^2)$

$$w_k = \frac{2.672 \times (1.40 \times 0.50) \times 0.161 \times 10^9}{7.70 \times 10^9} = 0.0391$$

$$g_k = \frac{14.208 \times 10^3}{(0.465 \times 0.5) \times 10^6} \qquad = 0.061$$

$$q_k = \frac{9.338 \times 10^3}{(0.465 \times 0.5) \times 10^6} \qquad = 0.041$$

Case 1

$1.4g_k + 1.6q_k = 1.4 \times 0.061 + 1.6 \times 0.041 \quad = 0.150$

Case 2

$1.4g_k + 1.4w_k = 1.4 \times 0.061 + 1.4 \times 0.0391 \quad = 0.140$

Case 3

$0.9g_k - 1.4w_k = 0.9 \times 0.061 - 1.4 \times 0.0391 \quad = 0.0002$

Case 4

$1.2g_k + 1.2q_k + 1.2w_k$
$\quad = 1.2 \times 0.061 + 1.2 \times 0.041 + 1.2 \times 0.039 = 0.169$

Pier is adequate in compression, but case 3 shows a small tension condition. The pier joints were all fully filled with mortar, and this pier is considered adequate.

Piers D and E were calculated in the same way and found to be adequate.

**Conclusion**

Based on the design stresses it was considered that the front elevation piers were adequate for all loading conditions. The minor cracking was raked out and re-pointed. The first floor was opened up, and 'strong point' bolted connections were made to those walls parallel to the floor joists. The floor joists were packed between the wall and the last joist, and the last two joist spacings were noggined out.

When considering the effects of wind on a building, it is important to ensure that walls that are parallel to floor joists are adequately tied into the floor diaphragm, and joists bearing onto the walls are also positively tied into the wall at regular intervals. These requirements are more important for low-rise buildings, as there is less preload available, and wind suction can have a greater effect on the wall.

Where in-situ or precast floors span onto walls, these provide sufficient restraint to the wall by virtue of their mass. With timber joists or purlin arrangements it may be necessary to supplement the built-in joists with metal straps. These straps should have a turned-down leg in the masonry, and be provided at suitable spacings. Where timber joists are supported on hangers, additional steel strapping is recommended.

## Case history 8.7

This insurance claim concerns a dwelling in Derbyshire that had developed cracking to the front and rear elevations. The building was inspected by the NHBC engineer, and the following structural report was submitted.

### Structural Report on 5 The Beeches, Derbyshire

#### 1.0 Introduction

This report has been prepared on the instructions of the insurance company following the development of cracking in the external walls. The inspection of the main walls was carried out on 5 August 1990, and was a visual inspection only. The walls were checked for plumb and bow using offsets and string lines. The hipped-end roof was also examined via the loft space and was seen to be constructed from prefabricated timber truss rafters.

Following the inspections to No. 5, the external walls to the remainder of the terrace block and other adjacent blocks were given a cursory examination. It was established that all the buildings had been built on similar ground, and standard strip footings had been used.

#### 2.0 Description of the dwelling

No. 5 is on the end of a five-block terrace and is quite exposed on the side of a valley configuration, facing open countryside. The dwelling was built using artificial stone 100 mm thick to the external leaf of a cavity wall, the internal leaf being 3.5 N concrete blockwork. The terrace had a duo-pitched roof with hipped ends, and the first floor construction was of timber joists on hanger shoes. The ground floor construction was 100 mm ground-bearing slab.

Movement joints had been provided in the front and rear walls to cater for contraction movements in the external artificial stone facings.

Internal inspections revealed all partitions to be non-loadbearing paramount board. First floor joists spanned from gable wall to the cross walls and were supported on joist hangers. On No. 5 there was a staircase running parallel to the gable wall, and the floor had been trimmed around the stairs using two joists nailed together.

No alterations had been carried out to the building since it was built in 1981.

#### 3.0 Details of inspection

##### 3.1 Front and rear elevations

Close to the gable end, diagonal cracking was visible on the front and rear return walls. This cracking was through both leaves of the cavity wall and extended from ground level to the underside of the window cills, and the cracks were measured at 1.50 mm. The front and rear walls were found to be in a sound condition.

### 3.1.1 Gable elevation

No vertical or diagonal cracking was evident, but there was a sign of fine horizontal cracking at first floor level with some mortar spalling. In addition there was a significant bow of 10 mm measured in the wall at this height.

### 3.1.2 Foundations

Overall examination of the terrace revealed minor cracking in the artificial stone facing as a result of thermal and contraction movements. The front elevation was south facing and had a movement joint at the centre of the terrace.

No cracking was observed to the brickwork below the dampcourse level. There were no signs of subsidence cracking despite the presence of several large mature trees on the boundary of the development. These trees were checked out using NHBC Chapter 4.2 and found to have no influence.

### 3.1.3 First floor construction

First floor construction was 22 mm chipboard on 220 mm × 50 mm SC3 timber joists on galvanised steel joist hangers. The staircase was trimmed at each end by two 220 mm × 50 mm joists nailed together. Opening up of the floor adjacent to the main staircase trimmer and the wall revealed that the joist shoes had not been properly fixed and were showing a 10 mm gap between the back of the shoe and the wall. Some of the floorboard joints had also opened up, and the ceiling was cracked in similar places.

### 3.1.4 Roof space

The hip end had been formed using a mixture of mono trusses with traditional loose fixings at the corners. All the necessary bracing was in place, and the trusses were all plumb. There was a lack of metal restraint straps into the gable and return walls at eaves level.

### 4.0 Discussion of findings

No foundation settlement was evident. The fine cracking on various parts of the terrace result from thermal and contraction movement, often observed in this type of material, and no further action is warranted as it is considered that they are insignificant.

Measurements along the gable have shown localised bowing outwards, and this has resulted in cracking internally and externally to the front and rear return walls. It is considered that the design and construction of the end terrace unit lacks adequate lateral restraint at first floor level. This has resulted in outward bowing, probably caused by wind suction on the exposed gable wall. The

gable wall has shed the load into the front and rear walls, which has caused the cracking in those elevations. Little buttressing was available from the internal partitions, and the supporting of the floor on joist hangers has negated the usual diaphragm action of the floor.

### 5.0 Conclusions and recommendations

It was clear that the gable end wall lacks proper restraint. This is most noticeable at first floor level, but the restraint at roof level also needs to be supplemented. The outward bowing is expected to become worse, and unless measures are taken to improve the wall's stability more serious damage could result in high winds. It is recommended therefore that the following remedial works are carried out as soon as possible:

1   First floor joist trimmers each side of staircase to be extended onto the inner wall and securely fixed using metal angle cleats at the party wall.
2   A section of the gable wall is to be removed between the staircase trimmers, and this will require the wall

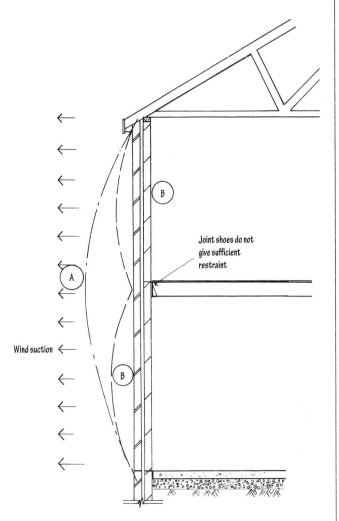

*Fig. 8.18* Deflection profiles due to wind suction on gable. A: wall poorly restrained at first floor level. B: after repairs wall fully restrained at first floor and roof levels.

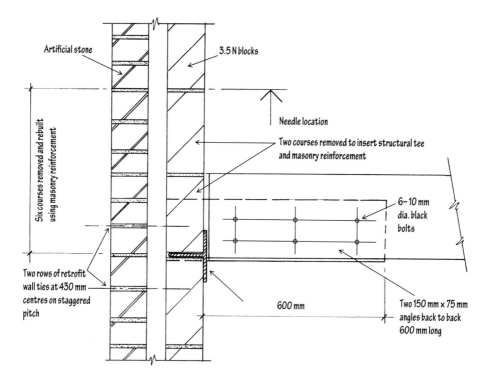

**Fig. 8.19** Gable strengthening at first floor level.

above to be needled and supported down to ground level using Acrow props. These temporary supporting works to be carried out by the builder and details submitted to the engineer for appraisal prior to commencement.

3 Provide for building in a structural tee section and reinforced masonry in the rebuilt section. All new masonry to be slate-pinned up to the existing, and this pinning to be installed from the ends of the masonry.

4 Provide additional galvanised steel restraint straps to roof trusses.

5 Following work on gable wall, provide additional retrofit wall ties from external stone leaf to internal blockwork.

6 Rake out external and internal cracking to a depth of 22 mm and repair using stitch bonding techniques.

7 On completion, re-board first floor ceilings, re-plaster repaired walls and redecorate.

The design of these works should be carried out by a chartered civil or structural engineer.

Figure 8.18 illustrates the suction mode on the gable and Fig. 8.19 details the gable-strengthening works.

## 8.3 Defects related to masonry materials

Serious structural defects can result if the basic components used in the construction of masonry walls are defective or inadequate for the conditions of use.

The basic materials are:

- bricks (made from clay, concrete or calcium silicate);
- concrete blocks (solid or hollow);
- mortar;
- additives;
- dampproof courses;
- wall ties, masonry reinforcement and other metal restraints.

### 8.3.1 Bricks

The properties of clay bricks depend on the types of clay from which they are made. The vast majority of clay bricks used in the UK are manufactured from the Oxford-type clays and the clays in the coal measure strata, i.e. Carboniferous clays. They are manufactured by pressing, extrusion (wire cut process), or hand made in moulds.

Bricks can be manufactured as solid units, perforated or with a frog, and are classified in BS 3921 (1985) in terms of ordinary quality and special quality.

Ordinary quality bricks are generally less durable than special quality bricks but are normally durable enough for use in the external leaf of a cavity wall.

Special quality bricks are defined as those bricks that are durable when built in situations of extreme cold weather and undergoing prolonged wet and freezing cycles of weather. They are used in such locations as garden walls, earth-retaining walls, and parapet walls. Special quality bricks are frost resistant, and there are stricter limits on their soluble salt content, as laid down in BS 3921.

For a brick to be so classified, the brick manufacturer must provide proof that the product has been in service under saturated and freezing conditions in the locations for its intended use for not less than three years, and evidence to show that inspections have confirmed its performance. Alternatively, bricks can be tested in sample panels by the manufacturer in exposed locations, and assessed by an independent testing laboratory over a three-year duration. Where no test data exist, a brick with a crushing strength exceeding 48.50 N/mm² or having a water absorption less than 7% shall be deemed to be frost resistant.

Compression strength is not necessarily an indication of frost resistance, owing to the variable porosity of some bricks and variations in the kiln firing process.

The strength of clay bricks is important, and Table 8.4 lists the compressive strengths for the various classifications.

### Frost resistance and soluble salt content

The two technical ratings for clay bricks relate to frost resistance and soluble salt content. Frost resistance is in three categories: F (high), M (medium) and O (poor). The soluble salts are split into two groups L: (low) and N (normal). Bricks that have poor frost resistance will soon spall, especially if they are used in a situation of high exposure and are subjected to frequent saturation and/or freeze cycles. The choice of brick will therefore depend on:

● the geographical exposure to which the building is to be subjected;
● the location of the brickwork within the building envelope, i.e. is it to be used in a normal external cavity wall or in a parapet wall or retaining wall?

Exposure is classified as sheltered, moderate or severe, and the height and distance that the site is from the sea will have a significant influence on the Driving Rain Index and wind speed factors.

Most of the UK has a moderate exposure rating, but in Wales, Lake District and west coast of Scotland the exposure rating is severe. These areas are shown in Fig. 8.20. The annual driving rain index is expressed in m²/s:

Driving rain index

$$= \frac{\text{Annual rainfall (mm)} \times \text{Wind speed (m/s)}}{1000}$$

Table 8.5 lists the allowable soluble salt contents for bricks. In the severe exposure areas it is advisable to avoid using bricks with an N rating, as special sulphate-resistant mortars may be required to avoid sulphate attack. In persistently wet areas, sulphates react slowly with tricalcium aluminate (a constituent of ordinary Portland cement) and hydraulic lime, and this causes the mortar to expand. Often the soluble salts wash out of the brickwork and crystallise on the face of the wall as a white deposit known as efflorescence. While not harmful, it gives the wall an unsightly appearance, and it can take many years to disappear.

The locations where brickwork is likely to remain saturated for long periods are:

● close to ground level, below dampcourse level and below ground;
● in free-standing walls, single-skin garages, earth-retaining walls, parapets and chimney stacks;
● in wall cappings, chimney cappings, copings and sill locations.

The effects of soluble salts and freezing can therefore be reduced substantially by good brick selection, good choice of mortar and good detailing to copings, flashings and dampproof requirements.

### Calcium silicate bricks

These bricks, generally referred to as sand limes, are classified in BS 187 according to their compressive strength. Their durability is related to the compressive strength of the unit. Calcium silicate bricks with a compressive strength of 14.0 N/mm² (Class 2) or greater have good resistance to frosts and saturation. Class 3 bricks (20.50 N/mm²) have an excellent frost resistance in most locations, but in the exposed regions higher strength bricks are recommended.

Calcium silicate bricks can be affected by sulphates in the soils or groundwater, and they should not be used in situations where they are likely to be saturated by seawater or contaminated by road salts. Filled or natural ground containing sulphates will require special consideration in terms of the brick strength and designated mortar type, and reference should be made to Table 8.2.

Calcium silicate bricks have a high initial drying shrinkage and the manufacturer's advice on recommended joint spacings should be followed.

*Table 8.4* Compressive strengths (N/mm²) of clay bricks

| Type of brick | Strength classification of brick | | | | | | | | |
|---|---|---|---|---|---|---|---|---|---|
| | 15 | 10 | A | B | 7 | 5 | 4 | 3 | 2 | 1 |
| Engineering bricks | | | 69 | 48.5 | | | | | | |
| Loadbearing bricks | 103.5 | 69 | – | – | 48.5 | 34.5 | 27.5 | 20.5 | 14 | 7.0 |

*Table 8.5* Allowable soluble salt contents for clay bricks

| Type of salt | Allowable salt content (%) |
|---|---|
| Sulphates | 0.50 |
| Calcium | 0.30 |
| Magnesium | 0.030 |
| Potassium | 0.030 |
| Sodium | 0.030 |

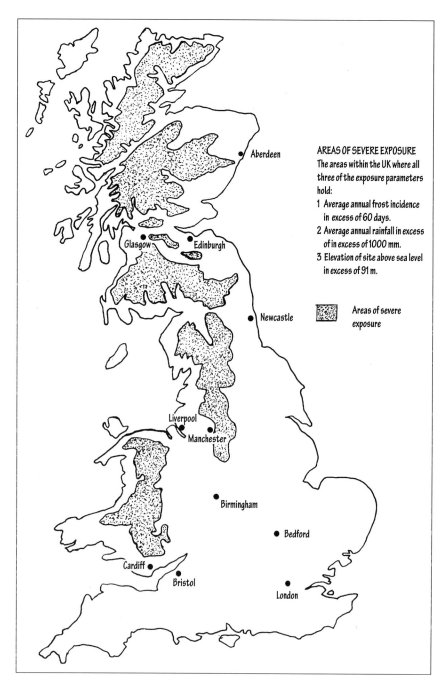

*Fig. 8.20* Areas of severe exposure in the UK.

### 8.3.2 Blocks

The two criteria for block selection are compressive strength and the insulating properties. Most concrete blocks meet the $3.50 \text{ N/mm}^2$ crushing strength required in most two-storey developments built with timber floors. Where concrete floors are specified the block strength required may be in excess of $3.50 \text{ N/mm}^2$, and may need to be checked by calculation. If in doubt use a 7.0 N block for ground to first floor and for blockwork below dpc level.

Clinker-type concrete blocks are very popular for the inside leaf of cavity walls, party walls and partition walls as they provide an excellent key for plaster finishes. Aerated autoclaved lightweight blocks are used mainly for their insulating properties, and can be obtained in a range of thicknesses. They have become very popular for use below ground, where they are known as Trench-blocks.

Quite a few of these blocks are suitable for use in ground conditions that would not be suitable for normal dense blocks. If the soils contain sulphates that would require the use of a Class 2/3 mortar in the proportions of 1:3 ordinary Portland cement:sand then ordinary blocks will not be suitable. If in doubt a check should be made with the block manufacturer's technical department as to the blocks' resistance to sulphates. Where sulphate-resistant cement is required it will be necessary to ensure that the blockwork has similar resistance.

The initial contraction shrinkage of autoclaved blocks is high, and the manufacturers require movement joints at a maximum distance of 6.0 m. If joints are not possible, then bed joint reinforcement should be used, and the mortar strength should be kept to less than the block strength.

In the UK most block manufacturers produce 215 mm high blocks. In Europe the construction industry uses a larger format block with thin joint bedding, which allows a greater wall height to be built. Several manufacturers are looking to promote this practice in the UK as it can show considerable savings in blocks and mortar. In Europe it is common practice, even on small projects, to have a mini-crane for moving materials about the site. It remains to be seen whether this type of blockwork will take off.

Table 8.6 lists the compressive strengths of Type A and Type B concrete blocks. Type A blocks are for general building use, including use below ground. Type B blocks are for general building use including use below ground, but for use below ground their compressive strength should not be less than 7.0 N/mm$^2$ unless the manufacturer can produce authoritative evidence as to the suitability of a lesser block strength.

### 8.3.3 Mortars

Mortar comprises a cement binder, fine sand and water. It can be thought of as a 'gap filling' glue, which allows masonry to be laid on an even bed to ensure uniform load distribution. The ideal mortar must:

- adhere completely and permanently to the masonry;
- remain workable during the laying period;
- harden sufficiently quickly to allow the building-up of the wall;
- resist rain penetration;
- resist the actions of frost and chemical attack;
- be able to accommodate the bedding-down of the structure as the ground settles.

A mortar must have sufficient strength when cured to support the masonry without cracking. It should, however, always be weaker than the bricks or blocks used, or cracking will result as the mortar dries out. Signs of over-strong mortar are recognised by the crack

pattern, which is usually vertical and passes through both the brick and the mortar.

Table 8.7 lists a selection of mortar groups for various wall constructions and materials. The following British Standards relate to the constituents in mortar:

- Ordinary Portland cement: BS 12 : 1996
- Sulphate-resistant cement: BS 4027 : 1996
- Lime (non-hydraulic and semi-hydraulic): BS 890 : 1995
- Sand: BS 1200 : 1976
- Plasticisers: BS 4887 : 1986
- Colouring agents: BS1014 : 1975

#### Ordinary Portland cement

This cement is used in most masonry construction in accordance with the recommended cement:sand ratio. The strength requirements of BS 12 are usually exceeded by a considerable margin, but cements from different plants have varying characteristics. While some plants produce cements that give high early strengths, these strengths do not necessarily become high strengths at later stages. Cement from other plants may be slow in gaining strength, but may reach a high ultimate strength within the required time scale.

Mortars made from ordinary Portland cement are susceptible to attack by sulphates and acids present in the bricks, soils and groundwater.

#### Sulphate-resistant cement

This cement is better able to withstand sulphate attack than ordinary Portland cement, but its mineral properties are not as good as those of high-alumina cement or super-sulphated cement, and it is also not resistant to attack by certain acids in the ground. It is used in the same proportions as ordinary Portland cement, and its ultimate strength is the same. Early hardening may be slightly slower than that of ordinary Portland cement.

Table 8.6 Minimum compressive strengths (N/mm$^2$) of concrete blocks

| Block type | Average of 10 blocks | Lowest individual block |
|---|---|---|
| A – 3.50 | 3.50 | 2.80 |
| A – 7.0 | 7.0 | 5.60 |
| A – 10.50 | 10.50 | 8.40 |
| A – 14.0 | 14.0 | 11.20 |
| A – 21.0 | 21.0 | 16.80 |
| A – 28.0 | 28.0 | 22.40 |
| A – 35.0 | 35.0 | 28.0 |
| B – 2.80 | 2.80 | 2.25 |
| B – 7.0 | 7.0 | 5.60 |

Table 8.7 Selection of mortar groups

| Type of masonry | Clay bricks | Concrete blocks, calcium silicate bricks |
|---|---|---|
| Internal walls, partitions | 5 | 5 |
| Internal leaf of external cavity walls | 5 | 5 |
| Backing to external solid walls above dpc | 4 | 4 |
| Backing to external solid walls below dpc | 3 | 3[a] |
| External cavity walls | | |
| outer leaf above dpc | 4 | 4 |
| outer leaf below dpc | 3[a] | 3[a] |
| Parapet walls, domestic chimneys | 3 | 3 |
| External free-standing walls | 3 | 3 |
| Sills, copings | 1 | 2 |
| Earth-retaining walls: well-drained fill | 1 | 2[a] |

[a] If sulphates are present in the soils or groundwater use mortar made with sulphate-resistant cement

Calcium chloride should never be used with this type of cement.

The resistance to sulphate attack provided by this cement is dependent on the production of a rich, well-mixed mortar.

## Masonry cement

These cements consist of ordinary Portland cement to which a proportion of ground chalk, limestone or similar filler has been added, together with an air-entraining

Table 8.8 Materials for dampproof courses

| Material and group | BS | Minimum weight (kg/m²) | Minimum thickness (mm) | Liability to extrusion | Durability | Other considerations |
|---|---|---|---|---|---|---|
| Group A: flexible | | | | | | |
| Lead | 1178 | 19 | 1.70 | Not in normal construction | Corrodes in contact with fresh lime or OP cement mortar | Apply thick coat of bitumen paint to mortar face and both lead surfaces |
| Copper | 1569 | 2.30 | 0.25 | Not in normal construction | Highly resistant to corrosion | Copper may stain external surfaces, especially stone |
| Bitumen, hessian base | 743 type A | 3.80 | | All these types are likely to extrude under heavy pressure, but this is unlikely to affect the water resistance | Decay of the hessian or felt may occur, but this does not affect the efficiency of the dpc if the bitumen is undisturbed | Materials should be unrolled with care. In cold weather the rolls should be warmed before use |
| Bitumen, fibre base | 743 type B | 3.26 | | | | |
| Bitumen, asbestos base | 743 type C | 3.80 | | | | |
| Bitumen, hessian base and lead | 743 type D | 4.34 | | | | |
| Bitumen, fibre base and lead | 743 type E | 4.34 | | | Types D, E and F are the most suitable for severe conditions | |
| Bitumen, asbestos base and lead | 743 type F | | | | | |
| Bitumen, sheet with hessian base | 743 type G | 5.43 | | | | |
| Polythene. Black, low density in the range of 0.915–0.925 g/ml | No BS | 0.475 | 0.457 | Not in normal construction | No evidence of breakdown in contact with other building materials | Thickness is that of a single sheet. Layers of thinner material are not recommended |
| Pitch polymer | No BS | 1.50 | 1.270 | As for polythene | As for polythene | |
| Group B semi-rigid type | | | | | | |
| Mastic Asphalt | 1097 1418 | | | Liable to extrude under pressure above 648 kN/m² | No deterioration | To provide key for mortar below next course of brickwork, grit should be beaten into asphalt immediately after laying and left proud of the surface |
| Brick | 3921 Maximum water absorption to be 4.50% | | | Not extruded | No deterioration | Brick dpc should consist of two courses of bricks laid to break joints bedded in 1:3 OPC mortar |
| Slate | BS 3798 | | Each slate to be at least 225 mm long and 4 mm thick, and be in accordance with tests in BS 3798 | Not extruded | No deterioration | Slate dpc should consist of two courses of slate laid to break joints. Bedded in 1:3 OPC mortar |

*Table 8.9* Selection of types of wall ties

| Least leaf thickness (mm) | Nominal cavity width (mm) | Permissible type of tie | | |
|---|---|---|---|---|
| | | Shape name | Type no. DD 140 Part 2 | Tie length (mm) |
| 90 | 75 mm or less | Butterfly, double triangle or vertical twist | 1, 2, 3 or 4 | 200 |
| 90 | 76–90 | Double triangle or vertical twist | 1, 2, 3 or 4 | 225 |
| 90 | 91–100 | Double triangle or vertical twist | 1, 2, 3 or 4 | 225 |
| 90 | 101 to 125 | Vertical twist | 1, 2, 3 or 4 | 250 |
| 90 | 126–150 | Vertical twist | 1, 2, 3 or 4 | 275 |

The strength and stiffness of masonry ties to DD 140 Part 2 ranges from the stiffest Type 1 to the least stiff Type 4
From BS 1243 the vertical twist type is the stiffest and the butterfly type is the least stiff

agent. This produces a mortar that is more plastic than that produced by ordinary Portland cement mortar. However, the strength of masonry cement mortars is substantially lower than that obtained from ordinary Portland cement mortars. This is due to the air content and to the dilution of the cement by the addition of fillers.

### Sand

Sand for mortars is often referred to as 'soft' sand, as opposed to sharp sand used for making concrete. The most important criterion for a building sand in terms of quality is its cleanliness. It is recommended that building sand should not have a silt content in excess of 7%. Poor-quality sand, containing deleterious matter, should never be used. Sand with a high proportion of fine grades has a greater surface area, and this has the effect of diluting the cement coating on the sand. This can result in a mortar that has less durability to the action of frosts and sulphate attack. Such mortars have very poor durability in coastal areas, where the driving rain index and exposure are higher than normal.

### Lime

This is used with ordinary Portland cement to improve the workability of the mortar, but in doing this there is a reduction in strength. In mortars the lime is usually non-hydraulic or semi-hydraulic. The hydraulicity of a lime denotes its capacity to harden off, and develop its full strength in a wet state. Lime is generally delivered to site in the dry hydrated form, often already mixed in with sand to the required proportions. This is a better method than site mixing as there is more control over the sand bulking factor. In addition the lime becomes more plastic when soaked overnight. BS 4721 *Specification for ready-mixed lime : sand mortars* gives the mix proportions for use when using on-site batching by volume.

### 8.3.4 Dampproof courses

These are used to prevent the upward or downward movement of moisture within a masonry wall. BS 743

specifies the quality and properties of the various types of dampcourse materials, and these are listed in Table 8.8.

Dampcourse materials such as slate, engineering (blue) bricks and some metals (with the exception of lead) do not have any problems related to load stressing, whereas the bitumastic and asphaltic types may give rise to problems of creep when they are subjected to exceptionally high stresses. BRE Digest 77 provides a useful reference for the applications of various dampproof course materials.

For situations where some flexural tension is desirable, such as free-standing walls or retaining walls, the best dampproof course is provided by three courses of engineering bricks in a good mortar. These bricks have a low water absorption, and very high compressive strengths of 48.50 $N/mm^2$ (Class B) and 69 $N/mm^2$ (Class A).

### 8.3.5 Wall ties

Wall ties are used to tie together unbonded leaves of masonry. In cavity wall construction the wall ties can be the galvanised butterfly type, galvanised fish tail type or stainless steel spiral type, and these should comply with BS 1243.

The standard spacing should be reduced to 750 mm horizontally when the cavity width exceeds 75 mm, and the ties should be the solid type.

With modern construction and the use of insulated cavities, stainless steel ties of the spiral type have become very popular. However, the butterfly tie is better when differential movements between the external and internal leaves are expected, as it is more flexible than the solid twist type.

Table 8.9 provides a selection of wall ties.

## 8.4 Workmanship

Masonry construction has been a long-standing traditional building method, employing craftsmen. For years these craftsmen built many buildings that had walls of generous proportions, based on empirical rules, which resulted in their having a greater degree of robustness. Buildings erected after 1950, following the introduction of structural codes of practice, now have walls that are thinner, and which have been calculated. These designed walls have had an allowance made for the degree of control and the standard of workmanship.

This workmanship factor for masonry construction is certainly higher than the factors of safety applied to reinforced concrete and steel-framed buildings.

Where reinforced masonry is used, it is essential that a greater degree of supervision is allowed for, if failures are to be prevented.

Workmanship defects in masonry can be avoided by good supervision. A good designer specifies the right mortar for strong, durable, rain-resistant and crack-free masonry, but they are reliant on the skills of the craftsman in achieving the required end product. The following items of bad workmanship can result in serious structural defects in masonry construction.

- **Mortars with inadequate ratios of cement:** The mortar mix proportions must be such that adequate strengths and durability are achieved for the particular type of masonry. This is very important in geographical locations with a high driving rain index.

  In addition the mortar mix must have adequate workability such that good adhesion to the bricks is achieved.

- **Sulphate-resistant cement mortars:** Sulphate attack on mortars is fortunately not very common, and takes a long time to develop. It can be avoided by good brick selection and the use of the correct type of cement and the correct mortar mix. Where it is obvious to the architect or designer that masonry will be in contact with sulphate-bearing filled ground, or it is in an area with a high driving rain index, they should specify the use of sulphate-resistant cement in the mortar.

- **Failure to build walls plumb:** If a wall is built out of plumb, the result can be that it is unable to carry its full potential load. It is therefore important to check the verticality of brickwork at frequent intervals. Blockwork goes up quicker than brickwork, causing fresh, soft, bed joints to squeeze out. As a result, blockwork tends to be built out of plumb more readily than brickwork.

- **Workability admixtures:** Avoid using unauthorised admixtures to improve the workability of a mortar. The use of calcium chloride in a mix is prohibited, as corrosion of embedded metals such as wall ties, masonry reinforcement can result.

- **Using incorrect ties and inadequate tie embedment:** Wall ties allow the two slender skins of a cavity wall to support each other, and hence create a stiffer wall than one in which each skin is independent. The minimum numbers of ties per square metre are 2.50 for 900 mm horizontal and 450 mm vertical spacing, and 3.0 for 750 mm horizontal and 450 mm vertical spacing for cavity widths exceeding 75 mm. All ties should be bedded in at least 50 mm into each leaf for strength. Position the ties such that the drip points downwards, and always ensure that the ties are level or slope to the outside face.

Do not bend ties to suit the coursing. This is common when using a natural stone external leaf, where the bed joints are rarely in line.

- **Wall ties:** Failure to provide an adequate number of ties, especially at window and door reveals. Ensure that ties are spaced at every block course or every fourth brick course if using bricks for both leaves at window and door reveals and at movement joint locations.

- **Laying bricks frog down:** Bricks walled frog up produce a stronger wall than when used frog down. This is because bricks laid frog down may not have a full bed of mortar below them. This is most important when heavy loads are to be carried or where piers are small.

- **Furrowing:** Furrowing is used by some bricklayers to spread the mortar, and can often lead to a bed joint not being fully filled. The practice of deep 'furrowing' should be avoided, as the strength of the wall can be substantially reduced. Light furrowing is permissible.

- **Excessively thick bed joints:** Using bed joints of excessive thickness, i.e. in excess of 10 mm, will produce an unsightly wall. It also makes the wall weaker, as the weight of the wall above will cause spreading of the mortar.

- **Protecting finished work:** Failure to protect finished work from the elements, especially frost, can lead to mortar being damaged and walls having to be rebuilt.

  In addition to protecting the finished work, the bricks and blocks in the stacks should also be covered.

  Bricklaying should stop when the air temperature falls to 2 °C, and should only resume when the air temperature has reached 1 °C and is still rising. The sands used in the mortar should be kept covered in cold weather, and the mixing areas should be protected accordingly from the elements.

  In warm weather, mortar should be laid in shorter lengths to limit the loss of moisture before the bricks are bedded. Bricks that are highly porous may need to be wetted up to improve the adhesion of the mortar. Care must be taken to avoid over-wetting, as this may lead to the brick 'floating' on the mortar bed.

- **'Knocking up':** This is the practice of adding water to a batch of mortar that has started its initial set. This is bad practice, as a weak mortar will result. Mortar should be used within 2 hours unless it is a special ready-mixed mortar that contains retarding agents.

- **Over-strong mortar:** Using over-strong mortar results in split bricks and excessive contraction shrinkage.

- **Recessed pointing:** The use of recessed pointing in locations that have a high driving rain index can lead to a reduced resistance to water penetration.

- **Poor lintel bearings:** The use of soft packing under lintels and not giving lintels a minimum bearing of 150 mm. Lintels should be bedded onto a full block, and for large-span openings suitable padstones should be used.

- **Lintels:** Failure to build in concrete lintels the right way up can result in cracked masonry above openings.
- **Inadequate bonding:** Not bonding in party walls, block partition walls etc. will result in a weaker construction with less resistance to wind forces.
- **Beam fill:** Failure to beam fill between floor joists will prevent the diaphragm action of the floor being effective, and prevent wind forces being transferred into the foundations.
- **Piers:** Failure to bond piers into the main body of the wall will produce a weaker wall.
- **Mixing blocks and bricks:** Mixing of aerated blocks and clay bricks in the same wall should be avoided otherwise cracking could result.
- **Using incorrect bricks in high-exposure locations:** Table 8.10, based on BS 3921, gives the correct brick classification according to their frost resistance and soluble salt content.

*Table 8.10* Durability classification of bricks

| Durability | Frost resistance | Soluble salt content |
|---|---|---|
| FL | Frost resistant (F) Durable in all building situations | Low (L) |
| FN | | Normal (N); no limit on soluble salt content |
| ML | Moderately frost resistant (M) Durable except when saturated and subject to repeated freezing and thawing | Low (L) |
| MN | | Normal (N); no limit on soluble salt content |
| OL | Not frost resistant (O) Liable to be damaged by freezing and thawing | Low (L) |
| ON | | Normal (N); no limit on soluble salt content |

## Points to remember

- Only use bricks that comply with BS 187.
- Store all bricks, blocks and cement under dry cover.
- Do not wet bricks prior to bedding.
- Use a mortar of the correct mix, appropriate for the strength of the masonry being used. Avoid mortar mixes that are too rich in cement.
- Always protect part-finished work from the weather.
- Ensure that movement joints are provided where walls exceed 7.50 m long when using blocks, sand lime bricks or artificial stone.
- Ensure that movement joints are provided if walls exceed 12.0 m long when using clay bricks.
- Materials affected by snow, ice or frost must not be used.
- Do not lay bricks when the air temperature is 2 °C and falling.
- At movement joints always provide additional ties on both sides of joint at 225 mm centres.

- Ensure that blocks and bricks have cured properly. Check that the blocks have been stored for at least 4 weeks prior to delivery to minimise the risk of cracking.
- Ensure that vertical chases in walls are restricted to one third of the wall thickness and horizontal chases to one sixth of the wall thickness.
- All wall ties should comply with BS 1243 : 1978.
- Ensure that wall ties have a 50 mm embedment in each leaf.
- Where cavities exceed 75 mm do not use butterfly ties. Solid vertical twist ties at 750 mm horizontal centres are required.
- Always lay bricks with a frog with the frog up.
- Always use special quality bricks for copings, free-standing walls and in areas where the driving rain index is high.

## References

*Building Research Establishment*

*Dampcourses*, BRE Digest 77 (1966).

*British Standards Institution*

BS 12 : 1996 *Specification for Portland cement.*
BS 187 : 1978 *Specification for calcium silicate (sandlime and flintlime) bricks.*
BS 743 : 1970 *Specification for materials for damp-proof courses.*
BS 890 : 1995 *Specification for building limes.*

BS 1014 : 1975 *Specification for pigments for Portland cement and Portland cement products.*
BS 1200 : 1976 *Specification for building sands from natural sources.*
BS 1243 : 1978 *Specification for metal ties for cavity wall construction.*
BS 3921 : 1985 *Specification for clay bricks.*
BS 4027 : 1996 *Specification for sulphate-resisting Portland cement.*
BS 4721 : 1981 *Specification for ready-mixed building mortars.*
BS 4887 *Mortar admixtures.*
  Part 1 : 1986 *Specification for air-entraining (plasticizing) admixtures.*

BS 5628 *Code of practice for use of masonry*
   Part 1:1992 *Structural use of unreinforced masonry*
   Part 3:1985 *Materials and components, design and workmanship.*
BS 5950 *Structural use of steelwork in building*
   Part 1:1990 *Code of practice for design in simple and continuous construction: hot rolled sections.*

BS 6073:1981 *Specification for precast concrete masonry units.*

W. Curtin, J.K. Beck, G. Shaw and W.A. Bray, *Structural masonry designer's manual* (1982).

# Chapter 9
# Foundations

## 9.1 Introduction

Defects that occur to a building because of foundation failure are caused by settlement or subsidence. On rare occasions they can be due to both. There is a distinction between the two causations, and therefore it will be useful if a definition is given for each.

**Settlement** is defined as movement in the structure caused by the weight of the building compressing the ground upon which the building stands. All buildings undergo some settlement, and the designer must provide a foundation that will ensure that the settlement is within an acceptable magnitude.

Excessive or differential settlement generally results in structural damage to buildings, especially masonry buildings. For example:

- foundations built on soft alluvium or peat beds that vary in thickness;
- foundations built on fill that has poor bearing capacity or has a potential for collapse settlement;
- foundations built over varied ground conditions.

**Subsidence** occurs when the foundation of a building moves down or up owing to the loss of support of the strata below the foundation or volumetric changes in the strata. Examples are:

- wash-out of soils caused by leaking drains;
- variations in groundwater due to leaking drains or a fluctuating water table;
- softening of clay strata due to water from leaking drains;
- landslip;
- mineral extraction, salt brine extraction etc.;
- clay heave or shrinkage due to rehydration of clay soils or frost action, which causes the clays to expand.

Any expert called in to advise on the cause of a serious structural defect must be able to identify whether the damage results from subsidence or settlement. When this has been identified, the owners of the property are then able to appoint a loss adjuster to pursue their building insurers or any warranty providers, or any other party who may have caused structural damage when carrying out adjacent works. Such an expert should have a detailed knowledge of building construction, damage assessment, soil mechanics and preferably some knowledge of how to remediate the defective building. He or she should also be able to write a clear and concise report on the observed defects, give a detailed account of the condition of the building, and advise on the need for any further investigation.

## 9.2 Causes of foundation movement to buildings

Section 9.3 will deal with the damage of buildings as a result of settlement of the bearing strata. This will cover a range of problems, including differential settlement, collapse settlements of filled ground, excessive settlements due to weak soils below foundations, and undersized foundations, which can result in bearing capacities greater than that which the soils can tolerate, leading to shear failure of the ground.

Section 9.4 will deal with damage caused by subsidence movements such as that from collapsed mine workings, erosion of granular soils by water action, subsidence caused by the removal of water from peat bands below a foundation, frost heave, chemical reactions in the ground, and problems due to the presence of trees in close proximity to buildings on clay soils.

## 9.3 Structural damage caused by settlement of the bearing strata

Settlement takes place as the weight of the building and its contents is transmitted downwards onto the ground via the walls and foundations. All buildings undergo an initial settlement, and provided the bearing capacity of the ground is not exceeded the settlements should be of a low order, i.e. up to 10–15 mm.

Clays have an initial settlement, but the total settlements take many years before they reach their maximum. This long-term settlement takes place as the porewaters in the clay are squeezed out under load until there is an equilibrium in the soil mass. With sands and gravel the total settlement will occur very quickly, and is usually complete by the time the roof goes on. It is therefore important to recognise this difference, because when a building is constructed on ground that varies from a clay stratum to a granular one, provision for the different settlement timescale must be allowed for. In such a situation total settlement would be the criterion, not bearing capacity, and the foundation should be reinforced along its length such that the variations of stress in the ground are accommodated as the foundation spans across the weaker strata onto the firmer strata without exceeding the allowable settlements of the firm strata.

### 9.3.1 Stratum variation

Where strip footing excavations reveal a variation in the soils at formation depth then the footing width should be based on the bearing capacity of the weaker soil. The foundation should also be reinforced along its length with light mesh in the bottom. Alternatively it may be possible to extend the excavation down deeper to keep the footing in the same firmer stratum. Before doing this, it is always prudent to excavate a trial pit off the building to confirm that the good stratum is present at an economic depth.

### 9.3.2 Soft spots

Where soft spots are encountered in a footing excavation, and trial holes have confirmed that there are good strata below at an economic depth, then the soft spot can be dug out locally and replaced with a GEN 3 mass concrete.

If the soft spot is found to be too deep to dig out then it may be more practical to reinforce the foundation to span over the weaker ground. When this method is used the foundation at each end of the reinforced beam section will need to be checked for the additional loading, and may need to be widened out at each end.

## Case history 9.1

### Introduction

An insurance claim was lodged by the owners of a bungalow following the development of cracking in the front and rear walls. The dwelling was situated on the outskirts of Northallerton, North Yorkshire, and faced onto open fields at the rear. The house was approximately 9 years old, and was built in traditional brick facings and concrete tiled roof.

### Investigation

An inspection of the house was carried out, and from the pattern of the cracks it was clear that some rotational movement had resulted along the gable wall. This movement had caused the gable foundation to rotate and transfer tensile stresses into the front and rear return walls.

A water level survey along the dampcourse showed that the gable brickwork was 16 mm lower than the same brick coursing 4 m from the gable end. The cracking on the rear wall was approximately 3 m from the gable wall, and was visible above and below the window. The cracks were wider at the top, confirming that some rotation was taking place.

The difference in level, while significant, could have been 'built in', and it was considered worthwhile to carry out a check on the plumb of the gable wall. This was carried out using heavy weights on a string line and measuring the offset. The gable was found to be 15 mm out of plumb from the dampcourse to eaves level.

The next step was to examine the foundations at the rear gable corner, and the hand-dug trial pit revealed that the gable foundation was on a firm clay but the footing was not concentric with the wall. The scarcement on the outside was measured at 55 mm, and the footing was 225 mm thick at a depth of 750 mm below footpath level.

This pit was used to progress the investigation further using a hand auger, and this revealed soft wet silty clays at a depth of 950 mm below footpath level. Shear vane readings in the soft clays recorded shear strengths of 15 kN/m$^2$, giving an equivalent bearing pressure of 30 kN/m$^2$. It was not possible to determine the total depth of this soft clay layer, owing to stones in the material, which jammed the auger at a depth of 1.80 m below footpath level.

A further trial hole excavated 4 m from the gable wall revealed good firm clay strata, and these were proved by hand augering for a distance of 1.50 m below the base of the footing.

### Conclusions and recommendations

The trial pits excavated revealed that the gable wall foundation was situated over a wet silty soft clay layer, and further long-term movement could not be ruled out. The bulb of pressure from the foundation clearly extended into this soft silty clay, and the perched groundwater was likely to fluctuate with climatic variations. To stabilise the foundations along the gable wall it was recommended that the gable wall foundations and return walls be underpinned using a mini-piles and in-situ needle beams system. The needle beams would be at close centres through the masonry above the foundation, and would be cast in situ with the inside and outside ring beams to form a ladder system of support. Figure 9.1 shows the beam and pile arrangement. To assist in the piling tender and to confirm the ground conditions at

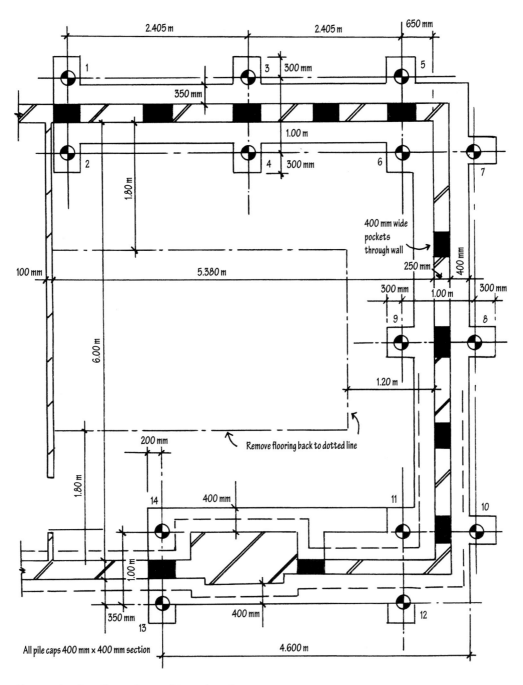

*Fig. 9.1* Plan of lounge showing piling and ground beam layout.

depth a single borehole was drilled, and the log of this borehole is shown in Fig. 9.2.

The piles used were 150 mm diameter, of thin gauge steel, and the ends of the piles were crimped into a cruciform. The piles were bottom driven using a Grundemat hammer onto a plug of gravel in the base of the pile, and the piles were driven to virtual refusal in accordance with the machine's driving characteristics. Once the piles had been driven to the required depths, the piles were concreted using a high-strength concrete grout placed in tremmie fashion from the base of the pile, withdrawing the tremmie pipe slowly as the pile filled up. The cross-needles and longitudinal ground beams were then constructed in one pour using RC 35 concrete. The

minimum pile length was 4.30 m and the maximum pile length was 7.50 m.

During the works it was necessary to rebuild sections of the external brickwork and realign the rear window frame. Part of the ground floor to the lounge was also taken up and the floor joists were relevelled after the sleeper walls had been rebuilt.

Internal cracking around the chimney breast and to internal corridor walls was repaired by raking out and stitch bonding across the cracks using stainless steel masonry reinforcement. The house was inspected by the author 12 months later; all the new brickwork had weathered in well, and the remedial works had clearly been successful.

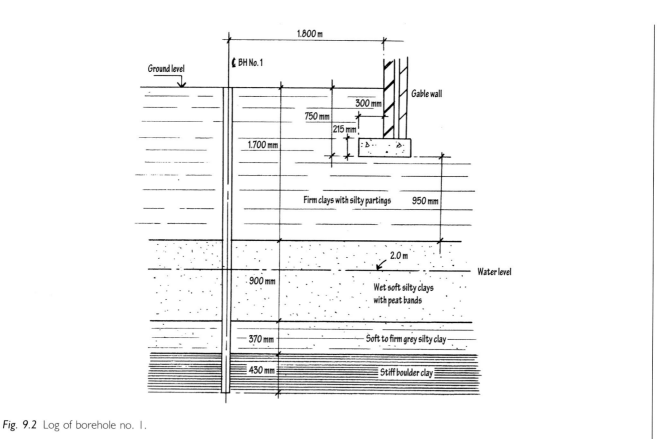

*Fig. 9.2* Log of borehole no. 1.

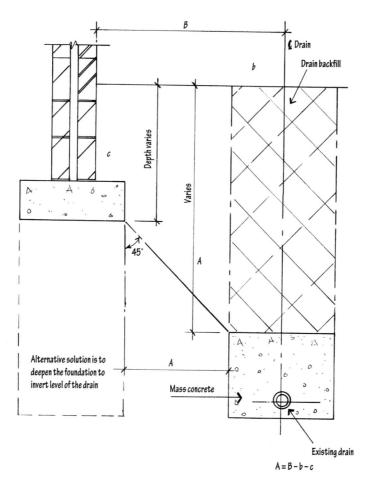

*Fig. 9.3* Deep drains parallel to a foundation.

### 9.3.3 Lateral displacement

It is a requirement of the Approved Documents that when drains are excavated close to a foundation then the foundation should be deepened to the drain invert, or the drain trench must be filled with concrete up to a depth at which a 45° line of repose from the footing edge will pass into concrete (Fig. 9.3).

Failure to observe this requirement often results in structural damage later as the soils under the foundation are subjected to a shear-slip type of failure. On sites where the ground conditions are non-cohesive, or where there is a high water table, such deep drainage should always be constructed before the house foundations.

This is the reason why, on sloping sites, it is preferable to construct houses from the lower level, working up the slope. This prevents previously constructed footings from being undermined.

## Case history 9.2

### Introduction

A claim was made against the warranty provider for structural damage to a bungalow in Sheffield, South Yorkshire. In the initial survey inspection it became clear that the total claim would have to be dealt with by both the warranty insurance provider and the house owner's insurance company. Following construction of the house, the owner had built an extension at the rear of the bungalow, which was suffering damage as a result of the effects of a large oak tree in an adjoining garden. In addition, the damage caused along the gable and front walls was thought to be the result of leaking drains, which had been damaged by tree root invasion. With the full cooperation of the loss adjusters the claim was investigated and remediated to the owners satisfaction. Figures 9.4 and 9.5 show the plan and elevations of the bungalow and external service runs.

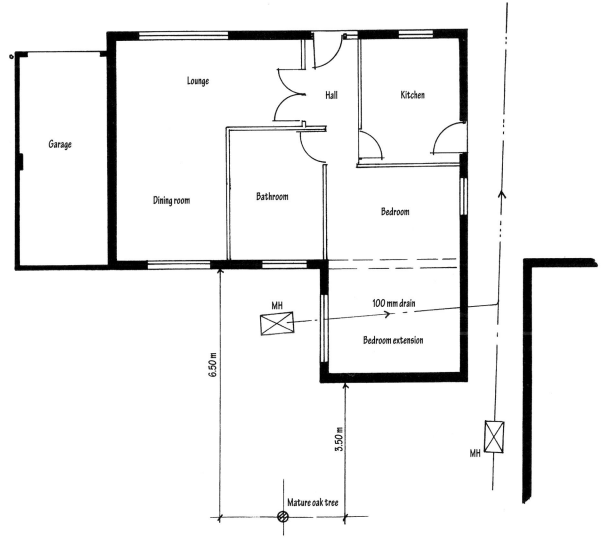

*Fig. 9.4* Location plan.

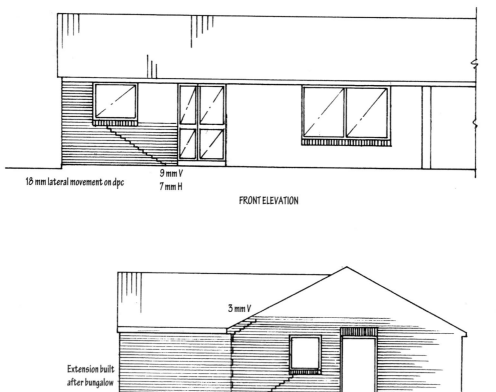

**Fig. 9.5** Front and gable elevations.

## Investigation

An external examination revealed that on the front, gable and rear elevations, cracking was evident in the substructure brickwork; and in some locations this extended up into the brickwork above the dampcourse level. On the front elevation there was cracking up to 9 mm at dpc level. On the gable elevation, there was cracking of approximately 9 mm at dpc level, and evidence that the substructure brickwork had rotated outwards slightly, showing a lipping of 18 mm on the dpc. At the right-hand side of the kitchen door the brickwork was out of plumb for about 1.0 m in height. On the rear elevation, hairline cracking was evident below the bedroom window, and there were 5 mm gaps between the rear wall fascia board and the extension fascia board.

Internally, the property was dry-lined. Within bedroom 1, the dry-lining at the junction to the original bungalow wall had a 15 mm bulge evident, with a 2 mm crack below ceiling level. Gaps 8 mm wide were noted between the gable wall skirting board and the kitchen entrance door frame sill.

Externally it was noted that a 100 mm foul drain ran parallel with the south-east gable wall at an approximate depth of 1.80 m. Two small *leylandii* trees in the adjacent garden were sited within 1.50 m of this drain.

At the rear of the property, a large mature oak tree was sited 4.0 m from the rear wall to the extension and 6.50 m from the rear wall of the bungalow. It was established that this tree was the subject of a tree preservation order.

A trial pit excavated along the south-east gable wall (Fig. 9.6) revealed the foundations to be at a depth of 900 mm below dampcourse level onto a firm yellow mottled clay. Tree roots were evident below the foundation, and there was a small gap below the foundation. The existing foul drain was 1100 mm from the wall, and a steel probe inserted into the drain trench revealed that the drain trench had no concrete backfill. A sample of the natural clay was taken for testing.

## Conclusions and recommendations

It was clear that the settlement of the bungalow gable wall and movement on the rear extension resulted from the combination of a badly filled drainage excavation with clay shrinkage due to the oak tree following a very dry summer. The two clay samples showed the plasticity index of the upper clays to be in the medium plasticity range, and that of the lower mudstones to be in the low plasticity range. Examination of the Atteberg limits indicated that the upper clays were desiccated.

To check on the potential influence of the oak tree, guidance was sought from the NHBC Chapter 4.2, and it

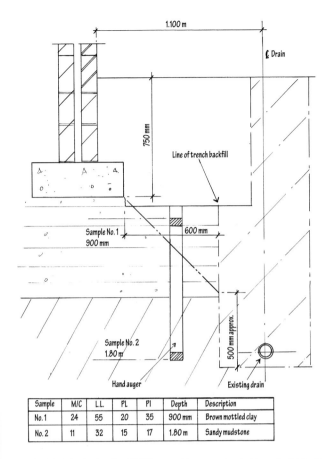

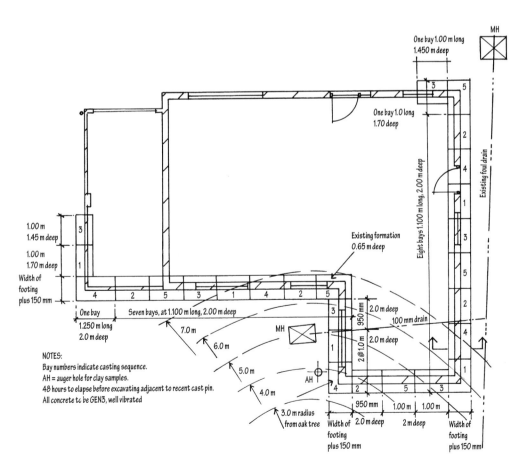

| Sample | M/C | L.L. | PL | PI | Depth | Description |
|---|---|---|---|---|---|---|
| No. 1 | 24 | 55 | 20 | 35 | 900 mm | Brown mottled clay |
| No. 2 | 11 | 32 | 15 | 17 | 1.80 m | Sandy mudstone |

**Fig. 9.6** Trial pit details. M/C = moisture content; LL = liquid limit; PL = plastic limit; PI = plasticity index.

was seen that the influence affected the whole of the rear elevation. The foundations to the extension clearly were not adequate at 650 mm depth below ground level and would require to be underpinned.

The gable wall foundation would also have to be underpinned down to a depth in accordance with Chapter 4.2, or the drain invert, whichever was the deeper. The rear wall and gable returns would also require deep underpinning.

Discussions were held with the local authority tree conservation officer, and it was agreed that removal of the tree could create more problems. It was agreed that the tree could be pruned back, subject to an arboriculturist's report and recommendations. It was therefore decided to base the underpinning depths on the low plasticity index and provide the maximum depth, based on the extension, across the whole rear elevation to act as a barrier wall. The deep underpinning along the gable would extend around the front corner beyond the influence line of the drain excavation.

Figures 9.7 and 9.8 show the extent of the works carried out. The underpinning was carried out in a sequential operation, with existing drainage sleeved at the dwelling exit points. The base of the working space was covered with a 100 mm concrete slab to prevent water from softening the mudstone formation.

Following the underpinning the cracked masonry was repaired, and the floor to the extension was built up with a levelling screed to its original level.

**Fig. 9.7** Plan showing sequence of underpinning.

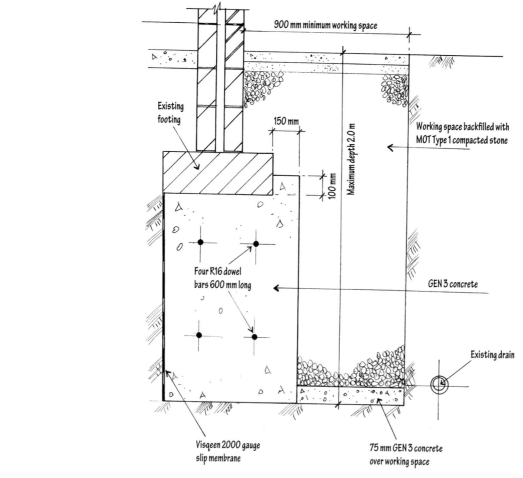

*Fig. 9.8* Typical section showing underpinning details.

### 9.3.4 Collapse settlement of fill

Any building constructed on filled ground may be at risk from differential settlement because of the manner in which the fills were deposited. Fills that are placed to an engineered specification and to a uniform thickness are unlikely to experience differential settlement. If the fills are placed without supervision, inadequately compacted and poorly graded, then there is a much greater risk of settlement due to:

- the weight of the building;
- long-term self-weight consolidation;
- collapse settlement of the fills.

It is therefore clear that when deciding to build on thick deposits of fill the risks cannot be fully quantified. Obviously the more money spent on site investigation, the less will be the risk, as more of the unknown factors will become evident.

Alternatively a type of foundation could be adopted that carries less risk of failure, such as piles or deep trench fill onto natural strata.

If investigating structural damage to a building that has been constructed on fill using strip footings or raft foundations, it is likely that collapse settlement of the fills has taken place. The author has dealt with many sites where the engineer acting on behalf of the builder considers that old fills will adequately support wide reinforced footings without excessive settlement. Often the site investigation has produced an excellent range of standard penetration test (SPT) results, and the fills have been described as being in a medium dense state of compaction. The engineer has failed to appreciate that SPTs are a very crude measurement of fill densities, and should always be treated with caution when used in strata other than a well-graded sand. Reliance on such testing in mixed fills containing hard material is, quite frankly, a risky strategy.

Regretfully, the author has dealt with such sites where reinforced footings have been adopted without any ground improvement and which have subsequently become major damage claims. Such claims are very difficult to investigate, and can also be difficult to remediate owing to the type of fill material present. Collapse settlement of fill is not time dependent, and can occur many years after filling has taken place.

## Case history 9.3

### Introduction

A two-storey house developed cracking in the main walls 8 years after it was constructed. The small development was built 4 miles west of Scarborough, North Yorkshire, and the major structural damage claim was being dealt with by the warranty provider. The builder had ceased trading, so no records were available. Enquiries at the local building control office revealed that wide reinforced strip footings placed on the fill had been agreed to by the local building inspector.

### Investigation

A detailed examination of the property was carried out to assess the scale of damage. The house had been built using traditional facing bricks and lightweight insulating blocks with a timber joist ground floor and a trussed rafter roof supporting interlocking concrete tiles. There was severe cracking to the ground floor ceilings and to several internal blockwork walls.

In addition to the cracking, externally, on the front and gable elevations, it was found that severe damage had taken place internally to the loadbearing walls supporting the staircase. This had caused a serious bulge in the wall below the mid landing. Figure 9.9 shows the ground floor layout and trial pit and borehole locations. A trial pit was excavated at the location of the major cracking at the front of the house. This revealed a wide concrete footing at a depth of 1.28 m to the top of the footing. The footing scarcement was 500 mm. Excavation down the side of the footing revealed massive pieces of limestone fill, very voided and difficult to excavate.

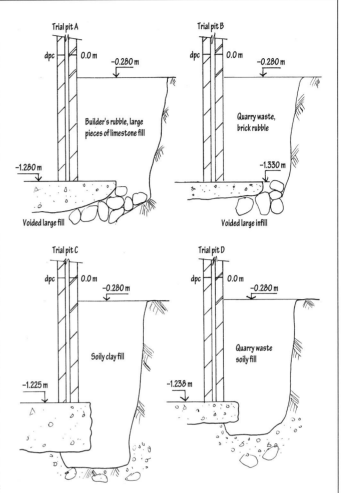

Fig. 9.10 Trial pit logs.

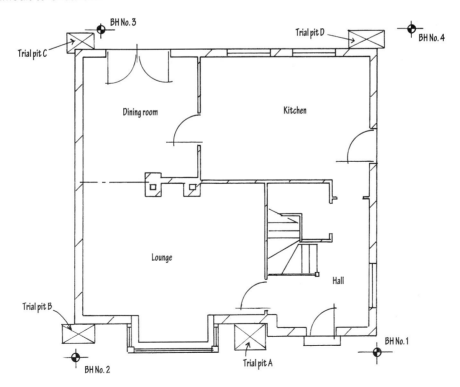

Fig. 9.9 Location of trial pits and boreholes.

In view of the type of fill encountered, a desk study was undertaken at the local library. This revealed that the development was in an area where limestone had been quarried, and to the immediate northern and western boundaries there were disused quarries. Three additional trial pits were excavated at the remaining corners of the house. The trial pit logs are shown in Fig. 9.10. These revealed similar fills at the front and rear of the house. At trial pit D the foundation was missing over the rear corner section. No groundwater was encountered, as the fills were very voided and the limestone was fissured.

To assist in the preparation of the remediation scheme four boreholes were drilled using a rotary percussive drill, and these encountered the natural creamy limestone rock at depths of from 2.20 m at the front to 3.0 m at the rear, and this rock was proven to a depth of 6.0 m.

## Conclusions and recommendations

From examination of the boreholes and trial pit data it was clear that the fines in the fill materials had been washed out into the highly voided limestone fills at a lower level, causing the gable and front corners to settle. This downward movement had rotated the loadbearing wall to the staircase, which was severely distorted.

It was decided to underpin the whole dwelling using steel mini-piles and in-situ double ground beams, as shown in Figs 9.11 and 9.12.

Because the fill contained oversize limestone boulders it was anticipated that difficulty would be encountered in driving piles beyond obstructions. Problems had been encountered during the borehole drilling, so the pile specification included for the use of pre-drilling using an Odex drilling bit to enable piles to be socketed into the limestone rock at a minimum pile length of 3.0 m.

Following piling, and during excavation for the ring beams, it was found that the rain water outlet from the small roof over the front bay window had not been connected into the site drainage system. This had obviously been flushing out the fines in the fill in the area of major damage at the front elevation. During the remediation works the staircase wall below the half landing was found to be in a dangerous condition, and it was needled and rebuilt.

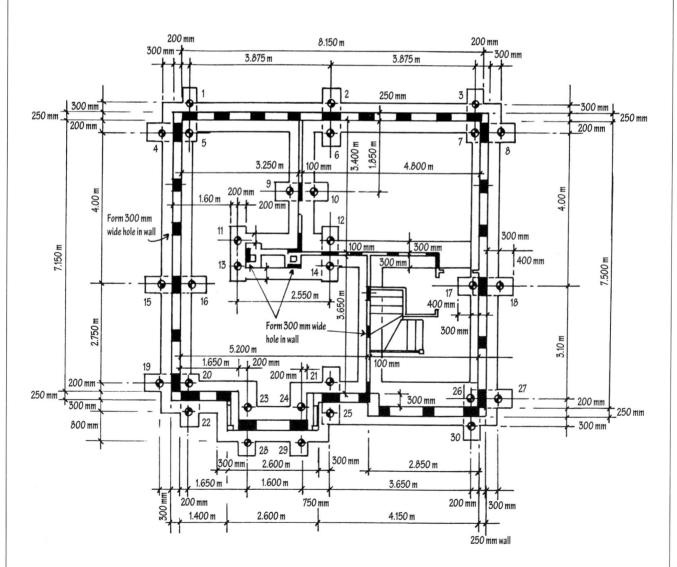

*Fig. 9.11* Plan showing pile layout and needle beams.

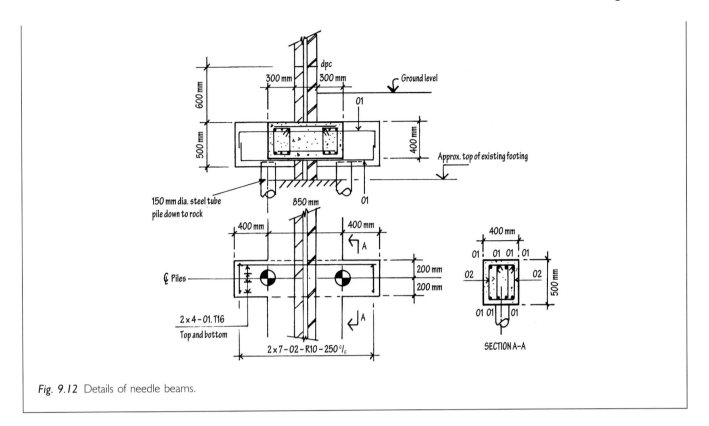

Fig. 9.12 Details of needle beams.

### 9.3.5 Inundation of fills

Where opencast sites, mineral quarries etc. are backfilled without proper compaction the fills will continue to settle as a result of self-weight consolidation for approximately 5–7 years after filling.

Figure 9.13 shows the magnitude of settlements as a percentage of the fill depth for various types of fill and different compaction methods.

A problem with such infilling in quarries is that the original groundwater levels can re-establish themselves from the bottom of the quarry upwards. This will result in successive episodes of collapse settlements as the fills are softened and flushed about by the action of the water. It has been established by testing in backfilled opencast workings at Corby, Northants, that water is a very effective medium for producing settlements in fill, and is comparable to surcharge pre-loading methods. It is therefore clear that this phenomenon must always be considered as a possibility when assessing the suitability of deep fills.

Even the use of ground improvement methods such as vibrocompaction can result in water being transmitted down the stone columns, which are as effective as vertical french drains. Such methods have been used in embankment construction to speed up the consolidation process, and have been seen to be very effective. The partial depth treatment of deep fills by vibrocompaction could well pose problems in the future.

### 9.3.6 Settlements in soft natural strata

In cohesive soils that are saturated, the effects of loading the soil are to squeeze out some of the porewater within

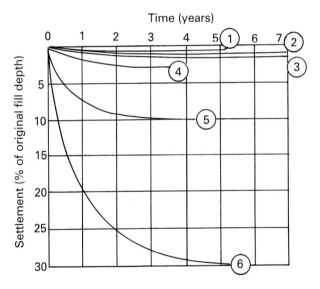

Fig. 9.13 Self-consolidation settlement of fills (Meyerhof, 1951). 1 well-graded soil, well compacted; 2 rock fill, medium compaction; 3 clay and chalk, lightly compacted; 4 sand, uncompacted; 5 clay, uncompacted; 6 mixed refuse, well compacted.

the soil mass. This is referred to as *primary consolidation*. A change of loading is required for this consolidation to take place, and it can be many years before the settlements cease. The most susceptible strata are the very soft silty alluvium deposits and beds of peat and organic silts. These soils are generally very soft or soft, have very high moisture contents, and have a potential for excessive settlements when loaded. It is not good practice to build on them using strip footings. In areas of soft alluvium containing peat bands it is safer to construct two-storey dwellings on piles, and for bungalows pseudo-rafts may be possible, subject to total settlements.

Raft foundations are generally suitable for ground-bearing pressures of 50 kN/m², i.e. shear vane reading of 25 kN/m².

For alluviums with shear vane values between 10 and 25 kN/m² piled foundations are generally recommended if the allowable settlements are not to be exceeded.

Where there is a stiff clay crust overlying the soft peats or silts, then, provided there is a cover of stiff clay of at least 1.50 m and the peat band does not exceed 150 mm, raft foundations have been shown to be suitable, subject to confirmatory trial pits.

Where the peat is below firm clays, and the thickness of the firm ground between the strip footings and the peat exceeds 4 m, the foundation bulb of pressure will not stress the peats. In this situation the footings should be kept as high as possible at a depth appropriate for the soil plasticity and climatic variation. If trees are within influencing distance of the building, foundations should not be deepened, and consideration should be given to the use of a raft foundation on a thick stone cushion or piled foundations taken below the peat.

When specifying rafts on marginal soils, reliable estimates of the likely settlements are essential for both total and differential. The pseudo-raft is based on empirical rules and there is no precise analytical solution for designing a raft. For a two-storey building there is no point in preparing sophisticated design solutions based on finite element analysis. Engineering judgement is a far better approach and more emphasis must be placed on the ground conditions supporting the raft.

## Case history 9.4

### Introduction

A claim for major structural damage had been submitted in 1982 to the insurance warranty provider by the owner of a large bungalow in a small village west of Malton, North Yorkshire. The claim was investigated and turned down, on the basis that the cracking evident was not considered to be serious, and it was recommended that cosmetic repairs be carried out by the owner. The report compiled by the investigator made reference to the possibility that a massive ash tree 19 m away could be the cause of the cracking.

In 1985 the owner lodged a further claim, alleging that the cracking was getting progressively worse, and the author was asked to investigate the claim. During the initial inspection the owner drew the attention of the author to cracking in the bedrooms. The owner had fitted some home-made tell-tales over the cracks, and had pencilled in the movements since 1982. It was clear from these cracks, and others observed, that differential foundation movements were taking place.

On examining the old file, it became apparent that no proper investigation had been carried out in 1982, and this was confirmed by the owner of the property. No trial pits had been investigated, and no desk study had been carried out. Owing to the progressive movements that were occurring, a full structural investigation was initiated. The positions of the trial pits and boreholes are shown in Fig. 9.14.

### Investigation

The bungalow had been architect designed for the owners, and was on a sloping site. The main walls were built in cavity construction using 215 mm random natural sandstone, 50 mm insulated cavity and a 100 mm clinker block inner leaf. The roof was built using purlins and collars supporting concrete tiles. The purlins spanned

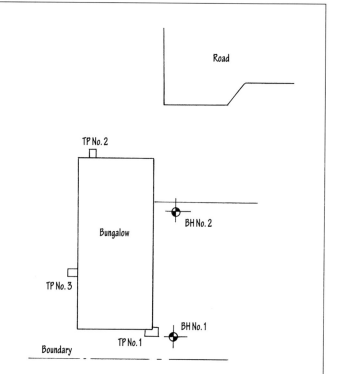

Fig. 9.14 Location of trial pits and boreholes.

onto the internal loadbearing cross-walls that divided the rooms. One trial pit was excavated close to the rear front gable wall down to the underside of the foundation. The stratum encountered below the footing at a depth of approximately 850 mm was a mixture of sandstone fragments within a loamy matrix. It looked like a highly weathered sandstone, and was in a loose state of compaction and was very moist. In view of the stratum encountered it was decided to commission a borehole investigation, and two 150 mm diameter boreholes were drilled using a light, cable percussion drill down to a depth of 5.0 m, terminating in a weathered sandstone.

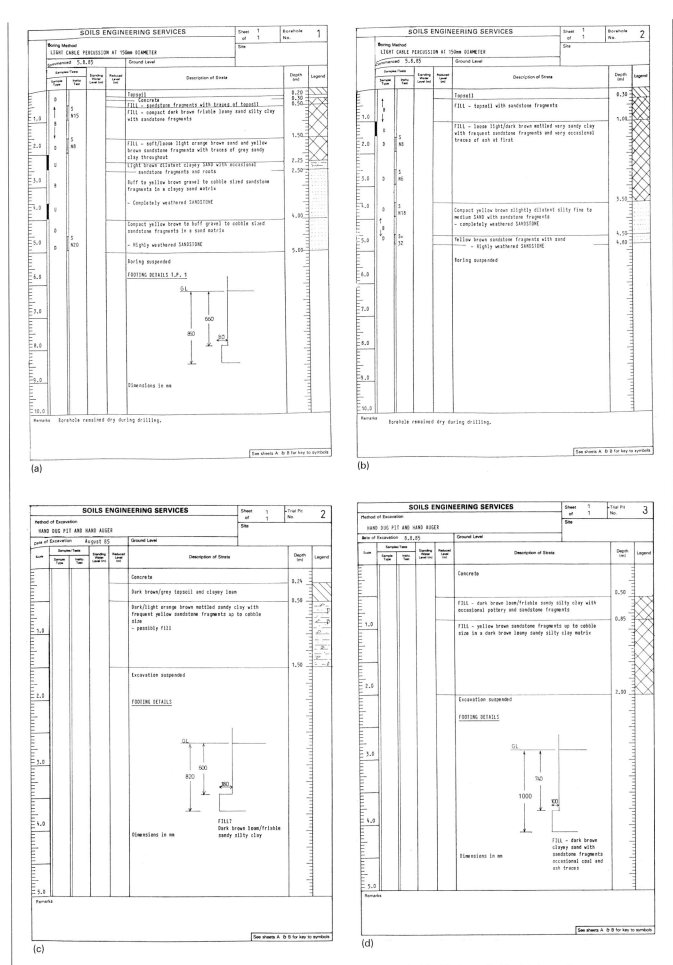

Fig. 9.15 Driller's logs of strata encountered: (a) borehole no. 1; (b) borehole no. 2; (c) trial pit no. 2; (d) trial pit no. 3.

## Findings

Borehole no. 1 revealed a loose loamy fill containing sandstone fragments down to 2.25 m, underlain by a relic topsoil followed by a buff to yellow gravel to cobble-sized sandstone fragments (a completely weathered sandstone) down to 5.0 m. Borehole no. 2 revealed similar fills down to 3.50 m, underlain by a completely weathered sandstone. Both boreholes were dry during the drilling operation.

Two further trial holes were excavated, and these confirmed that the foundations had been constructed on the filled ground at shallow depth. In addition, plastic drainage was encountered that had badly fitted joints, which when tested were found to be leaking. The borehole and trial pit locations are indicated in Fig. 9.14, and Fig. 9.15(a)–(d) shows the driller's logs of the strata encountered.

Because the dwelling was in a rural area, it was considered prudent to try to identify the source of the fills on the site. Discussions with local farmers revealed no clues, but a local whose hobby was treasure hunting with a metal detector suggested the area could have been part of a Roman complex.

A study of the various County Series geological maps and old Ordnance plans did make various references to a Roman camp adjacent to the house. A site walkover survey revealed the area in question to be composed of many undulating mounds, and it was considered possible that the dwelling in question was actually sited over an ancient Roman earthworks.

## Conclusions and recommendations

All the available evidence confirmed that the bungalow was built on an enigmatic fill, which was predominantly non-cohesive. The possibility of the large tree being responsible for the foundation movement was ruled out.

Despite the detailed investigation, the foundation movement could not be put down to a particular reason. It was considered that as the bungalow had been built in 1976, and had shown no damage until 1981, the movement must have arisen as a result of a change in the ground some time after construction, and it was concluded that the leaking drains had resulted in the fines being flushed out from below the footings, with inundation of the fills taking place in various areas.

It was recommended that the footings be underpinned down into the underlying natural strata using a mini-pile system with ground beams and needle pile caps.

Had this claim been investigated properly in 1982, the leaking drains would have been found and the minor damage repaired. The claim would also have been referred to the owner's house insurer as a subsidence claim. In view of the protracted timescale, and the house owner's understandable dissatisfaction, it was decided to accept the claim and not to involve the house insurer.

## Remedial scheme

A scheme was prepared by a specialist piling contractor, in which it was proposed to use a single ground beam below all loadbearing walls, spanning onto a twin pile needle beam, which was cast in situ with the beams. To enable the beam to be cast, the wall above was removed in short lengths and supported on sacrificial jacks at 750 mm centres. These are indicated in Figs 9.16 and 9.17.

There was a 75 mm gap between the top of the beam and the underside of the masonry to enable the concrete

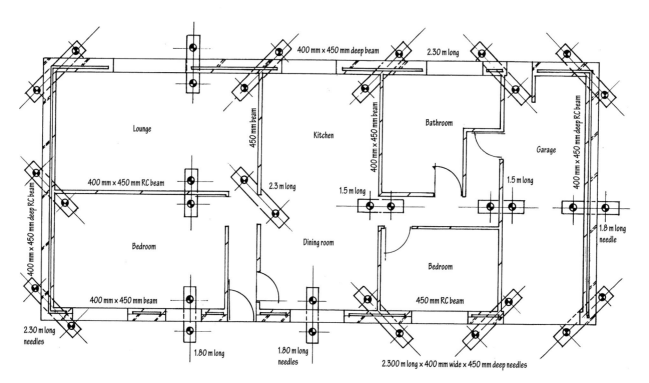

*Fig. 9.16* Plan showing location of needle beams.

to be placed into the shutters and to be vibrated. Following the striking of the formwork, this gap was made up using dry pack mortar stemmed in.

During installation of the beams down the garage gable, the masonry collapsed. Fortunately the site operatives were not injured. Further investigation revealed that the stone facing had not been tied very well into the blockwork. The bed joints in the random sandstone did not relate to the blockwork joints, and what few ties there were had not been given sufficient embedment. It was therefore decided to double up on

the jacks and place them in an alternate pattern at 500 mm centres. The garage wall was rebuilt with new retrofit ties installed. During installation of the ground beams construction joints were formed over needles or at one quarter of the beam's span.

The main beams were reinforced with three Y16 bars top and bottom, with R12 links at 250 mm centres. The cross needles were reinforced with four Y16 bars and R12 links in pairs at 250 mm centres. The steel piles were filled with concrete, and a single Y20 bar was placed in the top 2.0 m section to form a tie into the cross needle.

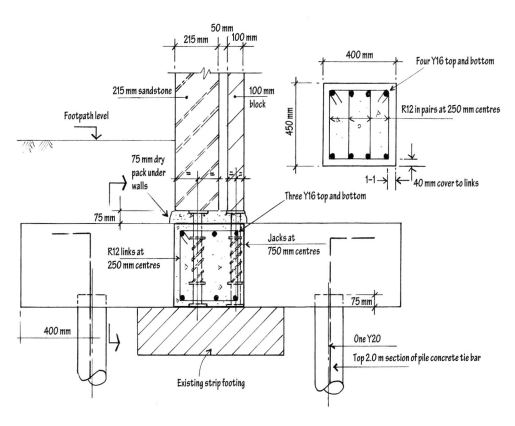

Fig. 9.17 Typical section through ground beams.

## Case history 9.5

### Introduction

The owners of a two-storey house 4 miles south of York had filed a claim with their household insurers for structural damage. The initial claim was rejected by the insurers on the grounds that the cracking evident was not major, and just required cosmetic repairs.

The building was then visually monitored for six months by the owners, who subsequently filed a claim

with the house insurance warranty provider, and the author was appointed to investigate the claim.

### Basic information

The house was a large, eight year old, detached two-storey building. The site had a stream running along the southern boundary, and according to records the stream was subject to extensive flooding.

Three large mature conifer trees were evident within close proximity of the house.

## Investigation

There was a large diagonal crack below the lower toilet window on the west gable, extending upwards to the landing window sill. Gaps were evident up the sides of the window frames up to eaves level. Corresponding cracking was mirrored on the internal leaf of the wall.

Diagonal cracking was evident above the window arch to the dining room on the west wall, with a smaller crack extending diagonally at eaves level above the top of the bay window frame in the dining room.

A large crack ran diagonally at eaves level on the eastern wall, and three further cracks were present beneath the window sills to bedroom 4, the utility room and down the side of the window to bedroom 4.

Diagonal cracking was visible in two places on the southern elevation wall, extending from the bay window frame up to the window sill above.

Extensive cracking was noted to all the ceilings, both upstairs and downstairs.

## Conclusions

The amount of structural damage observed in all the main walls of the property suggested that some form of differential foundation movement was taking place. It was necessary to carry out a detailed site investigation of the ground conditions, and of the type of foundations constructed.

## Site investigation

Initially six trial pits were excavated at various locations around the property, as shown in Fig. 9.18. The trial holes were used to examine the foundations and formation below, and were extended by hand augering down to 2.60 m. The hand augering proved to be relatively easy, owing to the soft materials encountered.

Trial pit no. 1 revealed the footings to be sitting on a black organic sand at a depth of 650 mm below ground level. From 1.60 m to 1.80 m a natural grey silty sand was logged, underlain by firm clays and very organic, very soft peaty silts at 2.45 m.

Trial pit no. 2 revealed the footings to be on a rubble fill at a depth of 650 mm underlain by damp sands to a depth of 1.90 m. From 1.90 m to 2.60 m a band of black

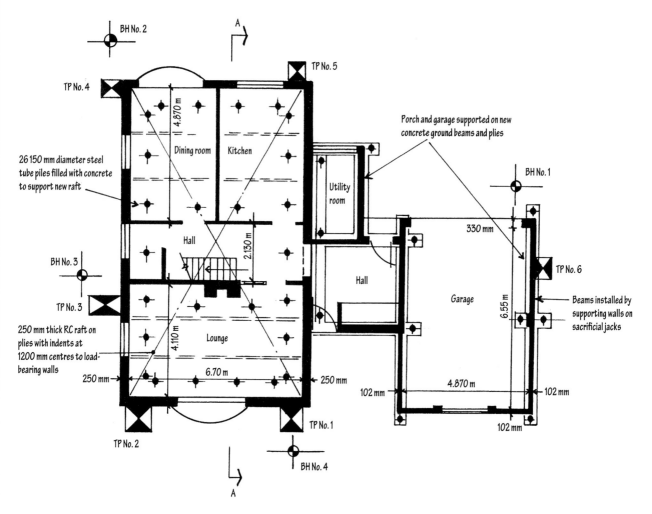

*Fig. 9.18* Location of trial pits and boreholes.

peaty silts was logged, underlain by very soft, grey, silty alluvium deposits.

Trial pit no. 3 was similar to trial pit no. 2, with the peat being evident at a depth of 1.90 m, extending down to 2.20 m and underlain by very soft, grey, silty alluvium to 2.60 m.

Trial pits no. 4 and no. 5 did not encounter peat, but from 900 mm to 1.40 m a soft, wet, silty clay was found, underlain by wet clayey sands to 1.80 m depth. From 1.80 m to 2.20 m firm clays were revealed, becoming stiff at the base of the hole.

Trial pit no. 6 encountered the peat at 1.90–2.30 m, underlain by very soft grey alluvium down to 2.60 m. The

footing was founded on a grey silty sand at 750 mm below ground level.

The cross-section A–A shown in Fig. 9.19 illustrates how part of the house is sitting on ground that has a reasonable bearing pressure, and part on a stratum that is very compressible. As the soft alluvium was too deep to consider traditional underpinning it was decided to provide a piled raft under the ground floor, with indents into the walls at close centres.

To assist in the piling design and tenders, four boreholes 10 m deep were drilled using a light cable percussion drilling technique. These boreholes had to be lined with steel casing tubes. Disturbed samples were

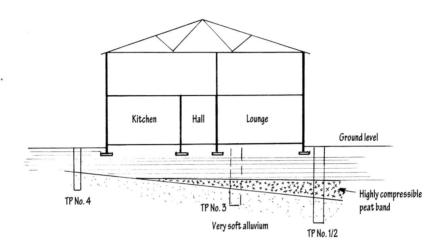

*Fig. 9.19* Section A–A.

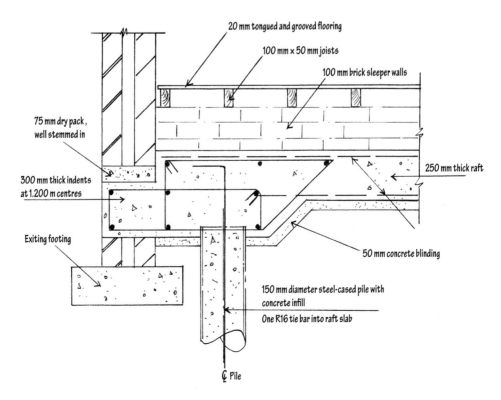

*Fig. 9.20* Section through raft indents.

taken for soil classification and chemical testing purposes. Undisturbed samples 100 mm in diameter were taken at regular intervals in the cohesive strata.

Standard penetration tests (SPT) were taken in the granular soils to indicate their densities and loadbearing characteristics.

Observations were made for groundwater entering the boreholes during drilling, and a check was made of the standing water level.

The boreholes indicated medium dense sands at 4.90–5.90 m underlain by firm to stiff clays to 8.80 m and medium dense sands and gravel down to 9.40 m. Water entered the bores at 3.10 m level and 8.80 m level, and stabilised at 2.80 m following withdrawal of the casings.

Two tenders were received for installing a piled raft under the ground floor using driven piles:

- Tender A was based on a 254 mm square precast pile, jointed in 2.0 m lengths and with a driven length of 6.0 m.
- Tender B was based on a 150 mm diameter steel tube pile filled with concrete grout and based on a driven length of 5.0 m.

The contract was awarded to Tender B. During the piling, three of the first five piles driven did not achieve their design set until they had reached 12 m. On completion of the piling, 36 piles were driven to various depths from 6.80 m to 13 m, with an average length of 7.0 m. The specialist piling contractor claimed for extra payment for the extra pile lengths driven, but this was rejected by the engineer.

Examination of the two types of pile showed that the end bearing component of the precast driven pile was 2.86 times greater than the steel tube, and it was considered that the steel tube pile had punched through the medium dense sands at 4.90 m and 8.80 m, and as a result had to go deeper to pick up the skin friction needed to carry the design load on the pile. It was the engineer's view that the piling firm had been provided with a comprehensive site investigation, and that the piling designer had provided a pile that was too small in diameter. As the firm in question was a piling specialist it was considered that their responsibility was to design a suitable pile to carry the load. They were given the opportunity to prove that the borehole information was inaccurate, but they declined.

Figure 9.20 shows a cross-section of the raft installed under the dwelling. The oversite concrete below the existing ground floor was the finished raft level. Indents were cast into pockets at 1.20 m centres, cantilevered from the piled raft, which was designed with top and bottom reinforcement to accommodate these.

### 9.3.7 Foundations on deep filled ground

Buildings constructed on made ground, i.e. mixed fills, may be at risk from differential settlements because of the nature of the fills or the condition of the fills, or for causes connected with the reason why the fills were placed in the first instance. When investigating serious structural damage caused by foundation failure, it is essential to determine the type of foundation used. Even buildings that have been constructed on piled foundations have been known to fail.

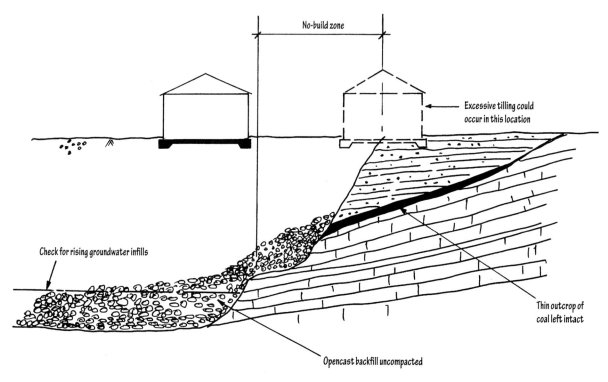

*Fig. 9.21* No-build zones in infilled opencast workings and infilled quarries.

Sites such as infilled railway cuttings, infilled mill dams and infilled quarries have the additional risk of buildings being constructed such that they straddle a high wall that has not been surveyed in. Figure 9.21 shows the no-build zone for such infilled situations, if differential settlements are to be avoided. On sites where the ground level has been raised above the original, the high wall line will probably be buried, and is generally difficult to locate using trial pits. The only definitive approach is to use deep boreholes at close centres, i.e. about 1.0 m apart.

For many such sites the risks can be reduced by obtaining a good-quality soil report, and by providing a robust foundation. In some circumstances it may even be cost-effective to reinforce the superstructure to strengthen the masonry.

However, failures can still occur, and these are generally as a result of changes in the composition of the fills, brought about by changes in the groundwater regime or the collapse or degradation of organic matter at depth.

The author has dealt with many such failures, and has reached the conclusion that strip footings are totally unsuitable on deep filled ground, even where the site investigation results indicate medium dense to dense fills. Experience has shown that even raft foundations cannot always provide the required stiffness to bridge over weak sections of fill.

It is essential when investigating such failures to carry out a desk study. Vital information can often be obtained from old maps or local archives that will assist in the investigation. This information is also relatively cheap to obtain.

## Case history 9.6

### Introduction

This case history illustrates the benefits of doing a desk study. If such a study had been carried out before

purchasing the site, it is likely that strip footings would not have been used by the builder.

The house that suffered structural damage was in Grimsby. Following the claim submission, a detailed

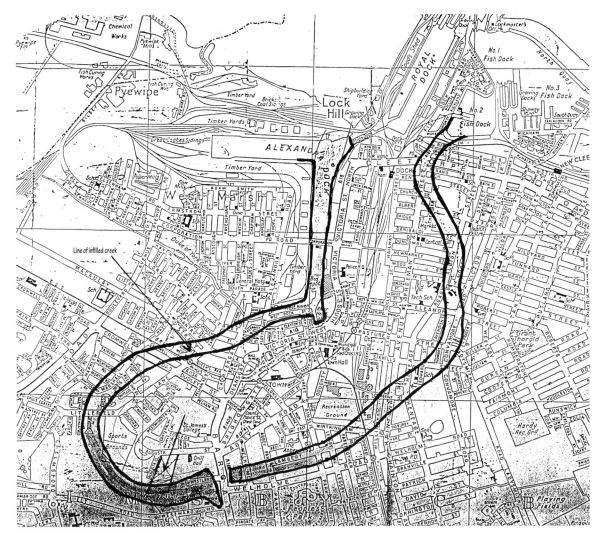

*Fig. 9.22* Infilled Grimsby Creek.

investigation was carried out. The initial inspections revealed that there had been extensive differential settlements. The owners of the house had commissioned a site investigation firm to carry out two boreholes to a depth of 8.0 m. These revealed the foundations to be sitting on sands and gravel underlain by dark brown silty peaty clay to a depth of 3.0 m, which was underlain by grey clayey sands containing pieces of rotting timber and other organic matter. These soft deposits were underlain by firm clays from 3.0 m to 8.0 m, which contained sandy partings.

The desk study revealed that the 1100 mm wide strip foundations had been installed by the original builder, who unfortunately had gone into liquidation. The foundations stood for a considerable time before the site was sold on to another builder, and he proceeded to build on the footings cast by the previous builder. This builder considered that as the original foundations had been inspected by the local authority building inspector it was in order to build on them. Part of the desk study involved the examination of old Ordnance plans, and it was noted that an earlier plan showed a large manor house in close proximity, which had a tree-lined road up to it. Later record plans showed that this tree-lined feature had disappeared, and it was decided to investigate this further. A local resident living in the area produced an old photograph of the large manor house, and in the photograph a small sailing boat with masts was in the background. This was considered strange, bearing in mind that the dwelling was several miles from the coast.

Following this line of investigation, a visit to the local library turned up an old plan, which revealed the line of the old 'creek'. This had been infilled in the past, and there was a history of dwellings suffering from subsidence in the Ainsley Street area of Grimsby and the surrounding streets that had been constructed over the creek. The infilling must have taken place 100–120 years previously. Looking at the scale of the maps, the creek ran from the Alexandra Dock and after going south returned in a loop, back into Fish Dock no. 2. It was likely

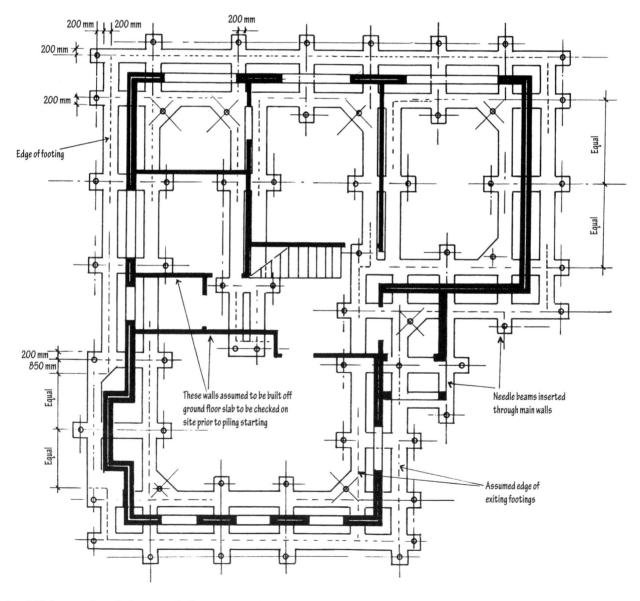

*Fig. 9.23* Layout of needle beams and piling.

that a one-way system operated, as it was not very wide. Plotting the large manor house and the house under investigation showed clearly that the house built in the 1980s was partly over the infilled creek. Figure 9.22 shows this infilled 'creek'. It was therefore clear that the house required to be stabilised, and major structural repairs would need to be carried out. The house was repaired by installing a pile and ring beam system under all the loadbearing walls. A total of 60 piles and 30 in-situ concrete needle beams were installed, and Fig. 9.23 shows the location of these. As can be seen from the plan, the remediation scheme for the foundations and associated superstructure works was fairly complex, and it took just over four months to complete the contract.

## Case history 9.7

### Introduction

This involved three dwellings forming a terrace built over an infilled railway cutting on a site in Rowlands Gill. Cracking had developed in several gable and party walls, and the housing association involved in the development had lodged an insurance claim. The housing association commissioned a consulting engineer, who carried out a trial pit investigation and concluded that the foundation movements resulted from inadequate stone sub-base below the rafts. While this was a possibility, it seemed unlikely in view of the age of the properties, and it was considered that poor compaction of the sub-base materials would have shown up shortly after construction.

### Investigation

Fortunately the site investigation for the housing project was available from the original consulting engineer. This showed the railway cutting to have been infilled with brick rubble, stone ash, clinker and soil. Examination of the 1858 Ordnance map showed the cutting to be quite wide, and plotting the terrace on the plan showed that the block was on the batter slope of the cutting, as shown in Fig. 9.24.

Along the rear of the terrace was a public walkway built over the line of the old railway line, and several large mature trees were evident. There was therefore a possibility that some of the clay fill in the cutting had become desiccated owing to tree roots, and this had occurred in the previous dry summer.

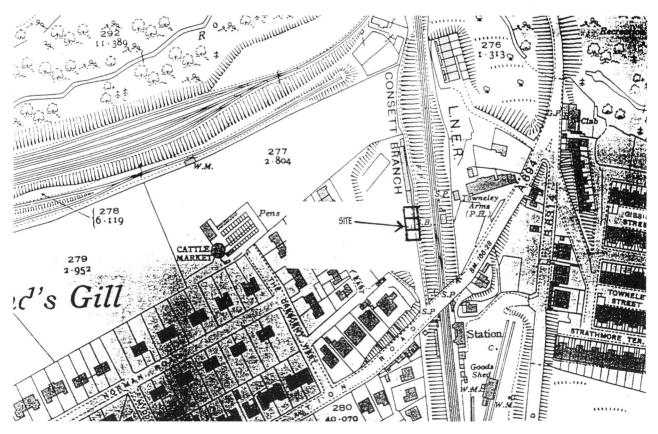

*Fig. 9.24* Site location.

Following an inspection of the structural damage it was decided that the raft foundation had to be stabilised prior to repairing the superstructure. Inspection of the raft details available showed that it consisted of a 180 mm concrete slab with 410 mm deep stepped edge beams and a 750 mm wide toe along the wall lines (Fig. 9.25). The slab was reinforced with A142 fabric reinforcement in the top and bottom of the slab, with additional larger reinforcement around the edge thickenings. The raft slab had been placed on a 600 mm cushion of DoT Type 1 sub-base.

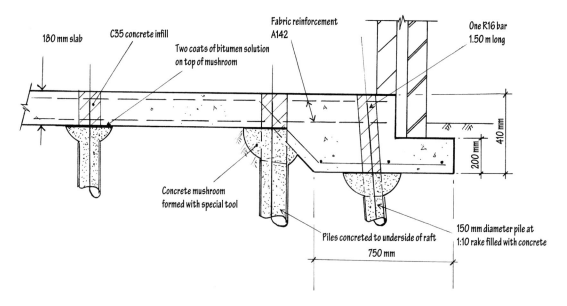

Fig. 9.25 Section through raft.

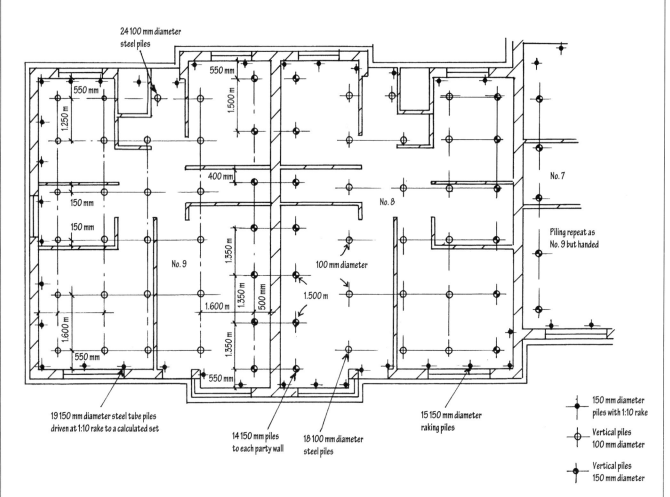

Fig. 9.26 Piling grid.

It was decided to stabilise the raft foundations by pin piling. This would require the raft to be cored, to enable 150 mm diameter steel tube piles to be driven.

Around the external walls inclined piles were used at 1200 mm centres, and infill piles at close centres were installed (Fig. 9.26). The grid dimensions were based on the ability of the slab and edge thickenings to span between the piles.

Once the piles were driven, the fill below the slab around the pile was removed using a special attachment fitted to a drill, leaving a mushroom-shaped void. The pile was then concreted and filled up to the underside of the slab. Once the concrete had set, the top of the pile was painted with two coats of bituminous compound before filling the cored hole with concrete. In some rooms the slab needed to be levelled using a thin latex cement screed followed by a new chipboard floor laid over polystyrene insulation.

The cracking to the party wall blockwork required the blockwork to be removed and rebuilt.

## Calculations

Line loads:

|  | Serviceability | Ultimate |
|---|---|---|
| Roof (total) say | 2.0 kN/m$^2$ | 3.0 |
| Main walls | 4.50 kN/m$^2$ | 6.30 |
| Ground floor $-0.18 \times 24$ | 4.32 | 6.07 |
| Finishes | 0.38 | 0.53 |
| Stud partitions | 1.0 | 1.40 |
| Imposed load | 1.50 | 2.40 |
| Total = | 7.20 kN/m$^2$ | 10.40 |

Front and rear walls:

| Roof | $= 5 \times 2.0$ | $= 10.0$ | 15.0 |
|---|---|---|---|
| Walls | $= 2.60 \times 4.50$ | $= 11.70$ | 16.40 |
| Floor slab | $= 1.0 \times 7.20$ | $= 7.20$ | 10.40 |
| Edge beam | $= 0.3 \times 24 \times 0.75 =$ | 5.40 | 7.60 |
| | Total | $= 34.30$ kN/m | 49.40 kN/m |

Party walls:

| | | kN | kN |
|---|---|---|---|
| Roof | $= 0.60 \times 2.0$ | $= 1.20$ | 1.80 |
| Walls | $= 3.50 \times 4.50$ | $= 15.80$ | 22.10 |
| Floor slab | $= 2.0 \times 7.20$ | $= 14.40$ | 20.80 |
| Thickening | $= 0.30 \times 24$ | | |
| | $\times 0.75$ | $= 5.40$ | 7.60 |
| | Total | $= 36.80$ kN/m | 52.30 kN/m |

Existing raft slab is 180 mm thick, reinforced top and bottom with A142 mesh. Design as a flat slab to determine the maximum pile spacing (using I. Struct. E Manual). Try 1.50 m pile spacing in two directions.

Span, $L = 1.50$ m therefore $F = 10.40 \times 1.50 \times 1.50$
$= 23.40$ kN

Maximum moment $= 0.083 \times 23.40 \times 1.50 = 2.92$ kN m

Using concrete grade 35 N/mm$^2$ and steel stress of 460 N/mm$^2$ and Chart No. 9.1 (Fig. 7.11 in Chapter 7):

Depth, $d_1 = 180 - 25 - 5 = 150$ mm

$$\frac{M}{b \times d^2} = \frac{2.92 \times 10^6}{1000 \times 150^2} = 0.129$$

From Chart No. 9.1:

$$\frac{100 A_s}{b \times d} = 0.05 \text{ therefore } A_s = \frac{0.05 \times 1000 \times 150}{100}$$

$$= 75 \text{ mm}^2/\text{m}$$

Minimum steel required is

$$0.13\% = \frac{0.13 \times 1000 \times 180}{100} = 234 \text{ mm}^2/\text{m}$$

Two layers of A142 provide 284 mm$^2$ in the slab section, and this is adequate with the piles spaced at 1.50 m maximum centres.

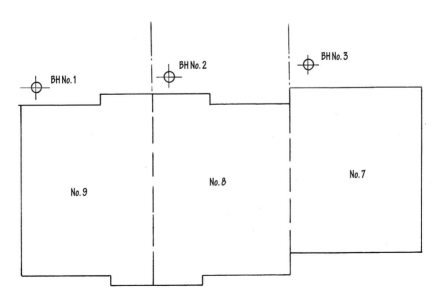

*Fig. 9.27* Location of boreholes.

## *Edge/party wall thickenings*

Using 410 mm thick edge beam and a single layer of A142 in 750 mm wide in the edge beam,

$$d = 440 - 40 - 5 = 395 \text{ mm}$$

then

$$\frac{100A_s}{b \times d} = \frac{100 \times 142}{750 \times 395} = 0.047$$

$$\frac{M}{bd^2} = 0.150$$

then $M = 0.150 \times 750 \times 395^2 = 17.55 \text{ kN m}$

Then moment of resistance $= 17.55 \text{ kN m}$

$$17.55 = \frac{wL^2}{10} = \frac{52.30 \times L^2}{10} \text{ therefore maximum } L = 1.83 \text{ m}$$

With two piles at 1.50 m spacing, pile loading $= 36.80 \times 1.50 = 55.20 \text{ kN}$

Use pile with a SWL of 60 kN

Figure 9.27 shows the location of the three boreholes drilled, and Figs 9.28, 9.29 and 9.30 detail the borehole strata encountered.

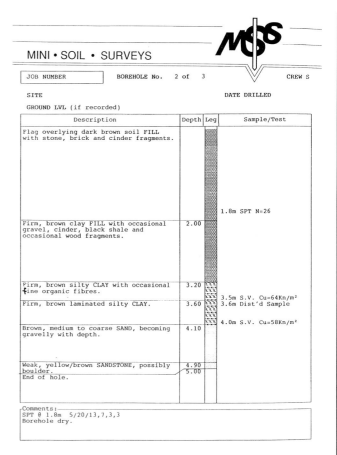

**Fig. 9.29** Log of borehole no. 2.

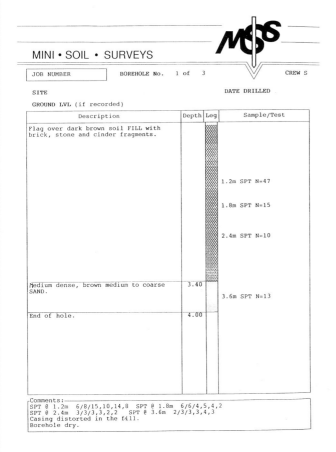

**Fig. 9.28** Log of borehole no. 1.

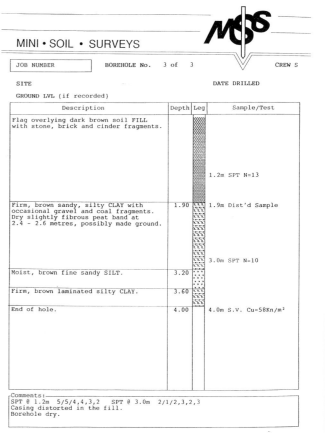

**Fig. 9.30** Log of borehole no. 3.

## 9.4 Subsidence-related defects

Defects can result in buildings as a result of the following types of subsidence:

- mining subsidence, from active and past mining of coal, ironstone and other minerals;
- brine pumping extraction and salt mining solution features;
- the formation of natural cavities, solution features in gypsum strata, swallow holes in chalks and carboniferous limestones;
- stone and mineral mining;
- desiccation or heave of cohesive soils by trees and seasonal variations in climate;
- slope instability and landslips;
- frost heave;
- spontaneous expansion of fills by oxidation and other chemical reactions.

### 9.4.1 Mining subsidence

#### Active mining

As the coalmining industry in the UK has reduced its capacity, active mining now takes place from fewer pits, which are now much larger and more complex. Selby coalfield is such an example of the mining of thick seams of coal by longwall extraction at depths of about 900 m.

The mining companies generally work on a 5 year plan of forward projection, and if any building is planned in an active mining area, the mining company should be consulted as to its plans. In addition a Coal Authority mining report should be sought. Any investigation of structural damage to buildings in mining areas should always be backed up with a Coal Authority search in regard to the past and present mining activity, either below the affected building or adjacent to it. If the damage is suspected to be due to mining activity a claim can be made to the mine operator for compensation, or – in the case of old coal workings – a claim should be submitted to the Coal Authority, as such mines are their responsibility.

Coal extracted by longwall mining methods produces subsidence waves that can damage buildings and services at the surface. Such damage can be exacerbated if the buildings are on or close to fault zones.

In general, the biggest problem arises where a building is on the fringe of a subsidence area, as this can result in severe differential movements.

Such a situation occurred in the County Durham coalfield, where different pits in the Seaham locality maintained a boundary of unworked coal between neighbouring pits. The effect of this was to create a situation such that a fault-type differential movement took place on the unmined side of both pit boundaries. This differential movement at the surface caused severe structural damage to buildings, and several houses had to be demolished because they had become unsafe. It is important to remember that active mining can also damage drainage services, which in turn can weaken the surrounding soils below foundations, resulting in secondary subsidence leading to structural damage.

#### Past mining

In coalmining areas it has been traditional to grout up pillar and stall workings with PFA and cement grouts where the coal workings have insufficient competent rock cover over them. The generally accepted rule of thumb is that where the thickness of competent rock exceeds 10 times the seam thickness, then no grouting up of the workings is required. Evidence gathered by Gerrard and Taylor based on observations of old mined coal seams subsequently opencasted has confirmed this rule of thumb to be an adequate guide for most old coal workings. However, there are situations where the rule does not apply: for example, where such thin pillars have been left, resulting in pillar crushing and subsequent collapse of the strata above. This is illustrated in Fig. 9.31.

Where properties are suspected of having been damaged as a result of the collapse of pillar and stall workings it is necessary to contact the Coal Authority at Bretby in Derbyshire, as they are responsible for old coal workings with the exception of a few private mines.

#### Mineshafts

Shafts for coal extraction belong to the Coal Authority. However, the responsibility for capping off and making such shafts safe rests with the developer of the land. In

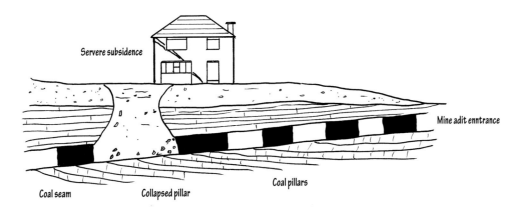

Fig. 9.31 Effects of pillar and stall coal workings.

terms of public safety the Coal Authority is obliged only to fence the shaft off. This generally requires the shaft to be capped off at rockhead level, using a shaft twice the shaft diameter. If the rockhead is too deep, then it may be necessary to grout the shaft fill using a cementitious pressure-grouting technique.

A building constructed over a mineshaft, especially a residential property, generally proves to be very difficult to sell. Obviously the risk of mine gases migrating upwards is higher, and the Coal Authority and most statutory bodies advise against building directly over shafts, as fatalities have been recorded.

## Case history 9.8

### Introduction

A builder was engaged to build a large bungalow on a site on the outskirts of Bradford. During the foundation inspections the building inspector for the warranty insurers pointed out that shallow coal workings were suspected below the site.

Examination of the geological plans for the area (Fig. 9.32) showed that there were two coal outcrops at shallow depth, and shaft locations nearby confirmed the seam thickness and shallow depth. The house was below the Black Bed coal outcrop but above the Better Bed outcrop, which was the lower seam.

Unfortunately the foundations proceeded without any reinforcement, and the site was not test drilled for voids. Several months later 18 boreholes were drilled, and these revealed shallow coal workings. In addition two trial pit excavations broke through into old coal workings at about 3 m depth. These workings revealed old pit props and discarded mining equipment. As it was a condition of the building society that the bungalow had to have a 10 year structural warranty, the owner of the land had to

take legal action against the builder, and the builder eventually had to carry out major grouting and foundation remedial works. This involved drilling a pattern of grout holes on a 2.0 m grid below and beyond the bungalow footprint for a further 2.0 m, and placing under pressure a mixture of pulverised fuel ash and cement until the workings were filled.

To cater for any possible residual subsidence, the foundations had to be strengthened by the provision of an in-situ concrete ground beam inside and out, joined by in-situ concrete needles taken through the external walls, as shown in Fig. 9.33. Cracking of the masonry had taken place prior to the grout stabilisation, and this was repaired on completion of the foundation remedial works.

### Ground beam calculations

Design beams to BS 8110
Concrete mix to be C35 N/mm$^2$ using high-tensile reinforcement

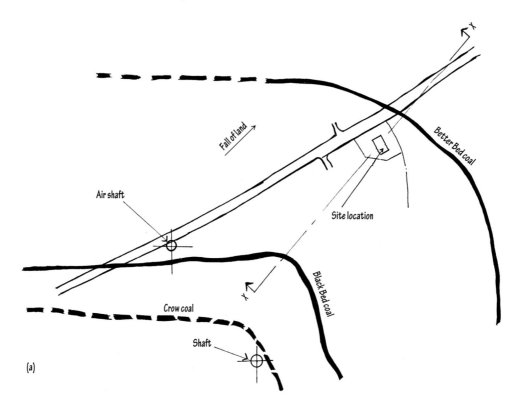

*Fig. 9.32(a)* Plan showing the coal seam outcrops.

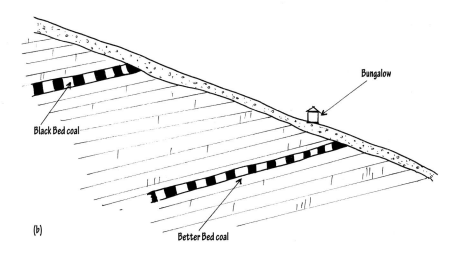

(b)

*Fig. 9.32(b)* Section X–X through the site.

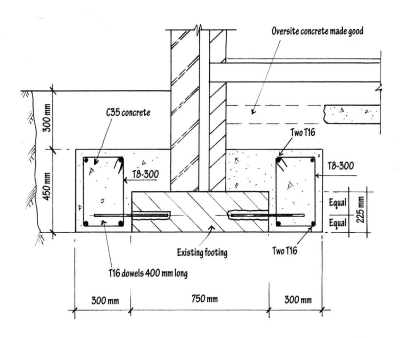

*Fig. 9.33* Foundation strengthening.

Concrete cover to links to be 40 mm

| Wall line loading: | (kN/m) |
|---|---|
| Roof loading – 2.0 kN/m² × 9.0 × 0.50 | = 9.0 |
| Main wall – 3.50 × 3.0 height | = 10.50 |
| Ground floor – 2.0 × 4.0 × 0.50 | = 4.05 |
| Foundations – 0.225 × 0.75 × 24 | = 4.50 |
| Total | = 27.55 |

With dowels at 450 mm centres the shear load per dowel

$$= \frac{27.50 \times 0.45}{2.0} = 6.198 \text{ kN}$$

From CP114, allowable shear stress to existing footing concrete $= 0.68 \text{ N/mm}^2$

Dowels cast in 200 mm, therefore shearing plane $= 158 \times 2 \times 200 = 63\,200 \text{ mm}^2$

Actual shear stress on concrete around dowels

$$= \frac{6.198 \times 10^3}{63\,200} = 0.100 \text{ N/mm}^2$$

Using a load factor of 1.50, the ultimate load/beam

$$= \frac{27.55 \times 1.50}{2.0} = 20.662 \text{ kN/m}$$

Beams designed to span 3.0 m and cantilever 1.50 m at corners

$$\text{Ultimate bending moment} = \frac{20.662 \times 3.0^2}{8} = 23.24 \text{ kN m}$$

$$\frac{M}{b \times d_1^2} = \frac{23.24 \times 10^6}{300 \times 390^2}$$

$$= 0.509 \text{ from Chart No. 9.1 (Fig. 7.11)}$$

$$\frac{100A_s}{b \times d_1} = 0.14$$

therefore $A_s = \dfrac{0.14 \times 300 \times 390}{100} = 164.0 \text{ mm}^2$

Use two T16 high-tensile bars top and bottom (402 mm²)

Shear,

$$v = \frac{V}{b_v \times d} = \frac{20.662 \times 3.0 \times 0.50}{300 \times 400} = \frac{30.99 \times 10^3}{300 \times 400}$$

$$= 0.26 \text{ N/mm}^2$$

$$\frac{100A_s}{b \times d} = \frac{100 \times 402}{300 \times 400} = 0.335 \text{ then } v_c \text{ is } < 0.46$$

where $V$ = maximum shear (ultimate), and $v_c$ = design concrete shear stress.

Provide minimum links using T8 bars. Minimum spacing = $0.75d$ = 300 mm

$$A_{sv} = \frac{0.40b_v s_v}{0.87 \times f_{yv}}$$

therefore $s_v = \dfrac{101 \times 0.87 \times 460}{0.40 \times 300} = 336$ mm

Use T8 links at 300 mm centres for all ground beams

### 9.4.2 Brine pump extraction and salt mining

Solution features can result in rock salt formations by natural water movement. Salt can be 180 times more soluble than natural limestone strata, and is therefore dissolved considerably faster. Because of this high solubility the rock salt rarely outcrops at the surface. In parts of Cheshire, rock salt beds have been dissolved away down to considerable depths below the existing ground levels as a result of mobile groundwaters. Below this wet zone the salt beds are unaffected: this zone is referred to as the *dry rockhead*, and is shown in Fig. 9.34. This zone is intact, as effective circulation of groundwater does not take place.

Between these two zones there is a buffer zone called the *wet rockhead*. In this intermediate zone the rock salt is partially dissolved, and consists of cavitated salt beds containing brine streams or brine runs known as *wild brines*.

A natural feature that is prevalent in areas of saliferous beds is the water-filled linear hollows known as *flashes*. These linear hollows are the result of sinkhole solution cavities collapsing to create a linear zone of sinkholes at the surface, which then become flooded to form a 'flash' feature.

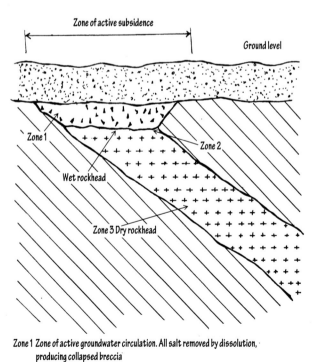

Zone 1 Zone of active groundwater circulation. All salt removed by dissolution, producing collapsed breccia

Zone 2 Zone of partially dissolved salt

Zone 3 In-situ saliferous beds. Groundwater circulation generally inactive

*Fig. 9.34* Formation of subsidence features in salt formations.

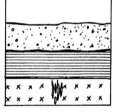

**Stage 1**
Solution-widened joints form by dissolution in the zone of groundwater circulation

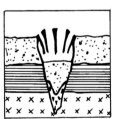

**Stage 2**
Solution-widened joints enlarge, coalesce and break down, causing progressive collapse of the cover deposits to form a pipe-filled breccia

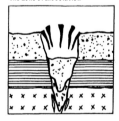

**Stage 3**
Continuing dissolution is deflected downwards, enlarging the zone of dissolution

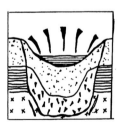

**Stage 4**
As the zone affected by dissolution enlarges, a greater area of cover deposits loses its basal support and subsides down

Many of the larger flashes evident in the Cheshire salt field have been caused by the wild brine pumping activities of private companies.

### Solution mining

Deposits that readily go into solution, notably salt, can be extracted by solution mining. This is known as brine pumping, and has been carried out in several areas of the UK. Cheshire and Cleveland were two areas, and the Cheshire salt fields were the most important. Consequently subsidence as a result of salt extraction still continues to affect large areas of Cheshire, owing mainly to the unpredictable nature of the groundwaters. It is very difficult to predict the amount of subsidence and its precise location. Soluble saliferous beds underlie large sections of Cheshire and Lancashire, and these have been exploited for centuries by brine pumping and salt mining. In the urban areas around Winsford, Middlewich, Northwich, Sandbach and Alsager major ground subsidence has taken place as a result of wild brine pumping.

Bridges, buildings, canals and railways have all suffered damage, and despite the cessation of brine pumping in the 1970s ground movements continue to take place. It is considered that these further movements are a result of the natural dissolution effects. In such areas it is prudent to construct buildings on a stiff raft foundation.

Various Acts of Parliament have been passed over the centuries to provide compensation funds for property owners who suffer damage as a result of brine subsidence. The funds were raised by the imposition of a financial levy on the salt producers. In 1952 the present Cheshire Brine Pumping Act was brought in, but following the cessation of brine pumping in the 1970s it is now more or less redundant.

### Salt mining

Rock salt was mined from shafts in the Marston and Winsford localities in the sixteenth and seventeenth centuries. The rock salt seams occurred in two beds, the Top bed and the Bottom bed, which were 20–30 m in thickness. Around Northwich the Bottom bed is about 100 m deep, and it is 150 m deep around Winsford.

During the period from 1870 to 1930 these worked mines flooded, and brine reservoirs were thus created. Subsequently these brine solutions were pumped out to remove the brine that had accumulated. This influx of clean water caused dissolution of the supporting pillars, and eventually the workings collapsed, resulting in catastrophic subsidence craters on the surface.

Most of the salt mining in the Cheshire salt field took place in the Top bed seam, and the worked seam thickness was 9–11 m. The shafts were sunk down to the seam, and salt was mined from radial adits, leaving pillars of rock salt to support the marl roof over. As the marl was not a very competent roof stratum, many sinkhole-type subsidence incidents resulted.

### 9.4.3 Natural solution features in gypsum, chalk and limestone

#### Gypsum

This stratum exists in a hydrous form and an anhydrous form, and can be 100 times more soluble in water than limestone. It is rare for gypsum to outcrop near the ground surface owing to the fast rate of dissolution by circulating water.

Beds of gypsum at shallow depth can be leached away by groundwater movement, and such beds can be tens of metres below ground surface. In Ripon, North Yorkshire, such gypsum beds are found 45–50 m below ground level, and it has been shown that the depth of the leached zone is integrally linked with the groundwater circulation. This dissolution of gypsum by water has resulted in a range of natural cavities and underground cave formations, and infilled breccia solution features. For many years subsidence incidents in and around the Ripon area have been monitored by the British Geological Survey, and several study papers have been written by Dr A.H. Cooper BGS. These subsidence features can result in very large craters at the surface. Those craters formed before the Ice Age are generally found to be filled with variable strata, and often thick beds of organic peat have resulted in the upper levels of the infilled subsidence hollows.

While the subsidence phenomenon in Ripon is of concern to developers, planners and other professions involved in building, it should be noted that in a recent study in the area of 150 properties, most recorded subsidence damage resulted from damaged drainage, soakaways too close to buildings and poor-quality foundations. Very few of those surveyed were related directly to gypsum dissolution.

In 1996 the DETR and Symonds Travers Morgan prepared a most useful report on the gypsum problems in and around Ripon. The aims of the report were to:

- assess the planning approval system to try and ensure a reduction of subsidence claims caused by gypsum dissolution;
- prepare a draft framework of advice suitable for use by developers, planners, engineers and insurance firms.

From a review of the historical information, and from work carried out by Dr Cooper, it appears that gypsum-related subsidence in the Ripon area has a very low frequency of occurrence, in the order of one incident per year over an area of more than 30 km$^2$.

In the Ripon area, many of the early subsidence features have filled up with peats and organic silts, and it is prudent not to build over them. Where the surface subsidence covers a large area, developments over peat bands may require the use of piled rafts. In such situations the piles should be designed to accommodate a pile redundancy, with the ground beams spanning over two beam spans.

Following the DoE report a development guidance map was produced for the Ripon locality, together with suggested procedures to be followed by engineers and surveyors in these areas. These areas are defined as development control areas A, B and C, and are shown in Fig. 9.35.

- **Area A:** No known gypsum present. This area lies to the west of the gypsum outcrop, and no potential for gypsum subsidence exists.
- **Area B:** Some gypsum present at depth. This is an area where gypsum is known to be present, but only at depths that appear to be below the influence of the groundwaters that are moving towards the buried valley of the River Ure. This area lies entirely to the east of the groundwater divide that is considered to lie beneath the crest of the Sherwood sandstone escarpment.
- **Area C:** Gypsum present and susceptible to dissolution. This area covers where the gypsum beds are present at depth and are within the effect of the circulating groundwater. It is in this area that there is therefore a greater likelihood of gypsum dissolution with associated surface depressions.

While many parts of area C have not been affected by recent subsidence, the implications of the Symonds Travers Morgan report are that subsidence features due to gypsum dissolution can be expected to occur at some time in the future. It is therefore area C that must have the most development control and the more thorough geophysical assessment when building in the area. If development is to be carried out in area C then an appropriate engineering balance must be achieved in the

overall design concept such that, if the building is large or prestigious, or highly sensitive to ground movements, then the only sure way of avoiding the effects of subsidence is to place the building on large-diameter piles taken down below the gypsum strata. Such solutions would most likely be uneconomic and unwarranted in terms of risk assessment.

Consideration could be given to incorporate design measures that would mitigate the subsidence effects and provide sufficient strength and flexibility for a building not to collapse suddenly. This approach would allow the building occupants to evacuate the building without harm, and would also leave a building that could be repaired. Such measures as masonry reinforcement in the superstructure walls, stiff edge-beam raft foundations, stiff heavily reinforced ground beams with a large spanning capacity and piling systems with built in redundancies can be adopted without too much expense, and in terms of risk assessment would be acceptable to most insurance companies. They would also meet the requirements of the Building Regulations Part A.

### Swallow holes in chalk and limestone formations

Limestones consist essentially of calcium carbonate, magnesium carbonate and other siliceous matter. Chalk is a soft white limestone. Both strata were formed as an accumulation of calcium carbonate from marine organisms or seawater.

When ordinary rainwater, containing carbon dioxide absorbed from the air, passes down through limestone

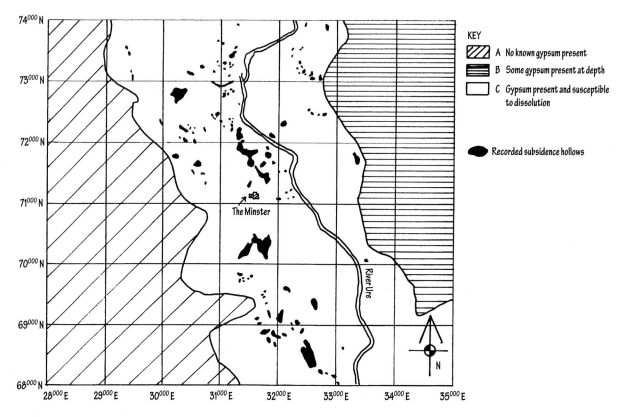

*Fig. 9.35* Development guidance map for the Ripon area.

strata, it can dissolve appreciable amounts of carbonate of lime. In passing through a bed of limestone or chalk, therefore, the percolating water removes small amounts of material in solution, and the effects, acting over a long period, are often significant. The water naturally takes the easiest route, and flows through cracks or faults where they are present. These cracks soon become enlarged by solution. A crack at the surface may in time be enlarged so much that it forms a wide opening sinking deeply into the ground. Such openings are known as *swallow holes*, and are quite frequent in limestone districts. Where the sinking has occurred below ground level, resulting in the ground above bridging over the swallow hole, such a feature can be difficult if not impossible to detect.

There was a building in the Chiltern Hills area that suffered severe damage as a result of a private swimming pool being pumped out, the water being discharged directly into the surrounding ground. The soils were very permeable, and it appears that the house had been constructed over a breccia pipe feature. This was a buried sinkhole filled with rock debris. The action of the water passing through the loose strata caused severe subsidence at the surface.

Although chalk is carbonate of lime, swallow holes are not formed to so great an extent as for example in limestones. The reason for this is that the joints in chalks are much closer.

### 9.4.4 Stone and mineral mining

Besides coal and metallic ores, other materials have been mined from shafts or adits, such as Elland flags, Permian sands for glass manufacture, chalk, fireclay for refractory products, limestone, lead, copper and slate. Most of these industries have now ceased, but they have left a legacy of ground instability.

In West Yorkshire mining for Elland flags was a major industry in the eighteenth century, and many mines sprang up around Halifax, Southowram, Idle, Bradford and Bingley. The sandstone flags were used as paving materials and, as some of the seams deepened, normal quarry operations were surpassed by deep mining from shafts. The workings were generally 25–30 m deep, with worked seams as thick as 3.50 m. Development over such workings often proves to be costly as most workings have insufficient rock cover above the seam. For coal workings the rule of thumb of 10 times seam thickness cover cannot be applied to Elland flag workings as the widths of the galleries are wider. It is therefore essential that a good site investigation be carried out that identifies the workings and confirms the bulking factor characteristics of the rock strata. This may require complete rock cores to be removed for testing. Once the bulking factor has been determined it should be possible to arrive at a safe depth for building over old unrecorded workings using raft foundations. Where the workings are found to be too shallow, it is unlikely to be economic to grout up such workings for development, and it would probably be more economic to quarry the pillars left.

### Sand mining

Mining for sand has been carried out in areas of the UK and one of the major uses was for glass manufacturing. The Permian sands were mined around Castleford and Pontefract from the early part of the twentieth century up until 1960. There have been quite a lot of subsidence incidents recorded in these localities, and the British Geological Survey carried out a detailed study and prepared a geological map showing the main mines and the affected areas.

### Chalk mining

This form of mining took place around the Bury St Edmunds locality and in areas around Norwich. Most subsidence incidents around Bury St Edmunds have been due to the use of soakaways for drainage purposes. Because of the problems in developing over such workings, large areas of land have been zoned as public open space as they are considered too costly and risky to develop.

### Limestone mining

The belt of country affected by the Silurian limestone working lies in the geological region of the South Staffordshire coalfield. The limestones that were deposited in this period were later folded and faulted as a result of tectonic activity, and subsequently were overlain by coal measures. The local succession for the Dudley and Walsall area is the Lower and Upper Wenlock Limestones, and these beds have been quarried or mined for their lime content between the first century and the twentieth century. The limestone has been used since Roman times for buiding stone and agricultural use. In the 1680s it became useful as a fluxing agent in the ironmaking process in the Black Country. Extraction was carried out on a similar basis to coalmining, and by the eighteenth century the pillar and stall method was commonplace. The heights of the stalls varied from 3 m to 10 m, which produced large caverns, later abandoned. Old limestone mining around Walsall and Dudley has experienced crownhole subsidence incidents from depths as great as 70 m, with even more spectacular subsidence following pillar collapses at deeper levels.

A lot of the extraction was incorrectly recorded to avoid paying tax revenues and this was achieved by pillar robbing, falsifying plans etc. Archival data must therefore be treated with some reservation, for the following reasons:

- Pillar robbing has left wider cavities.
- Some mines ceased working prior to the requirement to keep proper records.
- Many coalmining operations also extracted limestone as a bonus, and this was often undeclared to the Revenue.

A study of the abandoned limestone workings was commissioned in 1981 by the Department of the

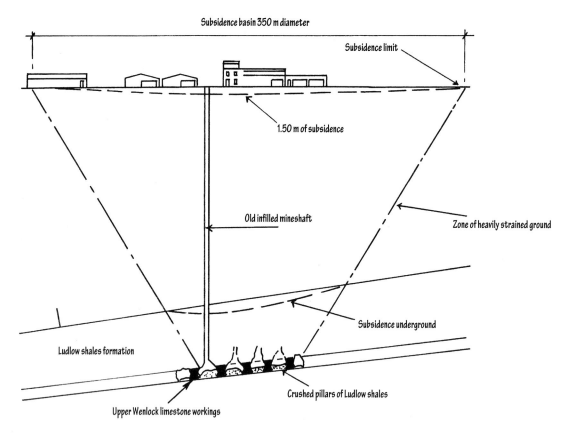

*Fig. 9.36* Subsidence at Wednesbury in 1978: pillar collapse in old limestone workings.

Environment and the local authorities of Dudley, Sandwell and Walsall. One of the worst subsidence incidents occurred in 1978 at Wednesbury, and took place over a period of six months. The surface subsided about 1.50 m at the centre of a 350 m diameter basin feature. This resulted in severe distortion of factory buildings in that area. It was considered that pillar collapse was responsible for the subsidence, as indicated in Fig. 9.36.

Most of these mines are flooded, and in locations where development is to proceed or where subsidence is affecting existing buildings the only practical method of treatment is to infill the workings. The normal methods of pressure grouting using cement and fly ash have been found to be very costly owing to the size of the cavities. Some areas have been treated using a *rock paste*, pumped in from the surface. The rock paste is generally old colliery spoil, which is in abundance in the West Midlands. The technique is to use sufficient water to make a paste that is fluid enough to pump. This allows it to flow easily below ground at distances up to 100 m from the injection point. While only about 90% of the mine volume is filled, this is sufficient to prevent progressive collapse of the roof of the mine. Infilling should always start at the deep end and work up the seam dip. In many cases the water from the mine can be used to mix the rock paste.

### Fireclay mining

Depending on its purity, fireclay was a much sought-after mineral, and was many times more valuable than coal, limestone or sandstone. It is well known that a coal seam may only have been mined as a bonus while the more economically worthwhile fireclay was mined.

Generally the fireclay was at a shallow depth, and was mined from adits.

### 9.4.5 Subsidence of clay soils caused by trees

Insurance statistics show that over 80% of subsidence claims for buildings on shrinkable clay soils are caused by the presence of trees or shrubbery close to the building. Quite often the offending trees are in a neighbour's garden or in a public open space.

Since the dry summers of 1947 and 1976 various organisations such as the NHBC and the BRE have produced authoritative guidance for building on clay sites where there are existing trees or on sites where tree planting is proposed in close proximity to buildings. In 1980 BS 5837 *Guide for trees in relation to construction* was introduced (new edition 1991), and this gave guidance on tree planting to new developments.

When carrying out an investigation of a building that has subsided, it should always be borne in mind that the cause of the damage may have been removed, possibly several years before the damage that resulted. When clays shrink as a result of excessive moisture depletion in the clay soils, the crack pattern is much different from those seen when clays swell as a result of

## Case history 9.9

A site for 40 houses was to be developed in Corbridge, Northumberland, on the site of an abandoned salt glaze-ware pipe manufacturing works. The works closed in 1960. The desk study for the site revealed that the fireclay had been extracted via several adits, and the workings were about 6 m below ground level, overlain by a good rock band. It was possible to gain access to parts of the mine via one adit, and the old abandonment plans were checked for accuracy by mining surveyors after the galleries had been provided with temporary support. The deepest section of the mine was found to be flooded and inaccessible.

A feasibility study was carried out based on excavating the remaining fireclay by opencast mining methods and replacing the overlying deposits to an engineered specification. This proved to be too costly, and after further investigation the developer decided to infill the workings with hydraulically pumped sand. Figure 9.37 shows the layout of the abandoned workings, with the stone barriers installed at various locations. These stone barriers were formed by drilling holes above a gallery junction and filling the gallery using a specially graded stone. The graded stone allowed the water in the hydraulically pumped sand to pass through, and the sand backed up behind. The water used for the pumping operation was pumped from the deeper, flooded mine section. The infilling was monitored using TV cameras inserted down the boreholes, and it proved to be a very successful filling operation. On completion of the filling the small gap at the roof of the workings was infilled with a PFA grout mixture pumped in under pressure.

The deeper part of the mine, which was flooded, was grouted up using traditional cement – PFA mixture, and the grout was placed using tremmie pipe apparatus. The sand used for the infilling came from a local source. It was considered unsuitable for concrete and the wrong grade for mortar, and was therefore purchased at an economic price. Following infilling, the housing development was built on stiff raft foundations.

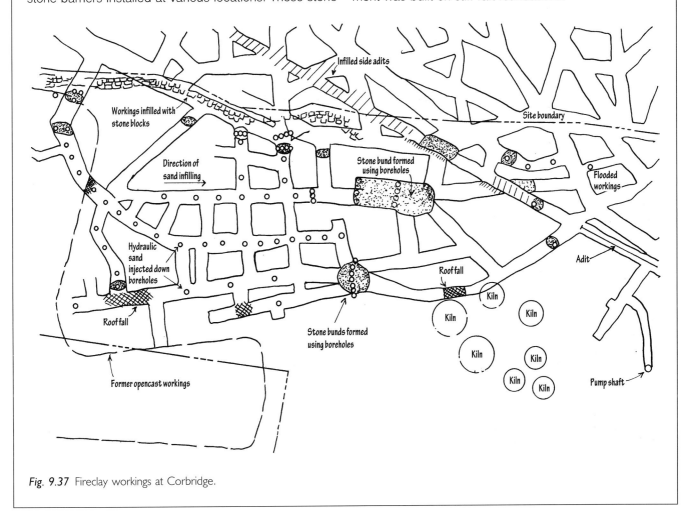

*Fig. 9.37* Fireclay workings at Corbridge.

rehydration, and Fig. 9.38 shows the nature of the movements due to shrinkage and heave of clay soils below a foundation.

The following guidelines and procedures should be adopted when investigating a major damage claim where trees may be the cause of the damage.

1  Try to determine the broad nature of the soil strata below the foundation. Is it cohesive or granular?
2  If trees are within close proximity, determine the tree species, the actual height, and their potential mature height.

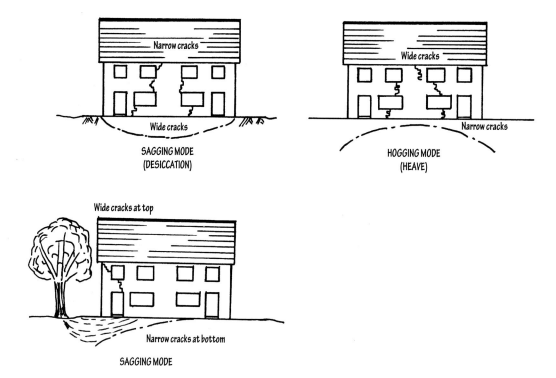

*Fig. 9.38* Subsidence and heave crack patterns.

3  If clay soils are found to be present below the foundations to a significant depth, then bulk samples should be taken for testing. The tests should include natural moisture content, liquid limit (LL), plastic limit (PL) and plasticity index (PI). The plasticity index is the LL minus the PL, and this can be used with the NHBC Chapter 4.2 tables to determine the depths of foundations. If clay tests are not available it is recommended to use high plasticity (PI > 40%) when using the tables.

4  Always excavate a trial hole down to and below the foundation. This will enable you to examine the soils for the presence of tree roots and any evidence of desiccation. Desiccated clays are difficult to mould by hand, and when excavated they have the appearance of small cubes of clay, hard and lumpy. Figure 9.39 illustrates the cracking pattern when trees are too close to a building on a shallow foundation in clay soils.

5  If sands or gravel are found below a footing, always check to make sure that they are not underlain by clay strata at an influencing depth. For these situations the thickness of the sand should be at least 75% of the required depths in Chapter 4.2.

6  If the foundations need to be underpinned, check that the clay soils extend down to a suitable depth. This should be at least 1.50 times the foundation width below the foundation base. Figure 9.40 shows a typical mass concrete underpinning detail to a foundation.

7  When considering the effects of a group of trees, the largest tree may not be the culprit if there are smaller trees that have a higher water demand. The tree to be considered is the one with the potential to cause the most damage.

8  Make a suitable allowance for the geographical location of the building.

9  If you are in any doubt about the tree species, or are unsure of the likely effects that pruning or removing a tree would have on a building, it is advisable to consult a qualified arboriculturist and obtain his or her views.

10  Always check for the possibility that there may have been previous hedges or trees. Consult old Ordnance survey plans, aerial survey photographs or local archives and old planning records, which may contain records of applications to remove trees.

11  Avoid traditional underpinning in areas where the ground becomes softer with depth. Mini-piling may be a more appropriate solution.

Table 9.1 lists the tree species, water demand and mature heights of trees found in the UK.

One of the problems when dealing with tree-related subsidence is identification of the trees, especially in the winter period when there are no leaves to examine. Figures 9.41, 9.42 and 9.43 show the common tree shapes for summer and winter for low, medium and high water-demand species.

It must be noted that dealing with tree-related subsidence is not an exact science, as there are many variations that can make the exception to the rules. The NHBC Chapter 4.2 is the best available guidance, but if there are trees that are actually taller than those species

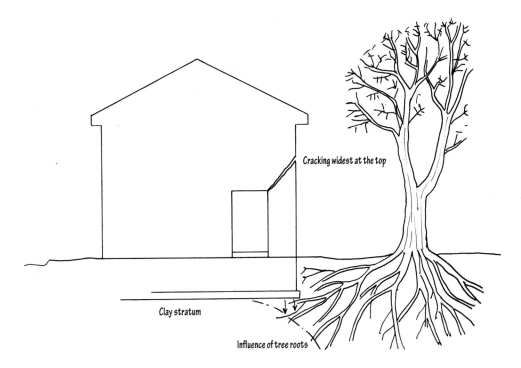

*Fig. 9.39* Tree root desiccation.

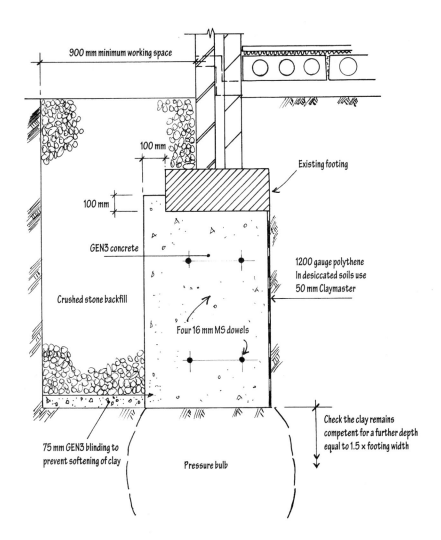

*Fig. 9.40* Typical underpinning detail.

*Table 9.1* Tree species

| Botanical name | English name | Water demand | Mature height[a] (m) |
|---|---|---|---|
| Acer campestre | Field maple | Moderate | 24 (13) |
| Acer negundo | Box elder maple | Moderate | 15 (12) |
| Acer platanoides | Norway maple | Moderate | 27 (18) |
| Acer pseudoplatanus | Sycamore | Moderate | 34 (22) |
| Acer saccharinum | Silver maple | Moderate | 31 (22) |
| Aesculus carnea | Horse chestnut | Moderate | 38 (20) |
| Alnus cordata | Italian alder | Moderate | 27 (18) |
| Alnus glutinosa | Common alder | Moderate | 26 (18) |
| Alnus incana | Grey alder | Moderate | 25 (18) |
| Betula pendula | Silver birch | Low | (14) |
| Betula verrucosa | Birch | Low | (14) |
| Carpinus betulus | Common hornbeam | Moderate | 30 (17) |
| Castanea sativa | Sweet chestnut | Moderate | 36 (20) |
| Cedrus spp. | Cedar | Moderate | (20) |
| Chamaecyparis lawsoniana | Cypress | High | 38 (24) |
| Crataegus monogyna | Common hawthorn | Moderate | 14 (10) |
| Cupressocyparis leylandii | Leyland cypress | High | (20) |
| Cupressus macrocarpa | Monterey cypress | High | (20) |
| Fagus spp. | Beech | Low | (20) |
| Fagus sylvatica | Common beech | Low | 41 (20) |
| Fraxinus excelsior | Common ash | Moderate | 45 (23) |
| Fraxinus ornus | Manna ash | Moderate | 24 (18) |
| Ilex aquifolium | Common holly | Low | 22 (12) |
| Juglans regia | Common walnut | Moderate | 23 (18) |
| Larix decidua | European larch | Moderate | 45 |
| Larix kaempferi | Japanese larch | Moderate | 37 |
| Magnolia spp. | Magnolia | Low | (9) |
| Malus aldenhamensis | Flowering crab | Moderate | 9 (7) |
| Malus tschonskii | Pillar apple | Moderate | 15 (9) |
| Morus spp. | Mulberry | Low | (9) |
| Picea abies | Norway spruce | Moderate | 43 (18) |
| Picea sitchensis | Sitka spruce | Moderate | 53 (20) |
| Pinus sylvestris | Scots pine | Moderate | 35 (20) |
| Platanus acerifolia | London plane | Moderate | 45 (26) |
| Populus alba | White poplar | High | 26 (22) |
| Populus canescens | Grey poplar | High | 35 (25) |
| Populus nigra | Black poplar | High | 30 (28) |
| Populus nigra 'Italica' | Lombardy poplar | High | 30 (25) |
| Populus tremula | Aspen | High | 20 (12) |
| Prunus avium | Wild cherry | Moderate | 30 (17) |
| Prunus dulcis | Almond | Moderate | (8) |
| Prunus padus | Bird cherry | Moderate | 15 (10) |
| Pseudotsuga menziesii | Douglas fir | Moderate | (20) |
| Quercus cerris | Turkey oak | High | 39 (24) |
| Quercus ilex | Holm oak | High | 28 (16) |
| Quercus robur | Common oak | High | 28 (20) |
| Robinia pseudoacacia | False acacia | Moderate | (18) |
| Salix alba | White willow | High | (24) |
| Salix caprea | Goat or pussy willow | High | (5) |
| Salix fragilis | Crack willow | High | (24) |
| Sorbus aria | Whitebeam | Moderate | (12) |
| Sorbus aria 'Lutescens' | Whitebeam | Moderate | 20 (13) |
| Sorbus aucuparia | Rowan | Moderate | 20 (9) |
| Sorbus commixta | Japanese rowan | Moderate | 16 (10) |
| Sorbus intermedia | Swedish whitebeam | Moderate | 15 (9) |
| Taxus baccata | Common yew | Moderate | 25 (12) |
| Tilia cordata | Small-leaved lime | Moderate | 30 (22) |
| Tilia euchlora | Crimean lime | Moderate | 18 (15) |
| Tilia europaea | Common lime | Moderate | 45 (22) |
| Tilia 'Petiolaris' | Pendent silver lime | Moderate | 27 (20) |
| Tilia platyhyllos | Large-leaved lime | Moderate | 31 (20) |
| Tilia tomentosa | Silver lime | Moderate | 27 (20) |
| Ulmus glabra | Wych elm | High | 29 (18) |
| Ulmus procera | English elm | High | 36 (24) |
| Ulmus 'Wheatleyi' | Wheatley elm | High | (22) |
| Ulmus spp. | Elm | High | (24) |

This table is produced with the permission of NHBC and is based on the Chapter 4.2 guidance.
[a] Figures in brackets are the NHBC Chapter 4.2 heights.

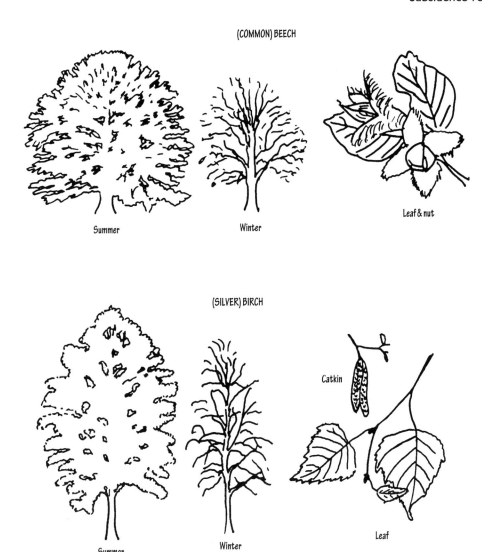

*Fig. 9.41* Tree identification: low water demand trees.

listed then the actual tallest heights should be used for assessing damage.

Special factors that must be considered when subsidence due to trees is suspected are as follows:

● Is the building on a site where previous heavy tree growth existed, and which has been removed prior to constructing the buildings? This includes such sites as former orchards, old fruit fields, and large houses with extensive landscaping.

● Always check on the types of shrubbery planted as part of landscape proposals. Generally such shrubbery is planted very close to the buildings, and species such as *Cotoneaster* can have a major effect on clay soils in a dry period.

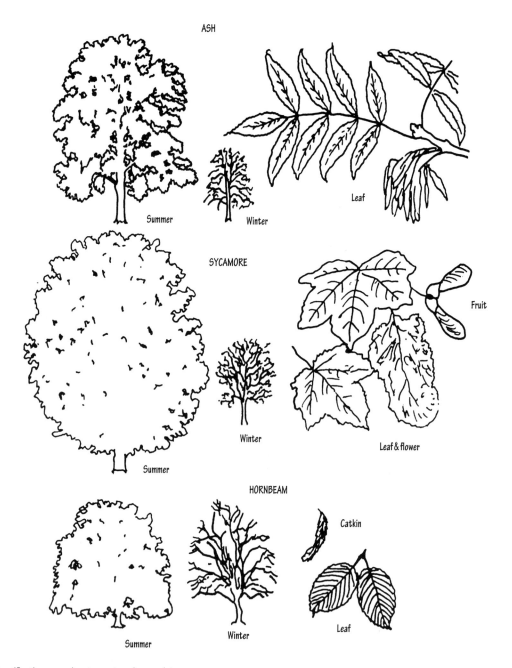

*Fig. 9.42* Tree identification: moderate water demand trees.

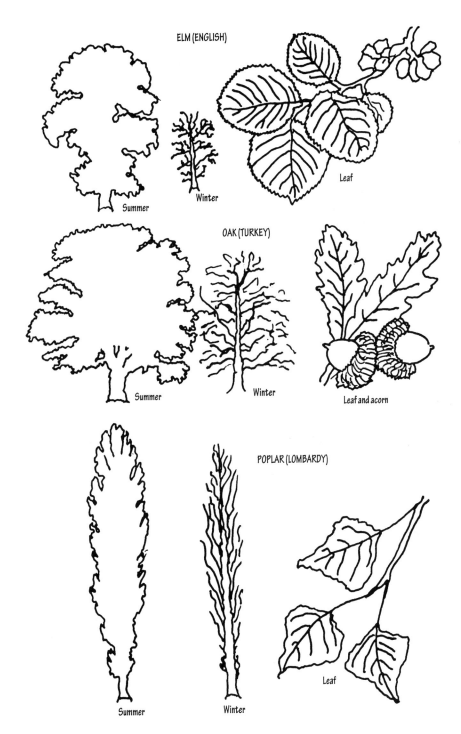

*Fig. 9.43* Tree identification: high water demand trees.

## Case history 9.10

### Introduction

A two-storey detached house in Leeds had suffered some cracking to the main walls internally and externally. The dwelling was 3 years old when the damage was reported to the insurance warranty provider.

### Investigation

The dwelling was architect designed to the owner's specification. The external walls were built in natural stone, with the inner skin built with lightweight autoclaved blocks. The main windows had dressed stone mullions and window heads. The internal walls were built in lightweight blockwork, and on the first floor timber stud partitions had been used. The roof was constructed using trussed rafters with Yorkshire slate tiles.

Examination of the exterior walls revealed movement down the sides of the main windows, and the central mullions had moved laterally about 5 mm.

Internally, most of the rooms had suffered some plaster cracking, and this was more evident at first floor level. Quite a lot of the ground floor cracking was attributed to contraction movement of the lightweight blockwork, which had been built without movement joints.

Tell-tales were placed over the serious cracks to assist in long-term monitoring. Examination of the plans submitted for Building Regulations approval showed that the house had been constructed on a pseudo-raft foundation, placed on an 800 mm cushion of compacted stone sub-base. This depth was checked by trial pits, and found to be accurate.

The original site investigation commissioned by the owner had revealed that the site was underlain by very shallow coal workings, and it was necessary to carry out a drill and grouting operation on a 3 m grid over part of the dwelling. The drilling carried out showed that the Churwell Thick coal seam outcropped below the house. Figure 9.44 shows the site layout and drilling pattern adopted and the coal seam outcrop. The workings were 3 m below ground level, and the Churwell Thick coal seam was approximately 1.0 m thick.

Examination of the rotary flush drilling logs revealed that cavities and broken ground were found in some boreholes, indicating that the roof over the workings had collapsed and migrated to the surface. There was only a thin mantle of rock above the coal seam. Subsequently the site had been stabilised by injecting into the workings a mixture of sand and cement grout.

The house was in a local conservation area, and the large sycamore trees and cherry trees were protected by Tree Preservation Orders. Following a precise level survey of the external paving and gardens it was clear that most of the movement was as a result of the effects

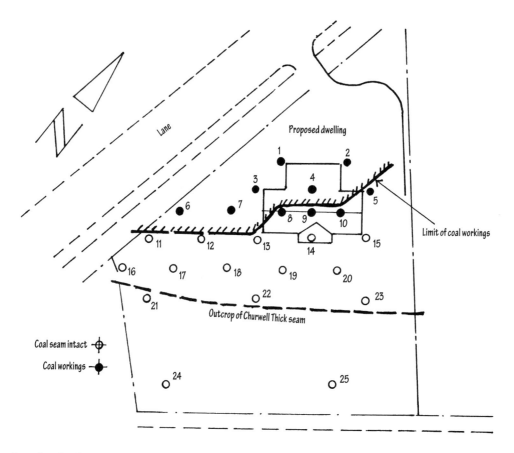

*Fig. 9.44* Location of coal outcrop.

of the extremely large cherry tree on the north-east boundary. It was noted that this tree was 850 mm lower than the ground level at the raft foundation edge, and it was clear that the stone cushion under the raft had been used to raise the level of the ground floor above the garden. This meant that the root growth was passing below the raft at a much lower level than that indicated in the NHBC Chapter 4.2 tables. Had this level difference been taken into account at the time of construction the depth of the stone cushion would have had to be 1200 mm. On the southern side of the dwelling the sycamore trees were also drying out the medium-plasticity clays.

Discussions took place with the local authority tree conservation department, and the request to remove some of the trees was rejected on the grounds that the house was in a conservation area. They were prepared to allow selective pruning if carried out under the guidance of a qualified arboriculturist. Figure 9.45 shows the existing trees.

It was then decided to provide a concrete root barrier, approximately 3.0 m deep, cast into the rockhead using a 25 N/mm$^2$ concrete strength. The location of the root barrier and its cross-section are shown in Fig. 9.46, and to prove the rock depth two boreholes were drilled

(Figs 9.47 and 9.48). This barrier was installed with a 2000 gauge polyethylene lining on both sides. Where the concrete needed to be jointed, the joint was half-lapped to provide a good seal. Prior to the installation of the root barrier the tell-tales were recorded, and monitoring continued for a further 12 months. The tell-tale measurements are shown in Table 9.2.

During the installation of the cut-off wall on the south side of the house, the old mine workings were broken into at very shallow depth. It was possible to look along the workings, and the previous grouting could be seen to have extended laterally beyond the house walls.

The intention of inserting the cut-off wall was to prevent further root invasion, to protect the house drainage, and to allow the clays between the cut-off wall and the house to rehydrate. This rehydration took place over approximately 18 months, and the cracks internally closed up approximately 2 mm.

The stone mullions in the rear windows had also moved back into their original position, and in February 1999 the internal walls were repaired and redecorated. This case history confirmed that great care must be exercised when trees are lower than the building level, and this change in level must be taken into account when using the NHBC Chapter 4.2 tables.

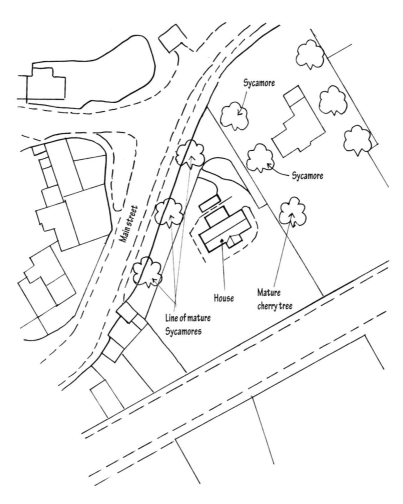

*Fig. 9.45* Location plan showing trees.

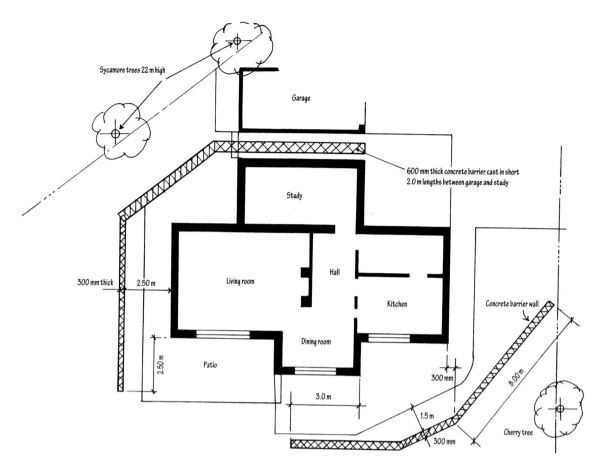

Fig. 9.46 Root barrier down to rock.

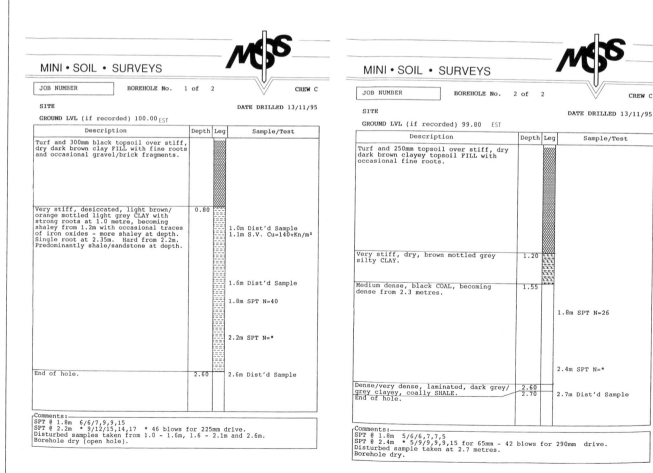

Fig. 9.47 Log of borehole no. 1.

Fig. 9.48 Log of borehole no. 2.

*Table 9.2* Tell-tale monitoring results

| Date of reading | A (mm) | B (mm) | C (mm) | Position of tell-tale over crack |
|---|---|---|---|---|
| *Tell-tale no. 1. Location: Bedroom facing road, left-hand side of window* | | | | |
| 15 Nov 95 | 43.54 | 38.58 | 42.30 | |
| 15 Feb 96 | 43.10 | 38.62 | 42.00 | Maximum reduction in |
| 5 Sept 97 | 42.00 | 38.20 | 42.00 | crack width = 1.54 mm |
| 26 Aug 98 | 42.00 | 38.24 | 42.40 | |
| *Tell-tale no. 2. Location: Gable wall, front bedroom* | | | | |
| 15 Nov 95 | 35.50 | 30.50 | 35.00 | |
| 15 Feb 96 | 35.59 | 30.20 | 35.39 | Maximum reduction in |
| 5 Sept 97 | 35.30 | 30.10 | 35.30 | crack width = 1.10 mm |
| 26 Aug 98 | 35.40 | 30.40 | 35.30 | |
| *Tell-tale no. 3. Location: On gallery landing* | | | | |
| 15 Nov 95 | 47.38 | 48.10 | 42.76 | |
| 15 Feb 96 | 47.10 | 47.70 | 42.56 | Maximum reduction in |
| 5 Sept 97 | 45.99 | 47.00 | 42.78 | crack width = 1.78 mm |
| 26 Aug 98 | 45.60 | 46.64 | 42.76 | |

Where a building has subsided because of clay shrinkage, the remediation generally entails underpinning down to a moisture-stable level. Provided some provision is made for the possibility of the clays on the inside of the dwelling rehydrating, then no problems should arise later. In certain circumstances it may also be wise to remove the offending tree, especially if it has not reached its potential mature height. Provided the heave effects can be accommodated, removing the tree removes the main cause of damage and the potential for further movement. Unfortunately some building owners get very attached to their trees, and the following case history illustrates the problems that can result if property owners do not accept the advice of professionals.

## Case history 9.11

### Introduction

A two-storey house built in 1987 in a village 5 miles north of York developed cracking in the external walls in 1988. The owners lodged a damage claim on the warranty insurance providers and the builder of the property.

### Investigation

The house had been built using traditional brick outer leaf and blockwork inner leaf cavity construction. The ground floor was a timber joist floor on sleeper walls. First floor construction was in timber joists with chipboard flooring, and the roof consisted of clay pantiles on trussed rafters. A single-storey study had been constructed off the lounge. The garage was under the first floor bedrooms, and this had a concrete floor. At the rear of the house there was a large mature Turkey oak tree, which was approximately 7 m from the study wall. The house layout is shown in Fig. 9.49.

Discussions with the builder and the local authority building inspector confirmed that 3 m deep foundations had been used at the rear of the property, and the depths were in accordance with the *NHBC Standards* Chapter 4.2 document. It was also established that no slip membrane had been used on the external face of the trench-fill concrete, and no low-density polystyrene had been used on the inside face.

The builder maintained that he had followed the guidance given by the local authority building inspector and had excavated to the depths laid down in the NHBC Chapter 4.2. Examination of site photographs taken by the owner clearly showed that many thick tree roots had been severed along the rear foundation excavations.

The foundations had clearly not been installed to NHBC requirements. These are illustrated in Fig. 9.50, and can be stated thus:

1  Where foundations are in excess of 1.50 m, the internal face of the foundations must be provided with a low-density compressible material to cater for heave of the internal dumpling.
2  The external face of the trench-fill concrete must be cast against a smooth excavation face.
3  The ground floor construction must be a fully suspended floor with a void, i.e. no sleeper walls.

The ground floor was opened up, and this revealed that the oversite concrete had heaved, and was badly cracked in several places. Examination of old Ordnance

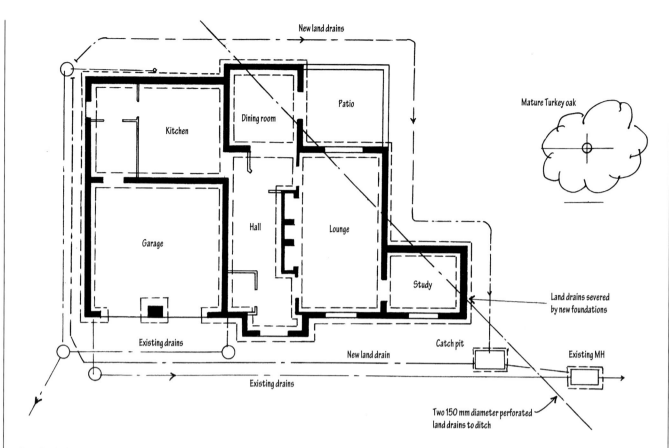

*Fig. 9.49* Site layout.

plans revealed that the land had been part of the adjoining garden, and there had been several ponds in the grounds. Discussions with the previous owners confirmed that these gardens were prone to flooding, and to alleviate this, several plastic perforated drainage pipes had been installed leading into the open ditch passing down the eastern boundary.

## Conclusions and recommendations

It was clear from the crack patterns to the walls and oversite concrete that the clays below the floor were heaving as a result of a rapid rehydration of the clays below the floor. In addition the clays on both sides of the foundation were locked in, and the lack of a slip plane to the deep foundations had also caused the foundations to heave. Trial pits were excavated, and these revealed the remains of the two 150 mm plastic perforated drains, which had been severed during the foundation works.

The excessive rainfall had drained into the ground, and the deep concrete footings along the north and eastern sides of the house had acted as a dam. This had caused a phenomenal increase in the moisture content of the clays, which must have been in a desiccated condition as a result of the root action of the Turkey oak.

These movements were vertical and lateral, and were exacerbated by the design of the house, which was lacking in strength in its centre. The main front hall went right through to the rear patio doors and in effect split the dwelling into two halves. In addition, the single-storey study, being lighter, was subjected to the most upward movement.

It was evident from the trial pits excavated that the foundations were at the correct depth. The clay at the base of the trial pits was also not desiccated at a depth of about 2.50 m. The damage that had been caused was therefore not considered to have resulted from the effects of the Turkey oak, but as a result of inadequate foundation construction and a failure to recognise that the severed land drainage should have been reconnected and diverted round or through the trench-fill concrete.

The initial recommendations were to remove the tree, remove the clays from below the ground floor and provide a fully suspended floor. The owner called in an independent building surveyor, who concluded that the foundations of the house were akin to a diving bell, and expressed the view that there must be an artesian pressure head causing the upward movement. He obviously had not dealt with many clay heave situations. Fortunately the consulting engineer advising the owner did not go along with this far-fetched theory. The owner, however, still refused to accept the removal of the tree, but did agree to the clays below the ground floor being removed and the land drainage that had been severed being reconnected into the ditch. The clays below the ground floor were removed to a depth of 2.50 m and replaced with pea gravel. The ground floor joists were increased in size to enable them to span between loadbearing walls. A new land drain was placed along the western boundary to intercept the perforated plastic

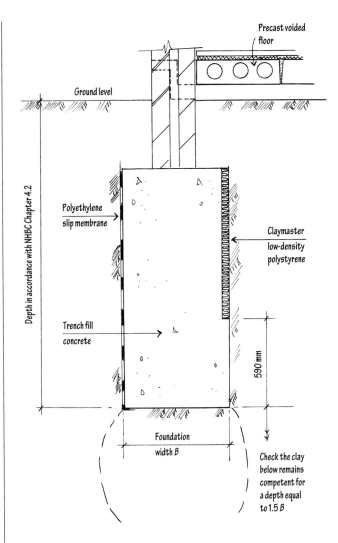

Fig. 9.50 Trench fill detail.

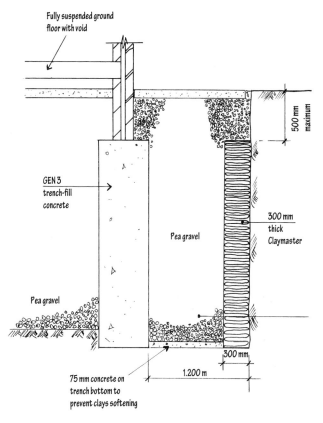

Fig. 9.51 External heave precautions.

land drains, and these were picked up into a catch pit, which discharged into the open ditch on the eastern boundary. During these works no evidence of ground-water under pressure was encountered.

In 1989 (the following year) the summer was extremely dry, and in September 1989 the owner reported that the house was on the move again. Examination of this movement showed it to be rotational, and this had resulted in lateral displacement on the dampcourse of about 10 mm.

The author was of the view that this movement had resulted from the Turkey oak having dried out the external clays. These clays were locked into the outside face of the concrete trench fill, and in moving down had imposed a drag on the external face of the trench fill.

This eccentric drag down had resulted in the rotational effect at the dampcourse, but it was also likely that the adhesion of the clays to the concrete had pulled the footing laterally as the clays shrank, exacerbating the drying-out effects.

Again the author strongly recommended that the Turkey oak be removed and a deep trench be excavated around the external trench-fill concrete to release the adhesion effects of the clay on the concrete trench fill.

The owner was not happy about losing the tree, and appointed another consulting engineer. This engineer unfortunately picked up on the original surveyor's diving bell theory, and in order to scotch this half-baked theory it was necessary to carry out further costly boreholes with piezometer monitoring.

The borehole report confirmed that there was no artesian water level, and the soil suction tests also showed there was no desiccation present below 2.50 m. The house owner was then presented with the following recommendations:

1   The Turkey oak tree be removed.
2   A 1200 mm wide trench be excavated down the side of the external face of the concrete trench fill, the excavation be backfilled with pea gravel, and a 300 mm thick layer of Claymaster be placed on the tree side of the trench, as shown in Fig. 9.51.

The owner finally agreed to this, and an application was made to the local authority to remove the tree, which was the subject of a tree preservation order. This was eventually granted, and the tree was removed. Following these works the brickwork was repaired, and 12 months later the owner reported that the removal of the tree had worked a treat.

With hindsight, the first repairs did not foresee the possibility of a very dry summer the following year, and predicting future weather patterns can be risky. Had the owner agreed to remove the tree initially a total and final solution would have resulted.

## Case history 9.12

### Introduction

This case history illustrates how much more difficult it is to resolve clay heave than subsidence caused by drying-out of clays. The problems with clay heave are that it is very difficult to control nature, and the heave is often time related. There may be situations where the clays can be rehydrated faster by installing cut-off walls, but generally such situations are rare. The recent succession of dry summers followed by dry winters had caused a cumulative moisture deficit, and it takes a long time for this dehydration to revert back to normal levels.

Following construction of a two-storey detached house on a development in York the owners reported cracking to the external walls and internal walls shortly after moving in. The claim on the warranty provider fell within the initial two-year period, and was therefore the builder's responsibility. Unfortunately, the builder went into liquidation, and in accordance with the warranty policy the claim was taken up by the insurance warranty provider.

### Investigation

Externally the damage to the front elevation wall was considered to be minor, but the cracking internally was considered to result from clay heave of the internal clay dumpling. Checks with the local authority building control showed that traditional strip foundations had been installed. In view of the movements occurring a desk study was carried out, and it was evident from old Ordnance plans and an aerial survey photograph that the house had been constructed directly over the line of a previous hawthorn hedge. It was not possible to confirm the height of the hedge when it was removed prior to construction of the house. Figure 9.52 shows the hedge location relative to the main walls.

### Conclusions and recommendations

The position of the hedge was established from old plans; it passed through the centre of the property. It was recommended that the property be underpinned, and that the requirements of NHBC Chapter 4.2 be complied with. This required the installation of a 50 mm thick layer of Claymaster low-density polystyrene on the inside of the external foundations, as shown in Fig. 9.53, and the provision of a fully suspended voided ground floor.

The height of the hedge was assumed to have been 4 m when removed, and it was established that it had

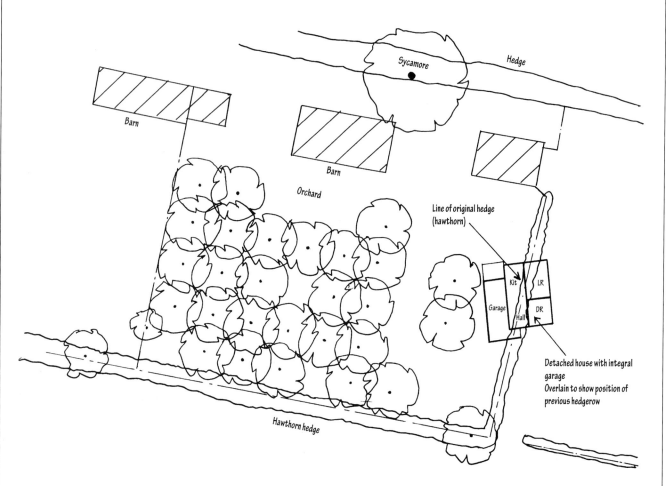

*Fig. 9.52* Old Ordnance plan.

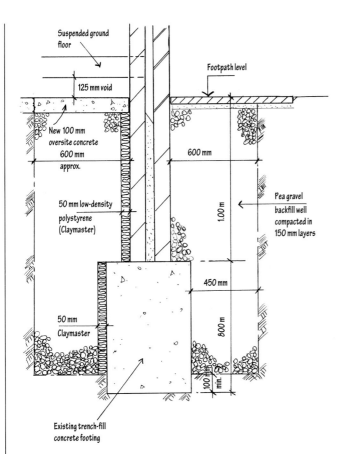

*Fig. 9.53* Modification to existing trench-fill footings.

been a hawthorn species. This required the foundations to be underpinned to a depth of 1.80 m. The internal footings also needed to be deepened to the same depth. These works were carried out using traditional underpinning techniques in 900 mm wide segments removed in a sequential order. During these underpinning works it was found that the cracking to the front elevation was due to the wall's having no foundation over the corner, and this was rectified by the underpinning. Despite these works the internal walls of the house suffered further movement, and it was considered likely either that the hedge had been higher than 4 m, or that it had been vigorous enough to cause desiccation to a greater depth. It was therefore decided that, before carrying out the internal repairs to the cracks, these be monitored until they showed signs of settling down. These monitoring results are shown in Table 9.3; they show that the clays took four years to rehydrate from removal of the hedgerow. Figure 9.54 shows the ground floor layout of the house.

*Table 9.3* Crack monitoring record. Claim Ref. No: 94/19591 Address: Richard Avenue, York

| Location of Avonguard pips | Date of reading | A | B | C |
|---|---|---|---|---|
| Dining room: Right-hand side of window sill | 30.01.96 | 37.66 | 41.69 | 36.00 |
| | 2.04.96 | 37.98 | 42.00 | 36.00 |
| | 10.09.96 | 36.00 | 41.80 | 36.00 |
| | 9.12.96 | 38.10 | 42.00 | 36.00 |
| | 12.08.97 | 38.00 | 42.00 | 36.10 |
| | 13.05.98 | 38.32 | 42.00 | 36.10 |
| Dining room: Top left-hand side of window | 30.01.96 | 28.33 | 39.00 | 46.20 |
| | 2.04.96 | 28.35 | 39.31 | 46.48 |
| | 10.09.96 | 28.20 | 39.20 | 44.68 |
| | 09.12.96 | 26.50 | 39.52 | 47.20 |
| | 12.08.97 | 28.50 | 39.40 | 46.30 |
| | 13.05.97 | 28.50 | 39.21 | 46.90 |
| Breakfast room: Top left-hand side of window | 30.01.96 | 35.90 | 33.10 | 36.28 |
| | 02.04.96 | 35.60 | 33.10 | 36.20 |
| | 10.09.96 | 34.10 | 33.30 | 36.28 |
| | 09.12.96 | 36.48 | 33.14 | 36.38 |
| | 12.08.97 | 36.18 | 33.50 | 36.00 |
| | 13.05.98 | 36.38 | 33.50 | 36.00 |
| Front bedroom: Left-hand side of window | 30.01.96 | 28.72 | 38.62 | 46.50 |
| | 02.04.96 | 28.80 | 38.54 | 46.39 |
| | 10.09.96 | 29.10 | 39.12 | 46.38 |
| | 09.12.96 | 28.64 | 38.14 | 46.58 |
| | 12.08.97 | 28.72 | 38.10 | 46.58 |
| | 13.05.98 | 28.60 | 38.80 | 46.49 |

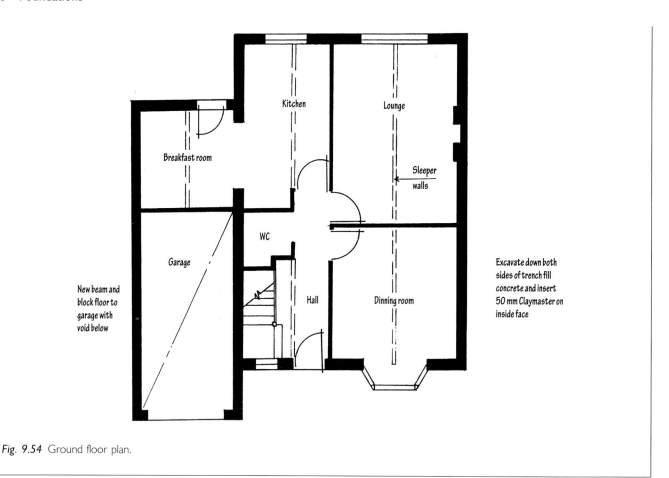

*Fig. 9.54* Ground floor plan.

## Points to remember

- Foundation movement can be caused by settlement or subsidence.
- In mining areas it is advisable to check the geological plans and obtain a Coal Authority mining report.
- When underpinning a foundation it is prudent to investigate the ground conditions below the new formation for a depth equal to 1.50 times the foundation width.
- Where there are variations in the stratum over the building footprint it is prudent to install some nominal reinforcement in the foundations.
- Where peaty soils exist at shallow depth, foundations should be placed below the highly compressible strata or be supported on piles.
- When carrying out an investigation of foundation movement, examine and possibly test the existing drainage.
- When dealing with foundation movement caused by trees or dry weather, check the clay plasticity index and consult the *NHBC Standards* Chapter 4.2 'Building near trees'.
- When carrying out an underpinning operation, ensure that no more than one-sixth of the foundation length is excavated and open at any one time.

- When investigating foundation movement on filled ground, do not place too much reliance on the standard penetrometer test results.
- In active coalmining areas, buildings should be constructed during a construction window. Always consult with the mine operators' subsidence engineer if mining damage has occurred.
- Avoid using soakaways in deep filled ground.
- In areas where chalks and gypsum deposits exist, avoid the use of soakaways.
- When investigating subsidence due to trees, always check the relative ground levels, especially if the trees are lower than the ground around the building.
- When investigating foundation movement due to trees, give careful consideration to the likely effects if the offending trees are removed. The potential for causing damage from ground heave must always be considered.
- Where shallow coal or other mineral workings have less than 10 times the seam thickness of rock cover, the workings must be grouted up.
- Where coal workings have less than five times the seam thickness of rock cover, raft foundations should be used following grouting-up of the workings.

# References

## Building Research Establishment

*Concrete in sulphate bearing soils and groundwater*, BRE Digest 363 (1991).

## British Standards Institution

BS 1377 : 1990 *Methods of test for civil engineering purposes.*
BS 5837 : 1991 *Guide for trees in relation to construction.*
BS 5930 : 1999 *Code of practice for site investigations.*
BS 8004 : 1986 *Code of practice for foundations.*
BS 8110 *Structural use of concrete*
    Part 1 : 1997 *Code of practice for design and construction*
    Part 3 : 1985 *Design charts for singly reinforced beams, doubly reinforced beams and rectangular columns.*

## Construction Industry Research and Information Association

Construction over abandoned mine workings, CIRIA Special Publication 32 (1984).

## Institution of Structural Engineers

*Subsidence of low-rise buildings* (1994)

## National House-Building Council

*NHBC Standards*: Chapter 4.1, Land quality – managing ground conditions.
Chapter 4.2, Building near trees (1992).

G.G. Meyerhof, Building on fill, *Structural Engineer*, **29** (2) (1951).
M.J.V. Powell, *House builders reference book*, Newnes Butterworth (1979)

# Index